INTERNATIONAL CENTRE FOR MECHANICAL SCIENCES

COURSES AND LECTURES - No. 303

ANALYSIS AND ESTIMATION OF STOCHASTIC MECHANICAL SYSTEMS

EDITED BY

W. SCHIEHLEN
UNIVERSITY OF STUTTGART

W. WEDIG
UNIVERSITY OF KARLSRUHE

SPRINGER-VERLAG WIEN GMBH

Le spese di stampa di questo volume sono in parte coperte da contributi
del Consiglio Nazionale delle Ricerche.

This volume contains 36 illustrations.

Originally published by Springer Verlag Wien-New York in 1988

ISBN 978-3-211-82058-2 ISBN 978-3-7091-2820-6 (eBook)
DOI 10.1007/978-3-7091-2820-6

PREFACE

Stochastic models and their analysis have reached a high amount of interest and research activities. In engineering environments, turbulent perturbations of mechanical systems cannot be excluded and they affect their dynamical behavior with increasing intensity. However, the hierarchy of second and higher moments time correlations and spectral densities are the tools to simplify the modelling and to summarize the complex analysis and simulation in nonlinear dynamics. Modern stochastic concepts are based on the theory of diffusion processes, Ito-calculus and stochastic differential equations leading to efficient matrix methods for linear time-invariant systems. For nonlinear systems and stability problems there is an ongoing research and increasing interest.

The present volume contains the lectures and the tutorials presented during a course at the International Centre for Mechanical Sciences in Udine, in summer 1987. It summarizes the latest developments in stochastic analysis and estimation. It presents novel applications to practical problems in mechanical systems. The main aspects of the course are random vibrations of discrete and continuous systems, analysis of nonlinear and parametric systems, stochastic modelling of fatigue damage, parameter estimation and identification with applications of vehicle-road systems and process stimulations by means of autoregressive models. The contributions will be of interest to engineers and research workers in industries and universities who want first-hand informations on present trends and problems in this actual field of engineering dynamics.

Finally, we would like to thank the authors for their efforts in presenting the lectures and preparing the manuscripts for publication. Many thanks are also due to Professors S. Kaliszky, Rector of CISM, to Professor G. Bianchi, Secretary General of CISM and the very capable CISM staff for advice and help during the preparation and the performance of the course on analysis and estimation of stochastic mechanical systems.

Werner Schiehlen, Stuttgart *Walter Wedig, Karlsruhe*

CONTENTS

Page

RANDOM VIBRATIONS OF DISCRETE AND CONTINUOUS SYSTEMS

I. Elishakoff
Naval Postgraduate School, Monterey, California, USA

ABSTRACT

The topic of random vibrations of discrete and continuous systems is treated in four papers as follows.

1. Correlation and Spectral Analysis
   A Brief Outline

2. Measurement of Characteristics of Stationary Random Processes

3. Random Vibration of Multi-Degree-of-Freedom Systems with Associated Effect of Cross-Correlations

4. Wide-Band Random Vibration of Continuous Structures with Associated Effect of Cross-Correlations

---

* On leave from Technion - I.I.T., 32000 Haifa, Israel

CORRELATION AND SPECTRAL ANALYSIS — A BRIEF OUTLINE

I. Elishakoff
Naval Postgraduate School, Monterey, California, USA

The first harbinger of the new discipline of "Random Vibration" was the book by that name, edited by Crandall[1], which summarized the state-of-the-art in terms of findings obtained on jet aircraft and rocket-propelled vehicles. During the last quarter-century, the probabilistic response and reliability of structures subjected to random excitation were widely investigated in a variety of contexts[1-17] : atmospheric turbulence, rough seas, earthquakes, roadway irregularities, and acoustic pressure generated by jet engines and rocket motors.

The discipline of random vibration derives in general from the kinetic theory, Brownian motion, statistical mechanics, and the fluid turbulence theory; discrete structures (linear or non-linear) are treated by methods borrowed from the communication noise theory.

1. BASIC NOTIONS

The principal notion of the random-vibration theory is that of the random process, X, defined as a family (ensemble) of random variables with its parameter (or parameters) belonging to an indexing set (or sets). Under this definition, each "outcome" has associated with it a real or complex time function $X(t,\zeta)$ or more concisely $X(t)$, which may thus be viewed in four different ways:

(1) a family of time functions (both t and ζ variable);

(2) a single time function (t variable, $\zeta = \zeta_i$ fixed);

*On leave from Technion - I.I.T., 32000 Haifa, Israel.

(3) a random variable ($t=t_j$ fixed, ζ variable);

(4) a single number (both t and ζ fixed).

The n-dimensional probability distribution function and density function, related as

$$F[x_1,x_2,\dots,x_n\ ;\ t_1,t_2,\dots,t_n] = \mathrm{Prob}[X(t_1)\le x_1,\dots,X(t_n)\le x_n] \tag{1}$$
$$= \int_{-\infty}^{x_1}\dots\int_{-\infty}^{x_n} f(x_1,x_2,\dots,x_n\ ;\ t_1,t_2,\dots,t_n)dx_1\dots dx_n$$

are functions of state variables $x_1,x_2,\dots,x_n$, as well as of time instants $t_1,t_2,\dots,t_n$. A random process may be regarded as specified, if all multidimensional probability densities are specified for arbitrary values of $t_1,t_2,\dots,t_n$. A random process can also be described by moment functions of different orders,

$$E[X^{i_1}(t_1)\dots X^{i_n}(t_n)]= \int\dots\int x_1^{i_1}\dots x_n^{i_n}f(x_1,\dots,x_n\ ;\ t_1,\dots,t_n)dx_1\dots dx_n \tag{2}$$

where $i_1 + i_2 + \dots + i_n$ is the order of the moment function. The first and second-order moment functions are of particular importance in applications. The first-order moment

$$m_X(t) = E[X(t)] \tag{3}$$

is called the <u>mean function</u>, and the second-order moment

$$R_{XX}(t_1,t_2) = E[X(t_1)Y(t_2)] \tag{4}$$

- the <u>autocorrelation function</u>. A counterpart of the autocorrelation function is the <u>cross-correlation</u> function

$$R_{XY}(t_1,t_2) = E[X(t_1)Y(t_2)] \quad , \tag{5}$$

where X(t) and Y(t) are a pair of generally different random processes.

For the variance we have

$$\sigma_X^2(t) = E\{[X(t) - m_X(t)]^2\} = R_{XX}(t,t) - m_X^2(t) \quad . \tag{6}$$

Consider a random process representing the sum of n constituent random processes

$$X(t) = \sum_{i=1}^{n} X_i(t) \tag{7}$$

Its autocorrelation function will be

$$R_X(t_1,t_2) = E[X(t_1)X(t_2)] = E[\sum_{i=1}^{n} X_i(t_1) \sum_{j=1}^{n} X_j(t_2)]$$
$$= \sum_{i=1}^{n} E[X_i(t_1)X_i(t_1)] + \sum_{\substack{i=1 \\ i \neq j}}^{n} \sum_{j=1}^{n} E[X_i(t_1)X_j(t_2)] \quad , \tag{8}$$

or

$$R_X(t_1,t_2) = \sum_{i=1}^{n} R_{X_iX_i}(t_1,t_2) + \sum_{\substack{i=1 \\ i \neq j}}^{n} \sum_{j=1}^{n} R_{X_iX_j}(t_1,t_2) \quad , \tag{9}$$

implying that the autocorrelation function of the sum equals the sum of the autocorrelation functions of the constituents plus that of the cross-correlation functions obtainable for all pair combinations of the former.

An important class of random processes is the stationary process. $X(t)$ is said to be strongly stationary if its probability density of arbitrary order does not change for any shift of the whole group of points $t_1,t_2,\ldots,t_n$ along the time axis; i.e., for any n and τ,

$$f(x_1,x_2,\ldots,x_n \,;\, t_1,t_2,\ldots,t_n) = f(x_1,x_2,\ldots,x_n \,;\, t_1+\tau,\, t_2+\tau,\ldots t_n+\tau). \tag{10}$$

In other words, the random process is stationary, when the expressions of probability densities of any order are invariant in terms of the time origin. Otherwise, the process is said to be nonstationary. If Eq. (10) holds for $k \leq n$ only, instead of for all n - the process is called stationary of order k. A process stationary of order 2 is called weakly stationary. A strongly stationary random process is also weakly stationary, but the converse is generally not true.

The foregoing definition implies that

a) the first-order probability density is time-invariant

$$f(x\ ;\ t) = f(x,\ t+\tau) = f(x) \quad , \tag{11}$$

b) the second order probability density depends on the time lag $t_2 - t_1$:

$$f(x_1, x_2\ ;\ t_1, t_2) = f(x_1, x_2\ ;\ t_2 - t_1) \quad ; \tag{12}$$

c) the n-dimensional probability density function depends only on the (n-1)th time parameter $t_i - t_1$, $i=1,2,\ldots,n$.

Since the first-order probability densities are time-invariant, so are the moments of stationary processes, in particular the mean and variance; and since the second-order probability density depends only on the difference $=t_2 - t_1$, so does the autocorrelation function:

$$m_X(t) = \int_{-\infty}^{\infty} x f(x)dx = \text{const} \quad ,$$

$$\sigma_X^2(t) = \int_{-\infty}^{\infty} [x - m_X^2(t)]^2 f(x)dx = \text{const} \quad ,$$

$$R(t_1, t_2) = \int_{-\infty}^{\infty}\int_{-\infty}^{\infty} (x_1 - m_X)(x_2 - m_X) f(x_1, x_2\ ;\ t_2 - t_1)dx_1 dx_2 \tag{13}$$

$$= R(t_2 - t_1) = R(\tau) \quad .$$

It is readily verified that the autocorrelation and cross-correlation functions of a real random process are symmetric with respect to their arguments

$$R_{XX}(t_1, t_2) = R_{XX}(t_2, t_1),\ R_{XY}(t_1, t_2) = R_{YX}(t_2, t_1) \quad . \tag{14}$$

For an arbitrary function g(t), the autocorrelation function satisfies the condition

$$\int_a^b \int_a^b R_{XX}(t_1, t_2) g(t_1) g(t_2) dt_1 dt_2 \geq 0 \quad , \tag{15}$$

i.e. is nonegative-definite.

For a stationary random process, symmetry reduces to evenness:

$$R_{XX}(\tau) = R_{XX}(-\tau),\ R_{XX}(\tau) = R_{YX}(-\tau) \quad , \tag{16}$$

and the nonnegative-definiteness becomes

$$\int_a^b \int_a^b R(t_2 - t_1)g(t_1)g(t_2)dt_1dt_2 \geq 0 \quad . \tag{17}$$

2. SPECTRAL DENSITY FUNCTION

According to Bochner's well known theorem, every nonnegative-definite function has a nonnegative Fourier transform. The Fourier transform of the autocorrelation function is called the spectral density

$$S_X(\omega) = \frac{1}{2\pi} \int_{-\infty}^{\infty} R_X(\tau)e^{-i\omega\tau}d\tau \quad , \tag{18}$$

whence, by the transform inversion formula

$$R_X(\tau) = \int_{-\infty}^{\infty} S_X(\omega)e^{i\omega\tau}d\omega \quad . \tag{19}$$

The pair of equations (18) and (19) are known as the Wiener-Khintchine relationships. Analogically, the cross-spectral density $S_{XY}(\omega)$ of a pair of jointly stationary random processes $X(t)$ and $Y(t)$ is the Fourier transform of their corsscorrelation function:

$$S_{XY}(\omega) = \frac{1}{2\pi} \int_{-\infty}^{\infty} R_{XY}(\tau)e^{-i\omega\tau}d\tau \quad , \tag{20}$$

$$R_{XY}(\tau) = \int_{-\infty}^{\infty} S_{XY}(\omega)e^{i\omega\tau}d\omega \quad .$$

If we set $\tau=0$ in (19), it then follows that

$$E[X^2(t)] = R_X(0) = \int_{-\infty}^{\infty} S_X(\omega)d\omega \quad , \tag{21}$$

that is, the mean-square value of a stationary random process equals the total area under the graph of the spectral density. Some autocorrelation functions and the associated spectral densities are listed below: (a) Exponentially-correlated random process

$$R(\tau) = d^2 e^{-a|\tau|}, \quad S(\omega) = \frac{d^2}{\Pi} \frac{a}{\omega^2 + a^2} \quad ; \tag{22}$$

(b) damped-exponentially-correlated random process

$$R(\tau) = d^2 e^{-a|\tau|} \cos b\tau, \quad a>0 \quad ,$$

$$S(\omega) = \frac{ad^2}{\pi} \frac{\omega^2 + a^2 + b^2}{(\omega^2 - a^2 - b^2)^2 + 4\, a^2 \omega^2} \quad ; \tag{23}$$

(c)

$$R(\tau) = d^2 e^{-a|\tau|} (\cos b\tau + \frac{a}{b} \sin b|\tau|), \quad a>0$$

$$S9\omega) = \frac{2d^2 a}{\pi} \frac{a^2 + b^2}{(\omega^2 - a^2 - b^2)^2 + 4\, a^2 \omega^2} \quad ; \tag{24}$$

(d) Ideal white-noise process

$$R(\tau) = 2\pi S_0 \delta(\tau), \quad S(\omega) = S_0 \quad . \tag{25}$$

That is, $X(t_1)$ and $X(t_2)$ for non-coinciding time instants are uncorrelated random variables. The variance of a white-noise process tends to infinity, as follows from Eqs. (21) and (25), since the spectral density is constant for all frequencies ω. White-noise is physically unrealizable, but is analytically useful in cases where the spectral density can be considered as constant in the operative frequency intervals of the systems in question.

(e) Band-limited white-noise process

$$R(\tau) = 4S_0(\sin\omega_0\tau)/\tau \quad ,$$

$$S(\omega) = S_0, \text{ for } \omega \le \omega_0, \text{ and } 0 \text{ otherwise} \quad . \tag{26}$$

3. CONTINUITY, DIFFERENTIABILITY AND INTEGRABILITY

A random process $X(t)$ is said to be continuous in the mean-square sense at t, if

$$\lim_{\varepsilon \to 0} E\{[X(t+\varepsilon) - X(t)]^2\} = 0 \quad . \tag{27}$$

A random process, continuous at every t on some interval, is called continuous on this interval. Consider the equality

$$E\{[X(t+\varepsilon) - X(t)]^2\} = R(t+\varepsilon, t+\varepsilon) + R(t, t) - 2R(t, t+\varepsilon) \quad . \tag{28}$$

Accordingly, a necessary and sufficient condition of continuity of a random process at t, is for its autocorrelation function to be continuous at $t_1=t_2=t$. For a weekly stationary process it follows from the following equality

$$E\{[X(t+\varepsilon) - X(t)]^2\} = 2[R(0) - R(\varepsilon)] \quad , \tag{29}$$

that continuity of an autocorrelation function of t=0 is a necessary and sufficient condition of continuity of the random process for any t. The random process X(t) is said to be <u>differentiable in the mean-square sense at t</u>, if there exist a random function $\dot{X}(t)$, called the mean-square derivative of process X(t) at t, such that

$$\lim E\{[\frac{X(t+\varepsilon) - X(t)}{\varepsilon} - \dot{X}(t)]^2\} = 0 \quad . \tag{30}$$

The mean function of the drivative $\dot{X}(t)$ is

$$m_{\dot{X}}(t) = \lim_{\varepsilon \to 0} \quad [mX(t+\tau) - m_X(t)] = \dot{m}_X(t) \quad . \tag{31}$$

By analogy

$$R_{\dot{X}\dot{X}}(t_1, t_2) = \frac{\partial^2}{\partial t_1 \partial t_2} R_X(t_1, t_2) \quad . \tag{32}$$

For a weekly stationary process we have

$$m_{\dot{X}}(t) = 0, \; R_{\dot{X}\dot{X}}(t_1, t_2) = \ddot{R}_X(\tau) = - \ddot{R}_X(\tau) \quad . \tag{33}$$

Differentiating Eq. (19) twice, we find

$$S_{\dot{X}}(\omega) = \omega^2 S_X(\omega) \quad . \tag{34}$$

The variance of the derivative is

$$R_{\dot{X}}(0) = -\ddot{R}_X(0) = \int_{-\infty}^{\infty} \omega^2 S_X(\omega)d\omega \tag{35}$$

A necessary and sufficient condition of differentiability of a weakly stationary process is finiteness of the variance $R_{\dot{X}}(0)$ of the derivative. As is seen from (35), this also indicates that

$$\int_{-\infty}^{\infty} \omega^2 S_X(\omega)d\omega < \infty \quad , \tag{36}$$

i.e. $S_X(\omega)$ decays faster than ω^{-3} at high frequencies. Random processes represented by Eqs. (22), (23) and (25) are nondifferentiable, whereas those represented by (24) and (26) are differentiable.

A necessary and sufficient condition of existence of the n-th derivative $X^{(n)}(t)$ is for a continuous mixed 2n-th order derivative to exist

$$R_{X^{(n)}}(t_1,t_2) = \frac{\partial^{2n}}{\partial t_1^{\,n} \partial t_2^{\,n}} R_X(t_1, t_2) \tag{37}$$

The autocorrelation function of the n-th derivative of weakly stationary process is

$$R_{X^{(n)}}(\tau) = (-1)^{(n)} R_X^{(2n)}(\tau) \tag{38}$$

and its spectral density

$$S_{X^{(n)}}(\omega) = \omega^{2n} S_X(\omega) \quad . \tag{39}$$

The n-th drivative exists, if the 2n-th derivative of the autocorrelation function is continuous at $\tau=0$ or, on the other hand, the spectral density decays faster than $\omega^{-(2n+1)}$ at high frequencies. The random process given in Eq. (24) is differentiable only twice, whereas the band-limited white-noise given in Eq. (26) has any number of derivatives. The integral, in the mean-square sense of a random process

$$Y(t) = \int_u^{\ell} g(t,u)X(u)du X(t) \tag{40}$$

[where g(t,a) is a deterministic (nonrandom) function, and "a" and "b" are some constants], is a random function, the mean-squre limit of a sequence interpreted as Yn(t) of random functions

$$Y_n(t) = \sum_{i=1}^{n} g(t, u_i)\, X(u_i)\, (u_i - u_{i-1}) \quad . \tag{41}$$

The mean function of the integral is obtained as

$$\begin{aligned} m_Y(t) &= E[\int_a^b g(t, u)\, X(u) du] = \int_a^b g(t, u)\, E[X(u)] du \\ &= \int_a^b g(t, u)\, m_X(u) du \quad . \end{aligned} \tag{42}$$

For a weakly stationary process X(t), we have

$$m_Y(t) = m_X \int_a^b g(t, u)\, du \tag{43}$$

The autocorrelation function of the integral is

$$R_{YY}(t_1, t_2) = E[Y(t_1)Y(t_2)]$$

$$= \int_a^b \int_a^b g(t_1, u_1)\, g(t_2, u_2)\, R_{XX}(u_2, u_2) du_1 du_2 \quad .$$

4. GAUSSIAN RANDOM PROCESS

A random process X(t) is said to be Gaussian or normal, of the random variables $X(t_1),\dots,X(t_n)$ are jointly Gaussian, i.e. the n-th order probability density is given by

$$\begin{aligned} &f(x_1,\dots,x_n;\, t_1, t_2,\dots,t_n) \\ &= [(2\pi)^n \det V]^{-1/2} \exp[-\frac{1}{2}\, (x-m)^T V^{-1} (x-m)] \end{aligned} \tag{45}$$

where

$$x^T = [x_1 \quad x_2 \dots x_n] \quad , \tag{46}$$

and

$$V = E[(X-m)\ (X-m)^T] = [v_{ij}]_{nxn} \tag{47}$$

is the variance-covariance matrix with elements

$$v_{ij} = E\{[X(t_i) - m_X(t_i)][X(t_j) - m_X(t_j)]\} = R_X(t_i,\ t_j) - m_i m_j\ . \tag{48}$$

Let a Gaussian random process be weakly stationary. Then, all mean values and variances are constant

$$m_i = m\ ,\ \sigma_i^{\ 2} = v_{ii} = \sigma^2 \tag{49}$$

and the autocorrelation function depends not on two separate variables t_i and t_j, but on their difference $t_j - t_i$. Then, the probability density function (45) does not change for any shift of the whole group of points $t_1, t_2, \ldots, t_n$ along the time axis. Thus, weak stationarity of a Gaussian random process implies its strong stationarity.

The first and second-order probability densities of a Gaussian stationary random process are given by

$$f(x) = \frac{1}{\sigma\sqrt{2\pi}} \exp[-\frac{(x-m)^2}{2\sigma^2}]$$

$$f(x_1,\ x_2\ ;\ \tau) = \frac{1}{2\pi\sigma^2\ 1-\rho^2(\tau)} \exp\{-\frac{1}{2\sigma^2[1-\rho^2(\tau)]} \tag{50}$$

$$[(x_1-m)^2 - 2\rho(\tau)(x_1-m)(x_2-m) + (x_2-m)^2]\}$$

where $\rho(\tau)$ is a correlation coefficient of the random variables $X(t)$ and $X(t+\tau)$,

$$\rho(\tau) = [R(\tau) - m^2]\sigma^{-2}\ . \tag{51}$$

The first-order probability density of the derivative $X(t)$ of a Gaussian process is given by

$$f_{\dot{X}}(y;\ t) = \frac{1}{\sigma_{\dot{X}}\sqrt{2\pi}} \exp[-\frac{(y-m_{\dot{X}})^2}{2\sigma_{\dot{X}}^2}] \tag{52}$$

where, in accordance with (31) and (32)

$$m_{\dot{X}}(t) = \dot{m}_{X}(t)$$
$$\sigma_{\dot{X}}^2(t) = R_{\dot{X}\dot{X}}(t,\ t) - [m_{\dot{X}}(t)]^2 \quad . \tag{53}$$

For a stationary Gaussian process

$$\dot{m}_X(t) = 0\ , \quad \sigma_X^2 = -\ddot{R}_X(0) \quad . \tag{54}$$

For the particular case $m_X=0$, we have

$$f_{\dot{X}}(y) = \frac{1}{\sigma_X \omega_1 \sqrt{2\pi}} \exp[-\frac{y^2}{2\sigma_X^2 \omega_1^2}] \quad , \tag{55}$$

where

$$\omega_1^2 = \int_{-\infty}^{\infty} \omega^2 S_X(\omega)d\omega / \int_{-\infty}^{\infty} S_X(\omega)d\omega \quad . \tag{56}$$

Analogically, for $m_X=0$ we have the second-order probability density of a derivative of a stationary Gaussian process

$$f(y_1,\ y_2;\ \tau) = \frac{1}{2\pi\sigma_X^2\{\omega_1^4 - [\rho_X(\tau)]^2\}^{1/2}} \times$$
$$\times \exp\{-\frac{y_1^2 + \rho_X(\tau)y_1y_2 + y_2^2}{2\sigma_X^2(\omega_1^4 - [\rho(\tau)]^2)}\} \tag{57}$$

ACKNOWLEDGMENT

The support by the Naval Postgraduate School Research Foundation, and Office of Naval Research are greatfully appreciated.

REFERENCES

1. Crandall, S. H. (ed.)., Random Vibration, Vol. 1, 1958; Vol. 2, 1963. M.I.T. Press, Cambridge, Mass.
2. Bolotin, V. V., Statistical Methods in Structural Mechanics, State Publ. House for Bldg. (English trans. Holden Bay, San Francisco, 1969).
3. Robson, J. D., An Introduction to Random Vibration, Edinburgh at the University Press, 1963.

4. Crandall, S. H. and Mark, W. D., Random Vibration in Mechanical Systems, Academic, New York, 1963.

5. Lin, Y. K., Probabilistic Theory of Structural Dynamics, McGraw-Hill, New York, 1967 (second edition, R. Krieger Publishers, Malabar, FL, 1976).

6. Sobczyk, K., Methods in Statistical Dynamics, Polish Scientific Publ., Warsaw, 1973 (in Polish).

7. Price, W. G. and Bishop, R.E.D., Probabilistic Theory of Ship Dynamics, Chapman and Hall, London, 1974.

8. Newland, D. E., An Introduction to Random Vibrations and Spectral Analysis, Longman, London, 1975 (second edition, 1984).

9. Svetlitskii, V. A., Random Vibrations of Mechanical Mechanical Systems, Vieweg, Braunschweig, 1978 (in German).

10. Heinrich, W. and Hennig, K., Random Vibrations of Mechanical Systems, Vieweg, Braunschweig, 1978 (in German).

11. Elishakoff, I., Probabilistic Methods in the Theory of Structures, Wiley-Interscience, New York, 1983 (solutions manual, Wiley-Interscience, 1986).

12. Bolotin, V. V., Random Vibrations of Elastic Bodies, "Nauka" Publ. House, Moscow, 1979; English Transl. Martinus Nijhoff, The Hague, 1985).

13. Ibrahim, R., Parametric Random Vibration, John Wiley, New York, 1985.

14. Yang, C. Y., Random Vibration of Structures, Wiley-Interscience, New York, 1985.

15. P. Kree and C. Soize, Mathematics of Random Phenomena-Random Vibration of Mechanical Structures, D. Reidel Publishing Company, Dortrecht, 1986.

16. K. Piszczek and J. Niziol, Random Vibrations of Mechanical Systems, Ellis Horwood Limited, Chichester, 1986.

17. Elishakoff, I. and Lyon R. H. (eds.), Random Vibration Status and Recent Developments, (The Stephen Harry Crandall Festschrift), Elsevier, Amsterdam, 1986.

MEASUREMENT OF CHARACTERISTICS OF STATIONARY RANDOM PROCESSES

I. Elishakoff
Naval Postgraduate School, Monterey, California, USA

This paper reviews measurement procedures for probabilistic characteristics of stationary random processes, such as the mean function, autocorrelation function and spectral density.

1. UNBIASED, CONSISTENT AND EFFICIENT ESTIMATORS

Let us consider first the case of the random variable. We need a random sample from the population of all possible values of such a variable X, and the process of obtaining it is visualized as an experimental sequence of observations where each observed value is independent of the others, while all elements of the population have - throughout the experiment - the same probability of turing up. Thus a random sample of size n contains n independent items $X_1, X_2, \ldots, X_n$ with identical distribution F(x).

The extimator $\hat{g}$ of a parameter g of X is a random variable associated with a random sample $g=g(X_1,X_2,\ldots X_n)$ and its quality is judged in terms of the following two properties: it is said to be unbiased if

$$E[\hat{g}]=g \quad , \tag{1}$$

(i.e., the parameter to be established should equal the expected values of the estimator) and is said to be consistent if

$$\lim_{n\to\infty} P[\hat{g}-g]<\varepsilon]=1 \quad , \tag{2}$$

(i.e., $\hat{g}$ converges in probability to g).

Cramer [1] shows that the sample mean

*On leave from the Technion I.I.T., 32000 Haifa, Israel.

$$\hat{X} = (X_1+X_2+\ldots+X_n)/n \tag{3}$$

and sample variance

$$S^2 = \frac{1}{n-1}\sum_{i=1}^{n}(X_i-\hat{X})^2 \quad , \tag{4}$$

are both unbiased and consistent. By contrast the apparent sample variance

$$\hat{\sigma}^2 = \frac{1}{n}\sum_{i=1}^{n}(X_i-\hat{X})$$

is inconsistent. To show this, let us add and substract the value of the true mean m_X, namely

$$\hat{\sigma}^2 = \frac{1}{n}\sum_{i=1}^{n}\left\{(X_i-m_X) - \frac{1}{n}\sum_{j=1}^{n}(X_j-m_X)\right\}^2 \quad ,$$

or

$$\hat{\sigma}^2 = \frac{1}{n}\sum_{i=1}^{n}(X_i-m_X)^2 - \frac{2}{n^2}\sum_{i=1}^{n}\sum_{j=1}^{n}(X_i-m_X)(X_j-m_X)$$

$$+\frac{1}{n^2}\left\{\sum_{j=1}^{n}(X_j-m_X)\right\}^2$$

The mathematical expectation of σ^2 is, due to independence of X_j

$$E[\hat{\sigma}^2] = \frac{1}{n}\sum_{i=1}^{n}E[(X_i-m_X)^2] - \frac{2}{n^2}\sum_{i=1}^{n}E[(X_i-m_X)^2]$$

$$+\frac{1}{n}\sum_{j=1}^{n}E[(X_j-m_X)^2] \quad ;$$

but

$$E[(X_i-m_X)^2] = E[(X_j-m_X)^2] = \sigma^2 \quad ,$$

so that finally

$$E[\hat{\sigma}^2] = E\left[\frac{1}{n}\sum_{i=1}^{n}(X_i-X)^2\right] = \frac{n-1}{n}\sigma^2 \quad .$$

Consequently, for the sample variance to be consistent we must take Eq. (4).

A third important property of an estimator is its efficiency. If two unbiased estimators $\hat{g}_1$ and $\hat{g}_2$ are compared, the one with smallest variance tends to be more efficient in the sense that it yields a better estimate of g for a given sample size. An unbiased estimator g is of minimum variance if $\text{Var}[\hat{g}] < \text{Var}[\hat{g}_1]$ where g_1 is any other unbiased estimator of g.

2. MEAN FUNCTION

Consider first the nonstationary case. Assuming that n realizations of the random function X(t) of the same length are available and fixing the time instance t, the estimate of the mean function is

$$\hat{m}_X(t) = [X_1(t) + X_2(t) + \ldots + X_n(t)]/n \quad . \tag{5}$$

It is readily shown that

$$E[\hat{m}_X(t)] = m_X(t), \quad \text{Var}[\hat{m}_X(t)] = R_X(t,t)/n \quad ,$$

i.e. the estimator is unbiased and consistent.

As regards the stationary case, the number of available realizations is usually small, and often there is only a single one. The question is then, whether time-averaging could be used instead of averaging over an ensemble, and in what circumstances. The answer is that such an approach is feasible for ergodic random processes, in which case we have an estimator

$$\hat{m}_X(t) = \frac{1}{T}\int_0^T X(t)dt \tag{6}$$

which is seen to be unbiased. It remains to check when it is consistent. We calculate $\text{Var}[\hat{m}_X(t)]$; we have (Ref. 2, Eq. 40.12)

$$\text{Var}[\hat{m}_X(t)] = \frac{2}{T}\int_0^T \left(1 - \frac{\tau}{T}\right) R_X(\tau)d\tau \quad . \tag{7}$$

For $\text{Var}[\hat{m}_X(t)]$ tending to zero with increase of T, estimator (6) is consistent. In other words, consistency requires

$$\lim_{T\to\infty} \frac{2}{T} \int_0^T \left(1 - \frac{\tau}{T}\right) R_X(\tau)d\tau = 0 \tag{8}$$

which is a necessary and sufficient condition for ergodicity in the mean of the random process X(t) [Ref. 3, Eq. 9.110]. For example, if

$$R_X(\tau) = d^2\exp(-a|\tau|) \tag{9}$$

(see Eq. 22 in Ref. 4), then

$$\mathrm{Var}[\hat{m}_X(t)] = \frac{2d^2}{aT} \{1 - \frac{1}{aT} [1-\exp(-aT)]\} , \tag{10}$$

and the time interval needed for $\mathrm{Var}[\hat{m}_X(t)] \ll d^2$ is $aT \gg 1$. For example [5], is we specify $\mathrm{Var}[\hat{m}_X(t)] = 0.01\ d^2$, we have $T \approx 200/a$.

3. CORRELATION FUNCTIONS

Consider first the nonstationary case. If the mean function $m_X(t)$ is unknown, we use its estimate as per Eq. 5 and put, in analogy with Eq. (4),

$$\hat{R}_X(t_1,t_2) = \frac{1}{n-1} \sum_{i=1}^{n} [x_i(t_1) - \hat{m}_X(t_1)][x_i(t_2) - \hat{m}_X(t_2)] , \tag{11}$$

again with n-1 rather than n required for consistency.

As regards the stationary case, we consider, for generality a joint pair of processes X(t) and Y(t)

$$\hat{R}_{XY}(\tau) = E\ \{[X(t) - \hat{m}_X][Y(t + \tau) - \hat{m}_Y]\} ; \tag{12}$$

for the estimator $R_{XY}(\tau)$ we can use the results of the preceding section in treating the mean of the stationary function

$$Z(t) = [X(t) - m_X][Y(t + \tau) - m_Y] \tag{13}$$

Eq. (6), with the time interval $[0,T-\tau]$, yields

$$\hat{R}_{XY}(\tau) = \frac{1}{T-\tau} \int_0^{T-\tau} [X(t)-\hat{m}_X][T(t + \tau) - \hat{m}_Y]dt \tag{14}$$

m_X and m_Y are usually unknown and estimated with the aid of Eq. (6).

where

$$\xi^2 = \lim_{T\to\infty} \frac{1}{T} \int_0^T X^2(t)dt, \quad \xi > 0 \quad . \tag{21}$$

The random variables A and B are jointly uniformly distributed on the unit circle $A^2+B^2 \leq 1$, and "p" is a control parameter.

Ref. 7 gives the following exact solution of Eqs. (19)-(21) for the mean and mean-square values of X(t):

$$E[X(t)] = \frac{1}{3p^2} [12p - 16 - (5p - 8)(4 - p)^{1/2}] \quad ,$$

$$E[X^2(t)] = \frac{1}{2p} [1 - \frac{2(p-2)}{p} \log (1 - \frac{p}{4})] \quad , \tag{22}$$

whereas the solution based on the assumption of ergodicity of X(t) reads

$$E[X(t)] = \frac{8-(4-p)^{3/2}}{p(8-p)} \, , \; E[X^2(t)] = \frac{3}{2(8-p)} \quad . \tag{23}$$

Comparison of Eqs. (22) and (23) shows that the approximate and exact solutions coincide for $p=2$ and are very close in the internal $0 \leq p \leq 2$, with an error of about half percent. However, as the control parameter p appraches four, the error due to the ergodicity assumption increases and reaches 25% for the mathematical expectation and 100% for the mean-square value. Another such example, demonstrating the need for extreme caution in using the ergodicity assumption, is given in Ref. 8. Usually however, (in "nonpathological" cases,) utilization of this assumption appears justified [9].

4. SPECTRAL DENSITY

A natural immediate reaction to this problem is to use the Fourier transform of the estimator of the autocorrelation function:

$$\hat{S}_X(\omega) = \frac{1}{2\pi} \int_{-\infty}^{\infty} \hat{R}_X(\tau)e^{-i\omega\tau}d\tau = \frac{1}{\pi} \int_{-\infty}^{\infty} \hat{R}_X(\tau)\cos \omega\tau d\tau \quad . \tag{24}$$

However, $\hat{R}_X(\tau)$ is usally known for some limited time interval (-T,T), and the calculation yields accordingly

$$\hat{S}_X(\omega,T) = \frac{1}{2\pi}\int_{-T}^{T} R_X(\tau)e^{-i\omega\tau}d\tau \quad , \tag{25}$$

which may lead to considerable distortion of the spectral density pattern at low frequencies. Indeed, consider the example

$$\hat{R}_X(\tau) = \exp(-a|\tau|) \quad .$$

Here we have

$$E[\hat{S}_X(\omega)] - S_X(\omega) = \frac{1}{\pi}\int_T^{\infty} e^{-a\tau}\cos\omega\tau d\tau \quad , \tag{26}$$

and recalling Eq. (22) of Ref. 4, we obtain [5]

$$E[\hat{S}_X(\omega)] - S_X(\omega) = \frac{1}{\pi}\frac{a}{a^2+\omega^2}[1-e^{-aT}(\cos\omega T\frac{\omega}{a} - \sin\omega T)] \quad . \tag{27}$$

Calculation shows substantial differences between the mean of the estimator and the true spectral density. In addition, it is seen that $E[\hat{S}_X(\omega)]$ may take on negative values, whereas the true spectral density is nonnegative.

Consider now the standard method of Blackman-Tukey [10]. A continuous signal x(t) with zero mean value, is sampled at equal intervals, Δt, to yield a finite series x_n for n=0,1,2..., N-1. Δt must be so chosen that the maximum frequency is the record be $f_{max}=1/2\Delta t$,Hz. We use, thus, the following expression for the autocorrelation function, instead of Eq. (15):

$$\hat{R}_r = \frac{1}{N-r}\sum_{n=1}^{N-r} x_n x_{n+r} \; , \quad (r=0,1,\ldots,m) \quad , \tag{28}$$

where r is the lag number, m its maximum, and $\hat{R}_r$ the estimate of the true value R_r at lag r, corresponding to time τ rΔt. Since this autocorrelation function is meaningless for $\tau > \tau_{max} = n\Delta t$, it must be replaced by another function suitable for any τ, namely,

$$\tilde{R}(\tau) = R_r D(\tau) \tag{29}$$

with $D(\tau)$ satisfying the following requirements

$$D(0) = 1;\; D(\tau) = 0 \text{ for } \tau > \tau_{max} \tag{30}$$

An example of $D(\tau)$ is the lag window, an application known as "Hamming":

$$D(\tau) = [0.054 + 0.46\cos(\pi\tau/\tau_{max})]U(\tau - \tau_{max}) \tag{31}$$

where U is a unit step function. For further discussion of this subject, consult Blackman and Tukey [10], Priestley [11], and Robson et al. [12].

Use of the FFT (Fast Fourier Transform) for estimating the autocorrelation function and spectral density is discussed by Stockham [13], Welch [14], Rader [15] and others; see also the monograph by Bendat and Piersol [6].

ACKNOWLEDGEMENT

The work was support by the Naval Postgraduate School Research Foundation, and Office of Naval Research is greatfully appreciated.

REFERENCES

1. Cramer, H., Mathematical Methods in Statistics, Princeton University Press, Princeton, New Jersey, 1946.
2. Sweschnikoff, A. A., Untersuchungsmethoden der Theorie der Zuffallsfunktionen mit praktischen Anwendungen, B. G. Teubner Verlagsgesellschaft, Leipzig, 1965.
3. Papoulis, A., Probability, Random Variables, and Stochastic Processes, McGraw-Hill, New York, 1965.
4. Elishakoff, I., Correlation and Spectral Analysis - a Brief Outline, in this volume.
5. Livshitz, N. A. and Pugachev, V. N., Probabilistic Analysis of the Systems of Automatic Control, "Soviet Radio" Publishing House, Moscow, 1963.
6. Bendat, J. S. and Piersol, A. G., Random Data: Analysis and Measurement Procedures, Wiley-Interscience, New York, 1971 (revised edition, 1986).

7. Scheurkogel, A. J., Elishakoff, I. and Kalker, J. J., On the Error That Can be Induced by an Ergodicity Assumption, ASME Journal of Applied Mechanics, Vol. 48, 1981, pp. 654-657.

8. Elishakoff, I., Probabilistic Method in the Theory of Structures, Wiley Interscience, New York, 1983, Chapt. 8.

9. L. Arnold, Personal communication, 1983.

10. Blackman, R. B. and Tukey, J. W., The Measurement of Power Spectra from the Point of View of Communications Engineering, Dover Publications, New York, 1958.

11. Priestley, M. B., Basic Consideration in the Estimation of Spectra, Technometrics, Vol. 4, No. 1, 1962.

12. Robson, J. D., Dodds, C. J., MacLean, D. B. and Paling, V. R., Random Vibrations, Springer Verlag, Vienna, 1971, Chapt. 16.

13. Stockham, T. G., Jr., High-speed Convolution and Correlation, in Digital Signal Processing, Rabiner, L. R. and Rader, Ch.M., eds., IEEE Press, New York, 1972, pp. 330-334.

14. Welch, P. D., The Use of Fast Fourier Transform for the Estimation of Power Spectra: A Method Based on Time Averaging Over Short, Modified Periodograms, ibid, pp. 335-338.

15. Rader, Ch.M., An Improved Algorithm for High Speed Autocorrelation with Application to Spectral Estimation, ibid, pp. 339-341.

RANDOM VIBRATION OF MULTI-DEGREE-OF-FREEDOM SYSTEMS WITH ASSOCIATED EFFECT OF CROSS-CORRELATIONS

I. Elishakoff
Naval Postgraduate School, Monterey, California, USA

We first expound here the theory of response of linear systems to stationary random excitation. We then consider the random vibration of single- and multi-degree-of-freedom systems. In the latter case, special emphasis is placed on the role of cross-correlations in random vibration of discrete structures. This effect, which is almost always disregarded in literature, plays a significant role when the natural frequencies happen to be equal or very close to one another.

1. RESPONSE OF LINEAR TIME-INVARIANT SYSTEMS

Let the behaviour of a stable dynamical system be described by a differential equation with constant coefficients

$$P_n(D)X(t) = F(t) \tag{1}$$

where D is the differentiation operator, D=d/dt

$$P_n(D) = a_nD^n + a_{n-1}D^{n-1} +\ldots+ a_0 \tag{2}$$

and F(t) is weakly stationary random process. Now, as is seen from Eq. (39) [1] the spectral density of k-th derivative is

$$S_{D^{(k)}X}(\omega) = \omega^{2k}S_X(\omega) \tag{3}$$

Corresponding result for the differential expression in Eq. (1) [4,5] reads

$$|a_n(i\omega)^n + a_{n-1}(i\omega)^{n-1} +\ldots+ a_0|^2S_X(\omega) = S_F(\omega) \tag{4}$$

or

$$S_X(\omega) =| H(\omega)|^2S_F(\omega), \qquad H(\omega) = P_n^{-1}(i\omega) \tag{5}$$

---

*On leave from Technion - I.I.T., 32000 Haifa, Israel

where

$$P_n(i\omega) = a_n(i\omega)^n + a_{n-1}(i\omega)^{n-1} + \ldots + a_0 \tag{6}$$

and $H(\omega)$ is the frequency response function. It is determined by measuring the steady-state response when the excitation is a unit sine wave of that frequency.

Now, consider instead of (1) a general case

$$P_n(D)X(t) = Q_m(D)Y(t) \tag{7}$$

where $P_n(D)$ and $Q_m(D)$ are the n-th and m-th order polynomials respectively, and $Y(t)$ is a weakly stationary random process. Then, the left-hand expression has a spectral density $|P_n(i\omega)|^2 S_X(\omega)$ and the right-hand one $|Q_m(i\omega)|^2 S_Y(\omega)$. Therefore

$$|P_n(i\omega)|^2 S_X(\omega) = |Q_m(i\omega)|^2 S_Y(\omega) \tag{8}$$

and

$$S_X(\omega) = \frac{|Q_m(i\omega)|^2}{|P_n(i\omega)|^2} S_Y(\omega) \tag{9}$$

with the corresponding frequency response function

$$H(\omega) = \frac{Q_m(i\omega)}{P_n(i\omega)} \tag{10}$$

Consider a single-degree-of-freedom system subjected to a foundation excitation $Y(t)$ producing a response $X(t)$. The differential equation of motion is

$$m\ddot{X} + c\dot{X} + kX = c\dot{Y} + kY \tag{11}$$

which can be rewritten as

$$m\ddot{X} + c\dot{X} + kX = \gamma k\dot{Y} + \delta kY \tag{12}$$

where the control parameters γ and δ are introduced so as to cover the forced vibration problem too: $\gamma=\delta=1$ we revert to the foundation excitation, whereas for $\gamma=0$, $\delta=1/k$ we have the problem of response due to the random excitation $Y(t)$.

For the frequency response function we obtain

$$H(\omega) = \frac{Q_1(i\omega)}{P_2(i\omega)} = \frac{\gamma c i\omega + \delta k}{m(i\omega)^2 + c(i\omega) + k} \quad , \tag{13}$$

or with notation

$$\omega_0{}^2 = \frac{k}{m} , \qquad \zeta = \frac{c}{2m} \quad , \tag{14}$$

where ω_0 is the natural frequency and the viscous damping factor:

$$H(\omega) = \frac{2\gamma\zeta i\omega_0\omega + \delta\omega_0{}^2}{\omega_0{}^2 - \omega^2 + 2\zeta i\omega_0\omega} \quad . \tag{15}$$

Y(t) is assumed to be band-limited white noise with intensity S_0, and ω_c - a cut-off frequency. For the mean-square response we have

$$E(X)^2 = S_0 \int_{-\omega_c}^{\omega_c} \frac{(4\omega^2\zeta^2\gamma^2\omega_0{}^2 + \delta^2\omega_0{}^4)d\omega}{(\omega_0{}^2 - \omega^2)^2 + 4\zeta^2\omega_0{}^2\omega^2} \quad , \tag{16}$$

where

$$\omega_0{}^2 = (k/m)^{1/2}, \qquad \zeta = c/2m\omega_0 \quad , \tag{17}$$

ω_0 is the natural frequency, and ζ the viscous damping factor.

Lengthy algebraic manipulations, which are omitted here, yield:

$$E(X^2) = S_0\omega_0\left(\frac{\delta^2}{2} - 2\gamma^2\zeta^2\right) \frac{1}{2\sqrt{1-\zeta^2}} \ \ln \frac{\omega_c/\omega_0 + \sqrt{1-\zeta^2} + \zeta^2}{\omega_c/\omega_0 - \sqrt{-\zeta^2} + \zeta^2}$$
$$+ S_0\omega_0\left(\frac{\delta^2}{2} + 2\gamma^2\zeta^2\right) \frac{1}{\zeta} \left[\tan^{-1}\left(\frac{\omega_c/\omega_0 + \sqrt{1-\zeta^2}}{\zeta}\right) + \tan^{-1}\left(\frac{\omega_c/\omega_0 - \sqrt{1-\zeta^2}}{\zeta}\right)\right] \quad . \tag{18}$$

For ω_c tending to infinty, the excitation appraches ideal white noise. In the case of foundation excitation Eq. (18) reduces to Weidenhammer's (Ref. 2, Eq. 36) and Piersol's (Ref. 3, Eq. 94) results

$$E(X^2) = \pi S_0\omega_0(2\zeta + 1/2\zeta) \quad . \tag{19}$$

In the case of forced excitation, $\gamma=0$, $\delta=1/k$ Eq. (18) reduces to Elishakoff's (Ref. 4, Eq. 9.55) result

$$E(X^2) = \frac{S_0 \pi}{2m^2 \zeta \omega_0^3} I_0 (\frac{\omega_c}{\omega_0}, \zeta) = \frac{\pi S_0}{kc} I_0 (\frac{\omega_c}{\omega_0}, \zeta) \tag{20}$$

where the factor preceding I_0 is nothing else but the response of a single-degree-of-freedom system to ideal white noise excitation, and the integral factor I_0 approaches unity with $\omega_c/\omega_0 \to \infty$. For $\omega_c/\omega_0 < 1$ this factor is always less than unity. However, for sharply resonant systems (low damping, $\zeta \ll 1$) this factor is only slightly less than unity, providing $\omega_c > \omega_0$. This implies that, although the ideal white noise is a mathematical "fiction", use of the ideal white noise hypothesis for quick evaluation would be justified under the foregoing circumstances.

2. MULTI-DEGREE-OF-FREEDOM SYSTEMS

Consider a linear stable system with n degrees of freedom. The associated equations of motion are

$$[m]\{\ddot{X}\} + [c]\{\dot{X}\} + [k]\{X\} = \{F\} \quad , \tag{21}$$

where m, c and k are constant (n x n) symmetric matrices called the inertia, damping and stiffness matrices, respectively. The n-dimensional vector $\{F(t)\}$ represents a real, joint stationary random vector process with given spectral density matrix

$$[S_{FF}(\omega)] = [S_{F_r F_q}(\omega)]_{nxn} \quad . \tag{22}$$

Our aim is to find the spectral density matrix $[S_{XX}(\omega)]$ of the response vector process $\{X(t)\}$

$$[S_{XX}(\omega)] = [S_{X_r X_q}(\omega)] \quad . \tag{23}$$

The natural frequencies of the system are denoted by ω_r $(r=1,2,\ldots,n)$ the corresponding normal modes are denoted by $\{v_r\}$, so that

$$([c] - \omega_r^2[m])\{v_r\} = \{0\} \tag{24}$$

The normal modes satisfy the following equations

$$\{v_r\}^T[m]\{v_q\} = \delta_{rq} \;, \qquad \{v_r\}^T[k]\{v_q\} = \omega_r^2\,\delta_{rq} \;, \tag{25}$$

where δ_{rq} are Kronecker deltas. The modal matrix is then written as

$$[v] = [v_1 \quad v_2 \quad \ldots v_n]$$

so that Eq. (25) can be rewritten as

$$[v]^T[m][v] = [I], \qquad [v]^T[k][v] = [\Omega^2] \;, \tag{27}$$

with $\Omega^2_{rq} = \omega^2_r \delta_{rq}$.

Now the transformation

$$\{X\} = [v]\;\{Y\} \tag{28}$$

yields the independent set of equations for the natural coordinates

$$\{\ddot{Y}\} + [u]\;\{\dot{X}\} + [\Omega]\;\{X\} = [v]^T\{F(t)\} \;, \tag{29}$$

with

$$[u] = [v]^T[c][v] \;. \tag{30}$$

Now we assume that proportional damping is the case under consideration, i.e.

$$[c] = \alpha[m] + \beta[k] \;, \tag{31}$$

where α and β are real constants. Then

$$[u] = \alpha[I] + \beta[\Omega] = 2[\zeta_r\omega_r] \;, \tag{32}$$

so that finally Eq. (29) becomes

$$\ddot{Y}_r + 2\zeta_r\omega_r\dot{Y}_r + \omega_r^2 Y_r = \sum_{j=1}^{n} u_{jr}F_j = Z_r(t) \;. \tag{33}$$

Then (see Elishakoff [4], formula 9.16) we have

$$[S_X(\omega)] = [v][H(\omega)][S_Z(\omega)][H(\omega)]^*[v]^T \;, \tag{34}$$

where $[H(\omega)]$ is a diagonal matrix with

$$H_r(\omega) = 1/(\omega_r^2 - \omega^2 + 2i\zeta_r\omega_r\omega) \;, \qquad r = 1,2,\ldots,n \;. \tag{35}$$

Then, due to Wiener-Khintchine relationships, we get for the crosscorrelation matrix

$$[R_X(\tau)] = [v] \int_{-\infty}^{\infty} [H(\omega)][S_Z(\omega)][H(\omega)]^* e^{i\omega\tau} d\omega [v]^T \; . \tag{36}$$

At $\tau=0$ we get

$$[R_X(0)] = [v] \int_{-\infty}^{\infty} [H(\omega)][H(\omega)]^* d\omega [v]^T \quad . \tag{37}$$

As is shown in Ref. 4, this expression can be put as

$$E(X_r X_q^*) = \sum_{j=1}^{n} v_{rj} v_{qj} E_{jj} + \sum_{j=1}^{n} \sum_{\substack{k=1 \\ j \neq k}}^{n} v_{rj} v_{qk} E_{jk} \quad , \tag{38}$$

where

$$E_{jk} = \int_{-\infty}^{\infty} S_{Z_j Z_k}(\omega) H_J(\omega) H_k^*(\omega) d\omega \quad . \tag{39}$$

The first sum in Eq. (38) is interpreted as the result of interaction of identical modes, and is called modal autocorrelation; the second sum is the result of interaction of different modes, and is called modal cross-correlation. Consider, as an example a two-degree of freedom system governed by differential equations with the following matrices

$$[m] = \begin{bmatrix} m & 0 \\ 0 & 2\varepsilon m \end{bmatrix}, \qquad [c] = \begin{bmatrix} c(1+\varepsilon_1) & -c\varepsilon_1 \\ -c\varepsilon_1 & 2c\varepsilon \end{bmatrix} ,$$

$$[k] = \begin{bmatrix} k(1+\varepsilon_1) & -k\varepsilon_1 \\ -k\varepsilon & 2k\varepsilon \end{bmatrix}, \qquad F(t) = \begin{Bmatrix} F_1(t) \\ 0 \end{Bmatrix} , \tag{40}$$

where ε_1 and ε_2 are some nonnegative coefficients, $2\varepsilon=\varepsilon_1+\varepsilon_2$ and $F_1(t)$ is an ideal white noise with intensity S_0. When $\varepsilon_1=\varepsilon_2=1$ we have the system described by example 9.14 in the book by Meirovitch [5]. The natural frequencies are

$$\omega_{1,2}^2 = [2+\varepsilon_1 \pm \varepsilon_1 \sqrt{1+2\varepsilon^{-1}}] \; (k/2m) \quad . \tag{41}$$

The elements of the modal matrix read

$$v_{11} = 2\lambda_1 \quad , \qquad v_{12} = 2\lambda_2 \quad ,$$
$$v_{21} = \lambda_1(1-\sqrt{1+2\varepsilon^{-1}}) \quad , \qquad v_{22} = \lambda_2(1+\sqrt{1+2\varepsilon^{-1}}) \quad , \tag{42}$$

where

$$\lambda_1 = [2+\varepsilon-\varepsilon\sqrt{1+2\varepsilon^{-1}}]^{-1/2}/2\ \ m \quad ,$$
$$\lambda_2 = [2+\varepsilon+\varepsilon\sqrt{1+2\varepsilon^{-1}}]^{-1/2}/2\ \ m \quad . \tag{43}$$

For the viscous damping coefficients we have

$$\zeta_{1,2} = \frac{c}{2\ 2mk}(2+\varepsilon_1 \pm \varepsilon_1\sqrt{1+2\varepsilon^{-1}})^{1/2} \tag{44}$$

The calculations according to Eq. (38) yield

$$E(X_1{}^2) = \frac{4\pi S_o}{kc}\{[2+\varepsilon_1+\varepsilon_1\sqrt{1+2\varepsilon^{-1}}]^{-2}[2+\varepsilon-\varepsilon\sqrt{1+2\varepsilon^{-1}}]^{-2}$$
$$+ [2+\varepsilon_1-\varepsilon_1\sqrt{1+2\varepsilon^{-1}}]^{-2}[2+\varepsilon+\varepsilon\sqrt{1+2\varepsilon^{-1}}]^{-2} \tag{45}$$
$$+ \Psi(4+4\varepsilon_1-2\varepsilon_1{}^2\varepsilon^{-1})^{-1}(4+2\varepsilon)^{-1}\} \quad ,$$

where

$$\Psi = 2\bar{c}[\bar{c} + \frac{\varepsilon_1{}^2(\varepsilon+2)}{(2\varepsilon+2\varepsilon\varepsilon_1-\varepsilon_1{}^2)(2+\varepsilon_1)}]^{-1} \quad , \ \bar{c} = \frac{c^2}{km} \quad . \tag{46}$$

The first two terms in Eq. (45) represent the modal autocorrelation, whereas the third represents the contribution of modal cross-correlation. For $\varepsilon_1 \to 0$ we have the result for the single-degree-of-freedom system under ideal, white noise excitation (Eq. 20):

$$E(X_1{}^2) = \frac{S_o}{kc} \quad . \tag{47}$$

It is interesting to calculate the error introduced by omitting the cross-correlation term in the analysis:

$$\eta_1 = [1 - E(X_1{}^2)_{approximate}/E(X_1{}^2)_{exact}]\cdot 100\% \ . \tag{48}$$

For $\varepsilon_1 \to 0$ this error equals

$$\lim_{\varepsilon_1 \to 0} \eta_1 = \frac{200}{4+\varepsilon_2} \quad \text{percent} \quad . \tag{49}$$

So that for $\varepsilon_2=1$, we have a 40% error. For $\varepsilon_1=\varepsilon_2=1$, the case considered by Meirovich, the error is about 3%, which justifies omission of cross-correlation in Ref. 5.

Analogically,

$$E(X_2^{\,2}) = \frac{\pi S_O}{kc}\,[2+\varepsilon_1+\varepsilon_1\sqrt{1+2\varepsilon^{-1}}]^{-2}\,[2+\varepsilon-\varepsilon\sqrt{1+2\varepsilon^{-1}}]^{-2}\cdot$$
$$\cdot[1-\sqrt{1+2\varepsilon^{-1}}]^{2} + [2+\varepsilon-\varepsilon_1\sqrt{1+2\varepsilon^{-1}}]^{-2}[2+\varepsilon+\varepsilon\sqrt{1+2\varepsilon^{-1}}]^{-2}\cdot \tag{50}$$
$$\cdot[1+\sqrt{1+2\varepsilon^{-1}}]^{2} - \Psi(4+4\varepsilon_1-2\varepsilon_1^{\,2}\varepsilon^{-1})^{-1}(4+2\varepsilon)^{-1}(2\varepsilon^{-1}) \quad .$$

Now

$$\lim_{\varepsilon_1 \to 0} E(X_2^{\,2}) = 0$$

and correspondingly, the error η_2 on omitting the cross-correlation term here tends to infinity (!), since $E(X_2^2)_{\text{exact}} = 0$, and $E(X_2^2)_{\text{approximate}} \neq 0$.

This example indicates that extreme caution should be exercised in omitting the cross-correlations in random vibration analysis.

3. RANDOM VIBRATION IN SYSTEM WITH EQUAL NATURAL FREQUENCIES

Consider now the dynamical system described by the following differential equations

$$m\ddot{X}_1 + c\dot{X}_1 + kX_1 + k(X_1-X_2) + k(X_1-X_3) + k(X_1-X_4) = F_1 \quad ,$$
$$m\ddot{X}_2 + c\dot{X}_2 + kX_2 + k(X_2-X_1) + k(X_2-X_3) + k(X_2-X_4) = 0 \quad ,$$
$$m\ddot{X}_3 + c\dot{X}_3 + kX_3 + k(X_3-X_1) + k(X_3-X_2) + k(X_3-X_4) = 0 \quad , \tag{52}$$
$$mX_4 + c\dot{X}_4 + kX_4 + k(X_4-X_1) + k(X_4-X_2) + k(X_4-X_3) = 0 \quad ,$$

$F_1(t)$ being an ideal white noise with intensity S_o. The natural frequencies of the system are

$$\omega_1^2 = k/m, \qquad \omega_2^2=\omega_3^2=\omega_4^2=5k/m \quad . \tag{53}$$

The modal analysis yields the following formula for the mean-square displacement of the first mass

$$E(X_1^2) = \sum_{r=1}^{4} \sum_{q=1}^{4} R_{rq} \quad , \tag{54}$$

where R_{rr} describes the contribution of autocorrelation and R_{rq}, $r \neq q$ - that of cross-correlation. Calculations give

$$s = \pi S_o/cm, \qquad R_{11} = s/16, \qquad R_{22} = s/4,$$

$$R_{33} = s/144, \qquad R_{44} = s/36$$

$$R_{12} = R_{21} = (1 + \frac{4}{3} g^{-1})^{-1} \; s/8 \quad ,$$

$$R_{13} = R_{31} = (1 + \frac{4}{3} g^{-1}) \; s/48 \quad , \tag{55}$$

$$R_{14} = R_{41} = (1 + \frac{4}{3} g^{-1}) \; s/24 \quad ,$$

$$R_{23} = R_{32} = s/24, \qquad R_{24} = R_{42} = s/12, \qquad R_{34} = R_{43} = s/72, \; g=km/c^2 \; .$$

As is seen the contributions of R_{14}, R_{41}, R_{13}, R_{31}, R_{12}, R_{21} are negligible, the corresponding natural frequencies being far apart. However, R_{23}, R_{32}, R_{24}, R_{42}, R_{34} and R_{43}, corresponding to the equal natural frequencies, must be taken into account. Therefore

$$E(X_1^2)_{\text{exact}} = 0.625s \quad . \tag{56}$$

Omitting the cross-correlation results in

$$E(X_1^2)_{\text{approximate}} = \sum_{r=1}^{4} R_{rr} = 0.347s \tag{57}$$

the associated error reaching 44.5%.

ACKNOWLEDGEMENT

The support by the Naval Postgraduate School Research Foundation and the Office of Naval Research are gratefully appreciated.

REFERENCES

1. Elishakoff, I., Correlation and Spectral Analysis - a Brief Outline, in this volume.
2. Weidenhammer, F., Das Schwingungsfundament unter Zufallserregung, Ingenieur-Archiv, Vol. 31, 1962, pp. 433-443.
3. Piersol, A. G., The Measurement and Interpretation of Ordinary Power Spectra for Vibration Problems, NASA CR-90, 1964.
4. Elishakoff, I., Probabilistic Methods in the Theory of Structures, Wiley-Interscience, 1983, Chap. 9, Eq. 9.55.
5. Meirovitch, L., Elements of Vibration Analysis, McGraw-Hill, New York, 1975.
6. Lubliner, E. and Elishakoff, I., Random Vibration of System with Fintely Many Degrees of Freedom and Several Coalescent Natural Frequencies, International Journal of Engineering Science, Vol. 23, 1986, pp. 461-470.
7. Elishakoff, I., A Model Elucidating Significance of Cross-correlations in Random Vibration Analysis, in Random Vibration-Status and Recent Developments (The Stephen Harry Crandall Festschrift, Elishakoff, I. and Lyon, R. H., eds.) Elsevier, Amsterdam, 1986, pp. 101-112.

WIDE-BAND RANDOM VIBRATION OF CONTINUOUS STRUCTURES WITH ASSOCIATED EFFECT OF CROSS-CORRELATIONS

I. Elishakoff
Naval Postgraduate School, Monterey, California, USA

The paper deals with wide-band random vibration of continuous structures due to random space-time excitation. The normal mode method is described. Special attention is devoted to the question as to when the modal cross-correlations can be neglected.

1. DERIVATON OF BASIC RELATIONS

The structures treated here are taken to be linear, time-invariant dynamical systems of finite extent (for infinite structures, see e.g. Ref. 1). We will concentrate on the one-dimensional case, with the structure extending along a single spatial axis x, from x=0, to x=L. Generalization to two- or three-dimensional cases presents no difficulties in principle, but calls for additional notation. Let the displacement u(x,t) satisfy the following differential equation.

$$L(\partial/\partial x)u + 2\varepsilon\rho u_{,t} + \rho u_{,tt} = f(x,t) \quad , \tag{1}$$

where $L(\partial/\partial x)$ is a spatial operator, ρ mass density, ε a damping coefficient, f(x,t) the excitation function.

Eq. (1) is supplemented by the following (deterministic) initial and boundary conditions, respectively,

$$u(x,t_o) = u_o(x),\ u(x,t_o)_{,t} = u_o(x) \tag{2}$$

$$M_\alpha(\partial/\partial x)u(x,t) = 0 \ ,\ \alpha=1,2,\ldots \quad ;\ t\geq t_o \tag{3}$$

prescribed either at x=0 or at x=L.

Within the correlation theory, we will describe the random space-time

---

*On leave for Technion, I.I.T., 32000 Haifa, Israel.

process u(x,t) in terms of the mean

$$E[u(x,t)] = m_u(x,t) \tag{4}$$

and the space-time autocovariance function

$$E\{[u(x_1,t_1) - m_u(x_1,t_1)][u(x_2,t_2) - m_u(x_2,t_2)]\} = C_u(x_1,t_1;x_2,t_2) \ . \tag{5}$$

We will consider in this study only space/time random processes which are weakly stationary in time, i.e. in which the first- and second-order probability densities are independent of translation of the time origin. Therefore,

$$E[u(x,t)] = m_u(x) \ , \tag{6}$$

$$E\{[u(x_1,t) - m_u(x_1)][u(x_2,t+\tau) - m_u(x_2)]\} = C_u(x_1,x_2,\tau) \ . \tag{7}$$

Due to the weak stationarity of u(x,t), we can define the cross-spectral density function $S_u(x_1,x_2,\omega)$ through the Wiener-Khintchine relationships

$$C_u(x_1,x_2,\tau) = \int_{-\infty}^{\infty} S_u(x_1,x_2,\omega)e^{i\omega\tau}d\omega \ , \tag{8}$$

$$S_u(x_1,x_2,\omega) = \frac{1}{2\pi}\int_{-\infty}^{\infty} C_u(x_1,x_2,\tau)e^{-i\omega\tau}d\tau \ . \tag{9}$$

Our problem will be defined as follows: knowing the probabilistic characteristics of the weakly stationary excitation f(x,t), find those of the response u(x,t), dispensing with the initial conditions (2).

Central among the different methods of solution of stochastic boundary value problems is the normal-mode method, which will be described as follows. Consider first the free-vibration problem:

$$L(\partial/\partial x)u + \rho\, u_{,tt} = 0 \ , \tag{10}$$

for which we can put

$$u = U(x)\exp(i\omega t) \ . \tag{11}$$

Substitution in Eq. (1) and (3) yields

$$L(d/dx)U - \rho\omega^2 U = 0 \ , \ M_\alpha(d/dx)U = 0 \ , \ \alpha = 1,2,\ldots \ . \tag{12}$$

Denote by ω_i (i=1,2,...) the natural frequencies and $\psi_i(x)$ (i=1,2,...) the corresponding mode shapes

$$L(d/dx)\psi_i - \rho\omega_i^2\psi_i = 0, \qquad M_\alpha(d/dx)\psi_i = 0 \quad . \tag{13}$$

The modes shapes are orthogonal

$$(\rho\psi_i,\psi_j) = \int_0^L \rho\psi_i\psi_j dx = (\rho\psi_i,\psi_i)\delta_{ij} \quad , \tag{14}$$

where δ_{ij} is the Kronecker delta. Consequently

$$(L\psi_i,\psi_j) = \int_0^L L[\psi_i]\psi_j dx = \omega_i^2(\rho\psi_i,\psi_i)\delta_{ij} \quad . \tag{15}$$

Let us represent u(x,t) as a series in terms of mode shapes

$$u(x,t) = \sum_{i=1}^{\infty} U_i(t)\psi_i(x) \quad . \tag{16}$$

Substituting in Eq. (1) yields

$$\sum_{i=1}^{\infty} U_i(t)L[\psi_i] + \dot{U}_i(t)2\rho\varepsilon\psi_i + \ddot{U}_i(t)\rho\psi_i = f(x,t) \quad . \tag{17}$$

Multiplying this equation by $\psi_i(x)$, integrating over the length of the structure, and taking the orthogonality property (14) and Eq. (15), we have

$$\ddot{U}_i + 2\varepsilon U_i + \omega_i^2 U_i = Z_i(t) \quad , \quad i=1,2,... \quad , \tag{18}$$

where $Z_i(t)$ are the generalized forces

$$Z_i(t) = \frac{(f,\psi_i)}{(\rho\psi_i,\psi_i)} \quad . \tag{19}$$

The cross-variance function of the generalized forces $Z_i(t_1)$ and $Z_j(t_2)$ will be

$$C_{Z_iZ_j}(t_1,t_2) = E\{Z_i(t_1) - E[Z_1(t_1)]\ Z_j(t_2) - E[Z_j(t_2)]\}$$

$$= \frac{1}{(\rho\psi_i,\psi_i)(\rho\psi_j,\psi_j)} \int_0^L\int_0^L C_f(x_1,t_1;x_2,t_2)\psi_i(x_1)\psi_j(x_2)dx_1dx_2 \quad . \tag{20}$$

From Eq. (20) it follows that if $f(x,t)$ is weakly stationary in time, so are $C_{Z_iZ_j}$'s, that is

$$C_{Z_iZ_j}(\tau) = \frac{1}{(\rho\psi_i,\psi_i)(\rho\psi_j,\psi_j)} \int_0^L \int_0^L C_f(x_1,x_2,)\,\psi_i(x_1)\,\psi_j(x_2)dx_1dx_2 \quad . \tag{21}$$

Applying the Fourier transform to this equation, we have

$$S_{Z_iZ_j}(\omega) = \frac{I_{ij}(\omega)}{(\rho\psi_i,\psi_i)(\rho\psi_j,\psi_j)} \tag{22}$$

where $S_f(x_1,x_2,\omega)$ is the cross-spectral density and

$$I_{ij}(\omega) = \int_0^L \int_0^L S_f(x_1,x_2,\omega)\,\psi_i(x_1)\,\psi_j(x_2)dx_1dx_2 \tag{23}$$

is the cross-spectral density of the generalized forces in the modes i and j [2]. Note that $I_{ij}(\omega)$ is Hermitian, i.e. $I_{ij}(\omega) = I_{ji}{}^*(\omega)$.

Reverting to our original pattern, Eq. (18), we represent $Z_i(t)$ as the Fourier Integral

$$Z_i(t) = E[Z_i(t)] + \int_{-\infty}^{\infty} \tilde{Z}_i(\omega_1)e^{i\omega_1 t}d\omega_1 \quad ; \tag{24}$$

we find

$$C_{Z_iZ_j}(t_1,t_2) = \int_{-\infty}^{\infty}\int_{-\infty}^{\infty} E[Z_i{}^*(\omega_1)\tilde{Z}_j(\omega_2)]e^{i(\omega_2 t_2 - \omega_1 t_1)}d\omega_1 d\omega_2 \quad . \tag{25}$$

Due to weak stationarity

$$E[\tilde{Z}_i{}^*(\omega_1)\tilde{Z}_j(\omega_2)] = S_{Z_iZ_j}(\omega_1)\,\delta(\omega_2 - \omega_1) \quad , \tag{26}$$

and

$$C_{Z_iZ_j}(\tau) = \int_{-\infty}^{\infty} S_{Z_iZ_j}(\omega)e^{i\omega\tau}d\omega \quad . \tag{27}$$

Representing $U_i(t)$ in analogy Eq. (24), we have

$$\tilde{U}_i(\omega) = H_i(\omega)\tilde{Z}_i(\omega) \tag{28}$$

where

$$H_i(\omega) = 1/(\omega_i^2 - \omega^2 + 2\varepsilon i\omega) \quad , \tag{29}$$

(Note the similarity to Eq. (35) of Ref. 3). The stochastic orthogonality (26) yields then

$$S_{U_iU_j}(\omega) = H_i^*(\omega)H_j(\omega)S_{Z_iZ_j}(\omega) \quad . \tag{30}$$

Corresponding crosscorrelation function will be determined as

$$C_{U_iU_j}(\tau) = \int_{-\infty}^{\infty} H_i^*(\omega)H_j(\omega)S_{Z_iZ_j}(\omega)e^{i\omega\tau}d\omega \quad . \tag{31}$$

With Eq. (16) in mind, the autovariance function of u(x,t) will be

$$C_u(x_1,x_2,\tau) = \sum_{i=1}^{\infty}\sum_{j=1}^{\infty} C_{U_iU_j}(\tau)\psi_i(x_1)\psi_j(x_2) \quad , \tag{32}$$

with the corresponding cross-spectral density

$$S_u(x_1,x_2,\omega) = \sum_{i=1}^{\infty}\sum_{j=1}^{\infty} H_i^*(\omega)H_j(\omega)\frac{I_{ij}(\omega)}{(\rho\psi_i,\psi_i)(\rho\psi_j,\psi_j)}\psi_i(x_1)\psi_j(x_2) \quad , \tag{33}$$

and the displacement variance becomes

$$\mathrm{Var}[u(x,t)] = \sum_{i=1}^{\infty}\sum_{j=1}^{\infty}\frac{\psi_i(x)\psi_j(x)}{(\rho\psi_i,\psi_i)(\rho\psi_j,\psi_j)}\int_{-\infty}^{\infty} H_i^*(\omega)H_j(\omega)I_{ij}(\omega)d\omega \quad . \tag{34}$$

This is essentially Powell's [4] central result. The terms with $i=j$ are referred to as the <u>modal autocorrelations</u> and those with $i \neq j$, as the <u>modal cross-correlations</u>.

Note the special case when the modal cross-correlations vanish identically, and only the modal autocorrelations remain. Let f(t) be an ideal white noise in space, that is

$$C_f(x_1,x_2,\tau) = \delta(x_2-x_1)\psi(\tau) \quad . \tag{35}$$

The $C_{Z_iZ_j}(\tau)$ become, with Eq. (21)

$$C_{Z_iZ_j}(\tau) = (\psi_i,\psi_i)(\rho\psi_j,\psi_j)^{-2}\delta_{ij} \quad . \tag{36}$$

In this case the modal sums can be calculated in closed form. Such a calculation was performed for a string by Van Lear and Uhlenback [5], and for a beam by Eringen [6]. For this type of excitation a number of studies were recently performed by Wedig [7-9] using his covariance analysis method. Influence of the axial force was considered in the paper by Elishakoff [10].

Generally, however, the crosscorrelations do not vanish. We next take up the question of their contribution in such a general case.

2. EFFECT OF MODAL CROSS-CORRELATIONS

The effect of cross-correlations is almost totally neglected in the early literature on random vibrations of elastic structures; unfortunately, even some recent publications (see, e.g., book [11]) omit them.

To investigate the effect of cross-correlations, consider the case where the spacewise correlations is arbitrary, but f(x,t) is an ideal white noise in time

$$C_f(x_1,x_2,\tau) = 2\pi\ \chi(x_1,x_2)\ \delta(\tau) \quad . \tag{37}$$

Then the cross-spectral density becomes

$$S_f(x_1,x_2,\omega) = \frac{1}{2\pi}\int_{-\infty}^{\infty} C_f(x_1,x_2,)e^{-i\omega\tau}d\tau = \chi(x_1,x_2) \quad . \tag{38}$$

Consequently the cross-spectral densities (23) of the generalized forces are

$$I_{ij}(\omega) = \int_0^L\int_0^L \chi(x_1,x_2)\psi_i(x_1)\psi_j(x)dx_1dx_2 \equiv I_{ij} \tag{39}$$

and independent of ω. The integral term in Eq. (34) equals

$$\int_{-\infty}^{\infty} H_i^*(\omega)H_j(\omega)I_{ij}d\omega = \frac{8\pi\varepsilon I_{ij}}{(\omega_j^2-\omega_k^2)^2+8\varepsilon(\omega_j^2+\omega_k^2)} \quad . \tag{40}$$

Inspection of this expression shows that if the natural frequencies are not too close, then the terms with $i \neq j$ are smaller than those with $i=j$. The condition for omission of the crosscorrelation is [12]:

$$\omega_i \varepsilon \ll |\omega_i^2 - \omega_j^2| \quad ; \tag{41}$$

in the case where material damping is present instead of viscous damping, this condition is replaced by $\mu^2\omega_i^2 \ll |\omega_i^2 - \omega_j^2|$ as derived in Ref. 13.

In order to quantify the effect of the cross-correlations, consider the example [14] of axisymmetric random vibration of a cylindrical shell, simply supported at the ends; the operator $L(\partial/\partial x)$ in Eq. (1) reads:

$$L(\partial/\partial x) = D\partial^4/\partial x^4 + Eh/R^2 \tag{42}$$

where D is flexural rigidity, E Young's modulus, h thickness and R the radius. The boundary condition operators U_α in Eq. (3) are:

$$M_1(\partial/\partial x) = 1, \quad M_2(\partial/\partial x) = \partial^2/\partial x^2, \text{ at } (x = 0) ,$$

$$M_3(\partial/\partial x) = 1, \quad M_4(\partial/\partial x) = \partial^2/\partial x^2, \text{ at } (x = L) \quad . \tag{43}$$

Mode shapes are $\psi_i(x) = 2\sin(i\pi x/L)$ with following natural frequencies

$$\omega_j^2 = \frac{E}{\rho R^2} + \frac{Eh^2}{\rho L^4}\frac{(j\pi)^4}{12(1-\nu^2)} \quad . \tag{44}$$

The excitation is represented by a ring force, uniformly distributed along the circumference at x=a, so that

$$C_f(x_1,x_2,\tau) = \delta(x_1-a)\ \delta(x_2-a)\ \psi(\tau) \tag{45}$$

where $\psi(\tau)$ is the time-wise autocovariance function of the ring loading, taken as a band-limited white noise with cutoff frequency ω_c, so that the cross-spectral densities of the generalized forces become, using Eq. (23)

$$I_{ij}(\omega) = \int_0^1\int_0^1 \delta(x_1-a)\,\delta(x_2-a)\,[U(\omega+\omega_c) - U(\omega-\omega_c)]\,\psi_i(x_1)\,\psi_j(x_2)dx_1dx_2$$

$$= \psi_i(a)\,\psi_j(a)\,[U(\omega+\omega_c) - U(\omega-\omega_c)] \tag{46}$$

Let us seek the mean-square velocity, which coincides with the velocities variance for E[f(t)]=0. The final result is very similar to Eq. (34):

$$E[\dot{u}^2(x,t)] = \sum_{i=1}^{\infty}\sum_{j=1}^{\infty} \psi_i(x)\psi_j(x)\psi_i(a)\psi_j(a)\,\rho^{-2}\int_{-\omega_c}^{\omega_c} \omega^2 H_i^*(\omega)H_j(\omega)d\omega \quad . \tag{47}$$

It can be shown that in the above summation, a finite number of terms can be taken into account, corresponding to the number of natural frequencies N_c within the excitation band, i.e. $N_c=\max(j)$, so that $\omega_j \leq \omega_c$. Results of calculations through Eq. (47) are known in Figs. 1 and 2 reproduced from Ref. 14. For R/L ->∞ we have the case of a beam driven at the cross section x=a. The nondimensional mean-square velocity d_v^2 practically coincides with the sum associated with the modal autocorrelations (ξ denoting x/L). This is due to the fact that for a beam the natural frequencies are far apart and the modal cross-correlations can be neglected. An interesting corollary - first observed by Crandall and Wittig [15] - to the latter fact, is symmetry of the response about the midpoint of the beam, i.e. the same respone at x=a (the loaded section) as at x=L-a (loadless). For a shell with R/L=0.5, the situation is completely different (Fig. 2); the cross-correlations (denoted S_2) cannot be neglected any more, and their omission would yield a 60.8% error, for h/L=0.001; the symmetry characteristic of a beam is likewise absent.

ACKNOWLEDGMENT

The support by the Research Foundation of the Naval Postgraduate School and of the Office of Naval Research are gratefully appreciated.

REFERENCES

1. Pal'mov, V. A., The Correlation Properties of the Field of Vibrations of Plates Under a Broad-Band Random Load, Mechanics of Solids, Vol. 2, pp. 92-93, 1967.

2. Lin, Y. K., Probabilistic Theory of Structural Dynamics, McGraw-Hill, New York, 1967, p. 211 (second edition by R. E. Krieger Publishing Company, Malabar, FL, 1976).

3. Elishakoff, I., Random Vibration of Multi-degree-of-freedom Systems with Associated Effect of Cross-correlations, in this volume.

4. Powell, A., On the Fatigue Failure of Structures Due to Vibrations Excited by Random Pressure Fields, Journal of the Acoustical Society of America, Vol. 30, pp. 1130-1135, 1958.

5. Van Lear, G. A., Jr. and Uhjlenbeck, G. E., Brownian Motion of Strings and Elastic Rods, Physical Reviews, Vol. 38, pp. 1583-1598, 1931.

6. Eringen, A. C., Response of Beams and Plates to Random Loads, ASME Journal of Applied Mechanics, Vol. 24, pp. 46-52, 1957.

7. Wedig, W., Zufallschwingungen von querangestroemten Saiten, Ingenieur-Archiv, Vol. 48, pp. 325-335, 1979.

8. Wedig, W., Stationaere Zufallschwingungen von Balken-eine neue Methode zur Kovarianzanalyse, ZAMM, Vol. 60(6), pp. T.89-T91, 19.

9. Wedig, W., Typical Problems in Distributed Stochastic Systems and Results of the Integral Finite Element Method, Personal communication, 1981.

10. Elishakoff, I., Generalized Eringen problem: influence of axial force on a random vibration response of simply supported beam, Structural Safety, to appear, 1987.

11. Yang, C. Y., Random Vibration of Structures, Wiley-Interscience, New York 1985.

12. Bolotin, V. V., Statistical Methods in Structural Mechanics, Holden Day, Inc., San Francisco, 1969.

13. Elishakoff, I., On the Role of Cross-Correlations in the Random Vibrations of Shells, Journal of Sound and Vibration, Vol. 50, pp. 239-252, 1977.

14. Elishakoff, I. van Zanten, A. Th. and Crandall, S. H., Wide-Band Random Axisymmetric Vibration of Cylindrical Shells, ASME Journal of Applied Mechanics, Vol. 46, pp. 417-422, 1979.

15. Crandall, S. H. and Wittig, L. E., Chladni's Patterns for Random Vibration of a Plate, Dynamic Response of Structures, (Herrmann, G., and Perron, N., eds.), Pergamon, New York, pp. 55-71, 1972.

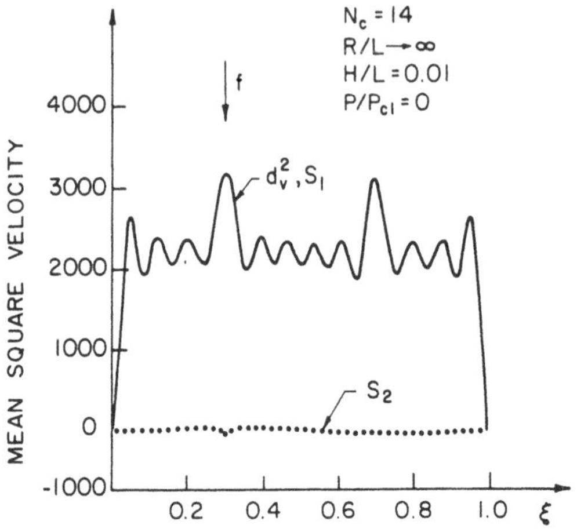
$N_c = 14$
$R/L \to \infty$
$H/L = 0.01$
$P/P_{cl} = 0$
f
d_v^2, S_1
S_2
MEAN SQUARE VELOCITY
4000
3000
2000
1000
0
-1000
0.2
0.4
0.6
0.8
1.0
ξ

Fig. 1

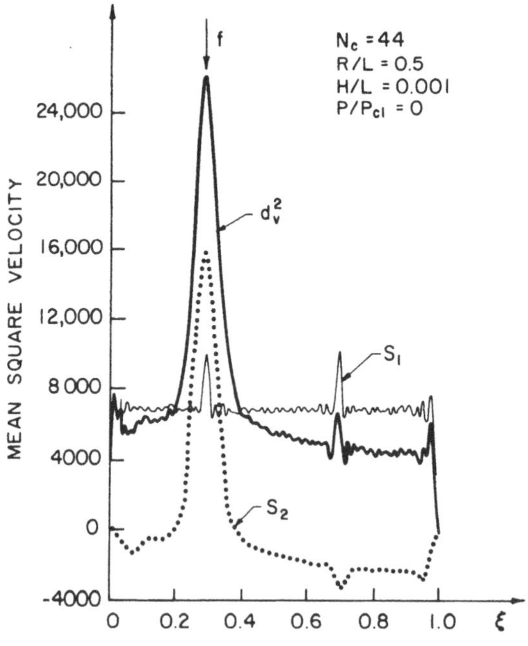
$N_c = 44$
$R/L = 0.5$
$H/L = 0.001$
$P/P_{cl} = 0$
f
d_v^2
S_1
S_2
MEAN SQUARE VELOCITY
24,000
20,000
16,000
12,000
8000
4000
0
-4000
0
0.2
0.4
0.6
0.8
1.0
ξ

Fig. 2

ANALYSIS OF NONLINEAR STOCHASTIC SYSTEMS

W. Kliemann
Iowa State University, Ames, Iowa, USA

TABLE OF CONTENTS

1. INTRODUCTION

The dynamics of a mechanical system are usually modeled as differential equations in Euclidian space $\mathbf{R}^d$ or on a smooth manifold M. For systems with lumped parameters and without memory an adequate model are ordinary differential equations (ODE) $\dot{x} = f(x,p)$, where the vector field f describes the directional field of the dynamics, depending on the systems parameters $p \in \mathbf{R}^m$. (Time dependent fields $f(x,t)$ fit into this framework by considering $(1, f(x,t)$ on $R \times M$.)

In general nonlinear ODE's are not explicitly solvable. The analysis of nonlinear mechanical systems therefore often involves the study of the qualitative behavior, followed by numerical computations. Qualitative theory includes concepts like invariant sets, limit sets, steady states, periodic solutions, stability, bifurcation, chaos, etc. Some of these concepts deal with the trajectories individually (like the first four mentioned above), while others compare the behavior of different trajectories, i.e. study the flows associated with the system.

If the system exhibits stochastic behavior, through parameter uncertainties, measurement errors, etc, an appropriate model is a stochastic differential equation (SDE) $\dot{x} = f(x,p,\xi)$, where ξ is a stochastic process on some probability space (Ω,F,P), describing the noise sources. An analysis of the stochastic system can be done using the same concepts as in the deterministic case. But since the solution $x(t)$ now is a random process, these properties have to be considered with respect to the measure P, e.g. almost sure, with positive probability, in p-th mean.

The purpose of this paper is two fold: First of all some techniques are outlined for the analysis of stochastic nonlinear mechanical systems that are modeled as SDE. These techniques include stochastic flows, associated deterministic control systems, ergodic theory, Lyapunov exponents, large deviations. Secondly we address the question: When does the behavior of a stochastic system differ considerably from the behavior of associated deterministic systems, e.g. the mean system or the "frozen" systems with fixed parameters $p \in \mathbf{R}^m$. We assume that the state and the parameters (either deterministic or their statistics in the random case) are known: The questions of identification and filtering are addressed in other papers in this volume.

In this first section we introduce some basic notation, consider a few examples and describe the range of problems to be solved in later sections.

1.1. DETERMINISTIC SYSTEMS

Throughout this paper systems are described as smooth ODE's

$$(1.1) \qquad \dot{x}(t) = X\big(x(t)\big)$$

on a paracompact, connected C^∞-manifold M of dimension d (in particular M may be the Euclidian space $\mathbb{R}^d$), where X is a C^∞ vector field on M. For background material on manifolds and vector fields consult the tutorial in this volume.

1. Example: The Linear System

Consider the linear ODE

$$(1.2) \qquad \dot{x}(t) = Ax(t) \qquad \text{in } \mathbb{R}^d.$$

For each initial value $x(0) = x_0 \in \mathbb{R}^d$ the solution is

$$x(t,x_0) = \Phi(t)x_0$$

where $\Phi(t) = e^{tA}$ is the fundamental matrix of (1.2). The spectral properties, i.e. the eigenvalues of A determine the behavior of (1.2): Let $\sigma_j = \lambda_j + i\mu_j$, $j = 1,\dots,d$ be the eigenvalues of A, then e.g.

- o (1.2) is exponentially stable (w.r.t. the origin $x \equiv 0$) iff $\max_j \lambda_j < 0$. Furthermore for each $\eta > \lambda$ there is a constant $c > 0$ such that

$$|x(t,x_0)| < c|x_0|e^{t\eta} \quad \text{for all} \quad t \geq 0, \quad \text{or}$$

$$\lim_{t\to\infty} \frac{1}{t}\log|x(t,x_0)| < \eta.$$

A specific example is the linear oscillator

$$\ddot{y} + 2b\dot{y} + (1+w)y = 0, \quad \text{or in the phase plane}$$

$$x = (x_1,x_2) = (y,\dot{y}) \in \mathbb{R}^2$$

$$\dot{x} = \begin{pmatrix} 0 & 1 \\ -1-w & -2b \end{pmatrix} x .$$

2. Example: A Periodic System

The motion of a helicopter blades during a flight can be descried by the linearized system

$$\ddot{z}(t) = C(t)\begin{pmatrix} \dot{z}(t) \\ z(t) \end{pmatrix} + H(t) ,$$

where $z \in \mathbb{R}^2$ describes position and torsion, $C(t)$ and $H(t)$ are 2×4 matrices with rotation period $T > 0$. Setting $x = (x_1, x_2) = (\dot{z}, z)$ we find

$$\dot{x}(t) = A(t)x(t) + F(t) , \tag{1.3}$$

where $x \in \mathbb{R}^4$, $A(t) = \begin{pmatrix} & C(t) & & \\ 1 & 0 & 0 & 0 \\ 0 & 1 & 0 & 0 \end{pmatrix}$, $F(t) = \begin{pmatrix} H(t) \\ 0 \\ 0 \end{pmatrix}$.

The periodicity and stability behavior of (1.3) follows from Floquet's theory: Let

$$\dot{y}(t) = A(t)y(t), \tag{1.4}$$

then the fundamental matrix $\Phi(t)$ of (1.4) has the form $\Phi(t) = P(t)e^{tR}$, where $P(t)$ is T-periodic and the eigenvalues $\rho_1, \ldots, \rho_d$ of e^{tR} are the characteristic roots of (1.4). The real numbers $\lambda_j = \frac{1}{T} \log|\rho_j|$ are uniquely determined by (1.4) and are called the characteristic exponents.

The system (1.3) now has for each T-periodic function $F(t)$ a periodic solution (with period T) if $\rho = 1$ is not a charateristic root of (1.4). If (1.3) has no periodic solutions with period T, then each solution is unbounded as $t \to \infty$ (resonance).

Setting $\Phi(t,s) = P(t)e^{(t-s)R}P^{-1}(s)$, we see that any solution of (1.3) with $x(0) = x_0$ is of the form

$$x(t,x_0) = \Phi(t,0)\left(x_0 + \int_0^t \Phi(0,s)F(s)ds\right),$$

while the periodic solution, in case it exists, is given by

$$x_p(t) = \Phi(t,0)[\mathrm{Id} - \Phi(T,0)]^{-1} \int_0^T \Phi(T,s)F(s)ds$$

$$+ \int_0^t \Phi(t,s)F(s)ds \, .$$

Therefore, if $\max_j \lambda_j < 0$, then all solutions of (1.3) converge towards the periodic solution.

3. Example: Van der Pol Oscillator

Van der Pol's equation

$$\ddot{y} + a(y^2-1)\dot{y} + y = 0 \tag{1.5}$$

is an example of a Liénard-type equation $\ddot{y} + f(y,\dot{y})\dot{y} + g(y) = 0$. We write (1.5) in the form for $x = (x_1,x_2) = (y,\dot{y})$ as

$$\dot{x} = X(x) = \begin{pmatrix} x_2 \\ -x_1 - ax_1^2x_2 + ax_2 \end{pmatrix} .$$

The steady state solution is (0,0) and linearization yields

$$DX(0,0) = \begin{pmatrix} 0 & 1 \\ -1 & a \end{pmatrix}$$

with corresponding eigenvalues $\sigma_{1,2} = \frac{a}{2} \pm \sqrt{a^2-4}$. Consider a as a bifurcation parameter:

for $-2 < a < 0$: $\mathrm{Re}\,\sigma_{1,2} < 0$ and $\mathrm{Im}\,\sigma_{1,2} \neq 0$,

for $a = 0$: $\sigma_{1,2} = \pm 2i$

and the system exhibits a Hopf bifurcation at $a_0 = 0$, i.e. the stable steady state (0,0) bifurcates into a stable limit cycle.

4. Example: Lotka-Volterra model

Consider the 2-dimensional Lotka-Volterra equation

$$\begin{pmatrix} \dot{x}_1 \\ \dot{x}_2 \end{pmatrix} = \begin{pmatrix} ax_1 - bx_1x_2 \\ bx_1x_2 - cx_2 \end{pmatrix} \tag{1.6}$$

, a,b,c positive constants.

The system has two steady states: $(0,0)$ and $(\frac{c}{b}, \frac{a}{b})$ in $[0,\infty) \times [0,\infty)$. The open quadrant $\mathbb{R}_+^2$ is an invariant set and we can consider (1.6) therefore on the manifold $M = \mathbb{R}_+^2$. On M the solutions are periodic orbits, described by $\frac{c}{b}(e^{x_1} - x_1) + \frac{a}{b}(e^{x_2} - x_2) \equiv \text{const} > 0$. In particular the system is marginally stable.

5. Example: The Brusselator

The Brusselator is a well-studied example of a system exhibiting a bifurcation from stable steady states to attractive nontrivial periodic solutions (Hopf bifurcation):

$$(1.7) \quad \begin{pmatrix} \dot{x}_1 \\ \dot{x}_2 \end{pmatrix} = \begin{pmatrix} p_1 - (p_2+1)x_1 + x_1^2 x_2 \\ p_2 x_1 - x_1^2 x_2 \end{pmatrix}, \quad p_1, p_2 \text{ positive constants.}$$

We use p_2 as a bifurcation parameter: The steady state is given by $(p_1, \frac{p_2}{p_1})$, linearization yields

$$DX(p_1, \frac{p_2}{p_1}) = \begin{pmatrix} p_2-1 & p_1^2 \\ -p_2 & -p_1^2 \end{pmatrix}$$

with corresponding eigenvalues

$$\sigma_{1,2} = \frac{1}{2}\left(p_1-(p_1^2+1)\right) \pm \sqrt{(p_2-(p_1^2+1))^2-4p_1^2} \,.$$

For $p_2 < p_1^2 + 1 - 2p_1 : \sigma_{1,2} < 0$ and real,

for $p_1^2+1-2p_1 < p_2 < p_1^2+1$: $\operatorname{Re} \sigma_{1,2} < 0$, $\operatorname{Im} \sigma_{1,2} \neq 0$, hence $(p_1, \frac{p_2}{p_1})$ is stable.

For $p_2 = p_1^2 + 1 : \sigma_{1,2} = \pm p_1 i$ and thus the system undergoes a Hopf bifurcation at $p_2^0 = p_1^2 + 1$.

The examples above cover the interesting concepts mentioned earlier for ODE systems. For stability considerations linear systems and their eigenvalues play a particularly important role, because linearizations contain information about the local behavior of nonlinear systems:

Let $\dot{x} = f(x)$ be a smooth ODE in $\mathbb{R}^d$ with steady state $x \equiv 0$. Denote by $A = D_x f(0)$ the Jacobian of f at the point $0 \in \mathbb{R}^d$, then the system in an open neighborhood of 0 looks like $\dot{x} = Ax + r(x)$.

If A is a stability matrix, i.e. $\max \operatorname{Re} \sigma_j = \max \lambda_j < 0$, then the system is asymptotically stable, locally around 0. If A is not a stability matrix we separate its spectrum $\Lambda = \Lambda_s \cup \Lambda_c \cup \Lambda_u$ according to

$$\begin{matrix} \sigma \in \Lambda_s \\ \sigma \in \Lambda_c \\ \sigma \in \Lambda_u \end{matrix} \quad \text{if} \quad \operatorname{Re} \sigma = \lambda \begin{cases} < 0 \\ = 0 \\ > 0 , \end{cases}$$

and denote by V_s, V_c, V_u the corresponding (generalized) eigenspaces. Then there exist invariant C^∞ manifolds M_s and M_u, tangential to V_s, and V_u respectively at 0, which collect the stable, and unstable points with respect to 0, i.e. $|x(t,x_0)| < C|x_0|e^{t\eta}$ for all $x_0 \in M_s$, $t > 0$ and any η greater than all λ in Λ_s (and $|x(t,x_0)| < c|x_0|e^{t\eta}$ for all $x_0 \in M_u$, $t < 0$ and any η smaller than all $\lambda \in \Lambda_u$). M_s and M_u are uniquely determined by the formulas above.

Furthermore there is an invariant C^∞ center manifold M_c, tangential to V_c at 0.

The exponential behavior of the system is described by M_s and M_u, while the systems equation on the center manifold M_c describes the asymptotic behavior of the solution up to exponential terms, see Carr (1981) for details. For smooth systems on compact manifolds see Ruelle (1979) for the cooresponding results.

1.2. STOCHASTIC SYSTEMS

A mechanical system with random components can often be described as a smooth ODE which depends on a stochastic process $\xi(t)$:

$$(1.8) \qquad \dot{x}(t) = X\big(x(t),\xi(t)\big).$$

We will restrict the class of noise processes $\xi(t)$ in several ways:

- o $\xi(t)$ has continuous trajectories and is Markovian, i.e. the future of $\xi(t)$ depends only on the present, not on the past - this is the stochastic analogue of the ODE property.

or

- o $\xi(t)$ is Gaussian white noise.

The white noise model is often easier to analyze, but its spectrum is flat, and the process is unbounded, while the Markovian model can handle bounded noises (e.g. on compact manifolds) and a wide range of spectral densities. To be more precise:

6. The white noise model

The mathematical formulation is the Itō or Stratonovic type SDE

$$\text{(1.9)} \qquad dx(t) = X_0(x(t))dt + \sum_{i=1}^{m} X_i(x(t)) \circ dW^i,$$

where we will use the Stratonovic integral (indicated by "$\circ$") throughout this paper, because it approximates physically realizable noises. Again the vector fields $X_0,\ldots,X_m$ are assumed to be C^∞ (and Itō's version is obtained by adding the correction terms to the drift X_0.) $(W^1,\ldots,W^m)$ denotes the standard, m-dimensional Wiener process, modeling the independent white noise sources of the system.

Solutions of (1.9), with initial variables independent of the increments of the Wiener process, are diffusions (in particular Feller processes with continuous trajectories) and the generator is $G = X_0 + \frac{1}{2}\sum X_i^2$. Note that the second order term $\sum X_i^2$ also contains first order fields.

7. The colored (or Markovian) noise model

Here the noise $\xi(t)$ itself is considered as the solution of an SDE

$$\text{(1.10)} \qquad d\xi(t) = X_0(\xi(t))dt + \sum_{i=1}^{r} X_i(\xi(t)) \circ dW^i \quad \text{on a manifold } N,$$

which then enters pathwise into the right hand side of (1.8). The solutions $x(t)$ of (1.8) are not diffusion processes on their own, but the pair process $(x(t),\xi(t))$, governed by

$$\text{(1.11)} \qquad d\begin{pmatrix} x(t) \\ \xi(t) \end{pmatrix} = \begin{pmatrix} X(x(t),\xi(t)) \\ X_0(\xi(t)) \end{pmatrix} dt + \sum_{i=1}^{r} \begin{pmatrix} 0 \\ X_i(\xi(t)) \end{pmatrix} \circ dW^i$$

is again a diffusion, now with generator $L = X + X_0 + \sum X_i^2$.

Note that the process (1.11) is degenerate in the sense that in the x-component there is no Wiener term present. This degeneracy and the fact that we have to analyse the pair $(x(t),\xi(t))$ in order to obtain information for the system $x(t)$ via the Markov machinery, constitute the difficulties in the colored noise analysis. If e.g. $x(t)$ is a one dimensional process and $\xi(t)$ is white, then the transition probabilities and invariant measures can often by computed explicitly from the corresponding one dimensional Kolmogorov equations, while the colored noise model always is at least two dimensional (and degenerate) so that PDE methods in general fail.

It is common to distinguish two types of noise sources in dynamical systems: additive (or external) and multiplicative (or internal, or parameter) noise. Additive noise enters into a system as

$$(1.12)\qquad \dot{x}(t) = X\big(x(t),p\big) + B\xi(t) ,$$

where B is a constant matrix of suitable dimension. Examples are measurement noise, transmission noise, environmental noise which does not affect the parameters p (and hence the vector field X), or a noisy input, e.g. in a vehicle system the road conditions.

Multiplicative noise affects the dynamics itself through randomness of one or more system parameters:

$$(1.13)\qquad \dot{x}(t) = X\big(x(t),\xi(t)\big) .$$

Examples are random restoring force in an oscillator due to vibrations, stochastic reaction rates in a chemical process, randomness due to variations in parts production, random damping in a vehicle system, or fatigue phenomena.

Of course (1.12) is just a special case of (1.13), but the qualitative analysis of the systems described above can be quite different:

8. Example.

Consider the linear system $\dot{x} = Ax$ in $\mathbb{R}^d$ with additive noise

$$(1.14)\qquad \dot{x}(t) = Ax(t) + B\xi(t)$$

and multiplicative noise

$$(1.15)\qquad \dot{x}(t) = A\big(\xi(t)\big)x(t).$$

The stability problem with respect to the origin $0 \in \mathbb{R}^d$, i.e. is $\lim_{t\to\infty} |x(t,x_0,\omega)| = 0$ w.p.1 does not make sense for (1.14) unless the process $\xi(t)$ itself tends to 0, because $\xi(t)$ drives the system permanently out of 0. The analysis here would take this route: Find conditions under which (1.14) admits a stationary solution (which would take the place of the steady state), and analyze the convergence of all other solutions towards the stationary one.

In the white noise case, i.e. $dx = Ax\,dt + B\,dW$, the answer is: Denote by $V = \sum_{i=0}^{d-1} \mathrm{Im}(A^i B)$ the reachability space of the matrix pair (A,B) and write the system equation as

$$(1.16)\qquad d\begin{pmatrix} x_1 \\ x_2 \end{pmatrix} = \begin{pmatrix} A_{11} & A_{12} \\ 0 & A_{22} \end{pmatrix}\begin{pmatrix} x_1 \\ x_2 \end{pmatrix} dt + \begin{pmatrix} B_1 \\ 0 \end{pmatrix} dW ,$$

with $x_1 \in V$, $x_2 \in V^{\perp}$. (1.16) has a stationary solution iff it has an invariant probability measure iff $\max_j \lambda_j(A_{11}) < 0$. The stationary solution is unique if $V = \mathbf{R}^d$, i.e. if rank $(B, AB, \ldots, A^{d-1}B) = d$, and then the invariant probability μ is Gaussian with support $\mathrm{supp}\,\mu = \mathbf{R}^d$. Therefore the process is uniquely ergodic, and all solutions tend towards μ in distribution. We see that the stability behavior of the deterministic system determines stationarity and ergodicity of the additive noise system. (Consult Ehrhardt and Kliemann (1982) for more details.) We generalize these findings to nonlinear systems in Sections 2 and 4.

For the multiplicative system (1.15) the origin is still a steady state. The corresponding stability problem is considerably more difficult to solve, as can be seen from the following preliminary discussion of the linear oscillator:

Consider the undamped linear oscillator $\ddot{y} + (1+w)y = 0$ or equivalently

$$(1.17)\qquad \dot{x} = \begin{pmatrix} 0 & 1 \\ -1-w & 0 \end{pmatrix} x \qquad \text{in } \mathbf{R}^2.$$

For $w = \pm\varepsilon$ the solutions live on ellipses in $\mathbf{R}^2$ with different major axis. If one can choose switchings between $+\varepsilon$ and $-\varepsilon$, it is possible to obtain solutions that grow to $+\infty$ or decrease to 0. If $\xi(t)$ is a stochastic process with values $\pm\varepsilon$ for w, which of the two patterns will the stochastic system follow? Or more generally: Determine the stability of $\ddot{y} + (1 + \xi(t))y = 0$ if $\xi(t)$ assumes positive and negative values.

These questions will be answered in Section 5, where several systems based on the deterministic Examples 1.1-1.5 will be examined.

To conclude these introductory remarks, we will define some concepts for the analysis of stochastic systems:

Given a probability space (Ω,F,P) with a filtration $\{F_t, t \in \mathbb{R}^+\}$ of σ-algebras in F. Let $\{x(t), t \in \mathbb{R}^+\}$ be a diffusion process with generator $G = X_0 + \frac{1}{2}\sum X_i^2$ defined on (Ω,F_t,P) with values in the state space $(M,\underset{\sim}{B})$, a d-dimensional connected, paracompact C^∞ manifold with its σ-algebra $\underset{\sim}{B}$ of Borel sets.

We denote by

$P(t,x,\cdot)$ the transition probabilities of $\{x(t), t \in \mathbb{R}^+\}$,

$P_t : b\underset{\sim}{B} \to b\underset{\sim}{B}$ the transition semigroup on the bounded $\underset{\sim}{B}$- measurable functions.

$A \subset M$ is an invariant set for $x(t)$, if for all $x_0 \in M$

$$P\{x(t,x_0,\omega) \in A \text{ for all } t \geq 0\} = 1.$$

The SDE (1.9) has a stationary solution $x^0(t)$, if there exists a random variable x_0^0 such that $x(t,x_0^0) = x^0(t)$ is a (strictly) stationary process, or equivalently, if there is an invariant probability measure μ for the transition probabilities of $x(t)$, i.e. $\mu(A) = \int P(t,x,A)\, d\mu(x)$ for all $t > 0$, all $A \in \underset{\sim}{B}$. (1.9) is uniquely ergodic, if μ is unique.

The SDE (1.9) has a periodic solution $x^p(t)$, if there is a random variable x_0^p and a time $T > 0$ such that for all $n \in \mathbb{N}$, and all time sequences $0 \leq t_1 < \dots < t_n$, the distribution of the random vector $(x(h,x_0^p),x(t_1+h,x_0^p),\dots,x(t_n+h,x_0^p))$ is periodic in h with period T. The SDE (1.9) has a steady state $x^s \in M$ if $x(t,x^s) = x^s$ w.p.1 for all $t > 0$, or equivalently if $X_0(x^s) = X_1(x^s) = \dots = X_m(x^s) = 0$. Steady states are in particular stationary solutions, with invariant probability $\mu = \delta_{x^s}$, the Dirac measure at the point x^s.

For the colored noise model (1.11) assume that $\xi(t)$ has a stationary (periodic,...) solution. Then (1.11) has a stationary (...) solution, stationarily (...) connected with $\xi(t)$, if there is a stationary (...) pair solution $(x^0(t),\xi(t))$ of (1.11).

Stability properties of stochastic systems will be studied in the context of Lyapunov exponents and large deviations. These concepts are introduced in Sections 3 and 4.

2. STOCHASTIC DIFFERENTIAL EQUATIONS: FLOWS AND SUPPORT THEOREMS

Let us first recall the concept of flows of smooth deterministic systems

$$(2.1) \qquad \dot{x}(t) = X(x(t)) \quad \text{on } M.$$

The solutions can be described from two different points of view:

- For each initial point $p \in M$ there is a unique open time interval $I_p = (\alpha(p),\beta(p)) \subset \mathbf{R}$ with $0 \in I_p$, and a unique C^∞ integral curve $\phi(\cdot,p) : I_p \to M$ with $\phi(0,p) = p$. The integral curve is the trajectory of (2.1) through p and satisfies $\frac{d}{dt}\phi(t,p) = X(\phi(t,p))$ for all $t \in I_p$. (Here $\frac{d}{dt}\phi(t,p)$ is just a shorthand notation for $\phi(\cdot,p)_*(\frac{d}{dt})|_t$.) If the vector field X is complete, i.e. the solution exists for all $t \in \mathbf{R}$, then of course $I_p = \mathbf{R}$.

- If X is complete, then there is a one parameter group of diffeomorphisms $\{\Phi(t),\ t \in \mathbf{R}\}$ on M, whose infinitesimal generator is X. (The infinitesimal generator of a C^∞ group action $\{\Phi(t),\ t \in \mathbf{R}\}$ on M is a vector field Y on M defined as $Y_p f = \lim_{h \to 0} \frac{1}{h}\left(f(\Phi(h,p)) - f(p)\right)$ for C^∞ functions f on M.)
Actually there is a 1-1 correspondence between one parameter groups of diffeomorphisms on M and complete vector fields (and similarly for local groups and all smooth vector fields). The integral curves are recovered from the group through $\phi(t,p) = \Phi(t)p$.
$\{\Phi(t),\ t \in \mathbf{R}\}$ is the flow corresponding to X.

The trajectory point of view emphasizes the solutions for initial value problems, while the flow describes the multipoint (in particular two point) motion of a system and allows us to compare the behavior of different trajectories.

1. Example:

For the linear system $\dot{x} = Ax$ in $\mathbf{R}^d$, the integral curves are given by $\phi(t,p) = e^{tA}p$ for $p \in \mathbf{R}^d$, and the flow is described as a group of diffeomorphisms through the fundamental matrix $\{\Phi(t) = e^{tA},\ t \in \mathbf{R}\}$.

For stochastic systems the standard construction of solutions of SDE's, reviewed briefly below, leads to diffusion processes described by their trajectories. To obtain associated groups of diffeomorphisms, additional techniques have to be developed. We present two approaches: groups of random diffeomorphisms as flows of SDE's, and groups defined through associated deterministic control systems via the support theorems.

2.1. STOCHASTIC DIFFERENTIAL EQUATIONS

Let (Ω,F,P) be a probability space and W_t a standard, m-dimensional Wiener process on Ω, with the σ-algebras $\underset{\sim}{F}(s,t) = \sigma\{W_\tau,\ s \le \tau \le t\}$ generated by W_t.

In this paper we will work on the canonical space, i.e.

$$\Omega = \{\omega, \omega : [0,\infty) \to \mathbf{R}^m\}$$

$$F = \text{Borel } \sigma\text{-algebra on } \Omega, \text{ completed}$$

$$P = \text{Wiener measure on } \Omega.$$

For notational convenience we identify $W_\cdot(\omega)$ and ω. On Ω we have the usual shift operation $\theta_t\omega(\cdot) = \omega(\cdot+t) - \omega(t)$.

Recall the definition of the Itō and Stratonovic stochastic integral: Let $\{\eta(t),\ t \ge 0\}$ be an a.s. continuous semimartingale, $\underset{\sim}{F}(0,t)$ adapted and define

$$\int_0^T \eta(t)dW_t = \underset{n\to\infty}{\text{prob-lim}} \sum_{k=0}^{n-1} \eta(t_k^n)[W(t_{k+1}^n) - W(t_k^n)]$$

$$\int_0^T \eta(t) \circ dW_t = \underset{n\to\infty}{\text{prob-lim}} \sum_{k=0}^{n-1} \frac{1}{2}(\eta(t_k^n) + \eta(t_{k+1}^n)[W(t_{k+1}^m) - W(t_k^n)]$$

where $t_k^n = \frac{k}{n}T$. The limits above are in the convergence in probability. If $\eta(t)$ satisfies $E\int_0^T \eta(t)^2 dt < \infty$, then for the Itō integral $I(\eta) = \int_0^T \eta(t)dW_t$ we have: There exists exactly one map

$$I : L^2([0,T] \times \Omega,\ dt \otimes P) \to L^2(\Omega,P) \quad \text{with}$$

I is linear, an isometry, i.e. $E|I(\eta)|^2 = E\int_0^T |\eta(t)|^2 dt$, and agrees with $\int_0^T \eta\, dW$ on random step functions. Furthermore, there is a version with continuous trajectories.

Stochastic integrals can be defined in more general situations and we were somewhat sloppy w.r.t. measurability requirements, but the point here is: The stochastic integral is defined in the L^2 set-up, which yields versions with continuous trajectories, but there is no immediate way to obtain flows of homeomorphisms, due to the inherent difficulties in the "backward" stochastic calculus.

Solutions of SDE, constructed usually via the successive approximation method, inherit these properties from the stochastic integral.

We first give an existence and uniqueness theorem for smooth SDE on manifolds in the Stratonovic set-up:

$$dx(t) = X_0(x(t))dt + \sum_{i=1}^{m} X_i(x(t)) \circ dW^i \qquad \text{on} \quad [0,T]. \tag{2.2}$$

2. Definition: A stochastic process $x(t,p)$, $p \in M$, $0 \leq t \leq \eta(p,w) \wedge T$ with values in M is a <u>local solution</u> of (2.2) with initial condition $x(0,p) = p$, if

- $\eta(p,w)$ is measurable, and for fixed $p \in M$ a stopping time w.r.t. $\sigma\{W_u - W_v, \ 0 \leq u \leq v \leq T\}$
- $x(t,p)$ is continuous in t and p,
- for any C^∞ function $f : M \to \mathbf{R}$, $f(x(t,p))$ satisfies for $t < \eta(p,w) \wedge T$

$$f(x(t,p)) = f(p) + \int_0^t X_0(f(x(s,p)))ds$$

$$+ \sum_{i=1}^{m} \int_0^t X_i(f(x(s,p))) \circ dW_s^i \, .$$

A solution is <u>maximal</u>, if $[0,\eta(p))$ is the maximal time interval, and strictly <u>conservative</u>, if $P\{\eta(p) = \infty \text{ for all } p \in M\} = 1$.

3. Theorem: For each $p \in M$ there is a unique maximal solution of (2.2). The solution is a strong Markov process with generator $G = X_0 + \frac{1}{2} \sum X_i^2$. The transition probabilities $P(t,p,\cdot)$ generate a probability measure P_p on $\underset{\sim}{C}_p = \{f : [0,\infty) \to M \cup \{\Delta\}$, continuous with $f(0) = p\}$ through $P_p\{f, f(t) \in A\} = P(t,p,A)$ for all Borel sets A in M. (Here $M \cup \{\Delta\}$ is the one point compactification of M in the noncompact case.)

4. Remarks:

- The proof of Theorem 3 uses local coordinates (U,ϕ) on M, the construction of solutions on $\phi(U) \subset \mathbf{R}^d$ and then patches these solutions together (see e.g. Kunita (1982)).

- If $M = \mathbf{R}^d$, the above set up reduces to the well known existence and uniqueness theorem.

- If $M = \mathbf{R}^d$, then the Itō case can be recovered from the above with the corresponding correction term: To (2.2) there corresponds the Itō equation $dx(t) = X_0^*(x(t))dt + \sum X_i(x(t))dW^i$, where

$$X_0^* = X_0 + \frac{1}{2} \sum_{i=1}^{m} \sum_{k=1}^{d} X_i^k \frac{\partial}{\partial x_k} X_i.$$

- If M is not a flat space $\mathbf{R}^d$, for the Itō set-up the coefficients of the SDE should not be regarded as vector fields. The solution should rather be constructed locally using Itō's formula.

2.2. STOCHASTIC FLOWS OF DIFFEOMORPHISMS

For simplicity we consider the case $M = \mathbf{R}^d$ first. From Theorem 3 we infer the existence of a unique maximal solution $x(t,p)$, which is continuous in t <u>and</u> p. Define for $t > 0$: $V_t = \{x \in \mathbf{R}^d, \eta(x) > t\}$, then the maximal solution defines a continuous map $x(t,\cdot,\omega) : V_t \to \mathbf{R}^d$ for almost all ω.

5. Definition: A family of maps $\Phi(t,\cdot,\omega) : M \to M$ is a <u>stochastic flow of diffeomorphisms</u>, if

- for all $t > 0$, all $p \in M$ $\Phi(t,p,\cdot)$ is $\sigma\{W_u - W_v, 0 \leq u \leq v \leq t\}$ measurable

- for almost all ω, $\Phi(\cdot,\cdot,\omega)$ is continuous in t and p, and satisfies $\lim_{t \downarrow 0} \Phi(t,p,\omega) = p$

- for almost all ω, $\Phi(s+t,p,\omega) = \Phi(s,\Phi(t,p,\omega),\theta_t\omega)$ for $s,t > 0$

- for almost all ω, $\Phi(t,\cdot,\omega) : M \to M$ is a (bijective) diffeomorphism.

6. Theorem: Under the above assumptions $x(t,\cdot,\omega) : V_t \to \mathbf{R}^d$ is a C^∞- diffeomorphism into $\mathbf{R}^d$. If furthermore the solution is strictly conservative, then it defines a stochastic flow of diffeomorphisms on

$\mathbb{R}^d$ iff the adjoint equation

$$(2.3) \qquad dy(t) = -X_0(y(t))dt - \sum_{i=1}^{m} X_i(y(t)) \circ dW^i$$

is also strictly conservative.

7. Remarks:

- The injectivity of the maps $x(t,\cdot,\omega)$ follows from Kolmogorov's criterion for the existence of continuous modifications of stochastic processes. These maps are homeomorphisms, because $x(t,\cdot,\omega)$ can be extended to a continuous 1-1 map on the one point compactification $\mathbb{R}^d \cup \{\Delta\}$. For the diffeomorphism property one has to look at the (linear) SDE for the Jacobian $D_x\, x(t,x,\omega)$, which has a nonsingular solution. This indicates the proof of the "only if" part, except for onto. The rest follows from backward stochastic analysis, which we will not get into here.

- Conservativity of (2.2) and (2.3) is crucial for the existence of stochastic flows. Sufficient conditions are e.g.: The vector fields $X_0,\dots,X_m$ on $\mathbb{R}^d$ are globally Lipschitz.

  In the one dimensional case, explicit conditions can be obtained from Feller's boundary classifications: Here (2.2) and (2.3) are strictly conservative iff $+\infty$ and $-\infty$ are natural boundaries, compare e.g. Gihman and Skorohod (1979).

Theorem 6 has a complete analogue if the state space of the sochastic system is a manifold M:

8. Theorem: Let $V_t = \{p \in M, \eta(p) > t\}$. Under the above assumptions $x(t,\cdot,\omega) : M \to M$ is a $(C^\infty-)$ diffeomorphism into M for all $t > 0$. If furthermore the solution is strictly conservative, then it defines a stochastic flow of diffeomorphisms on M iff the adjoint equation (2.3) is strictly conservative.

The proof again is obtained by patching together local solutions, i.e. solutions in $\mathbb{R}^d$. An immediate corollary is:

9. Corollary: If M is compact, then the solution of (2.2) defines a stochastic flow of diffeomorphisms on M.

10. Remark: The ideas outlined for the proof of Theorem 6 were developed by Kunita (see e.g. Kunita (1982)). Different routes were followed e.g. by Elworthy and Baxendale, who lifted the SDE (2.2) to the Hilbert manifold of diffeomorphisms on M and solved the corresponding equation directly (see eg. Elworthy (1978), or Baxendale (1980)). Or by Malliavin and Ikeda-Watanabe, who utilized an approximation of (2.2) by a sequence of ODE's whose flows will apppr̦oximate the desired stochastic flow (see e.g. Malliavin (1976), or Ikeda-Watanabe (1981)). We will make use of this last idea in the next paragraph in a slightly different set-up.

A stochastic flow of diffeomorphisms, $\Phi(t,\omega)$, as presented here, is defined for $t \geq 0$, and therefore is no group of diffeomorphisms, but only a semigroup. Use stochastic forward and backward differential equations on the space

$$\Omega = \{\omega, \omega : \mathbf{R} \to \mathbf{R}^m \text{ continuous}, \ \omega(0) = 0\}$$

with the Wiener measure P on Ω now given by the distribution of

$$W(t) = \begin{cases} W_1(t), & t \geq 0 \\ W_2(-t), & t \leq 0 \end{cases}$$

(where W_1 and W_2 are two independent standard Wiener processes), to obtain the solution $\{x(t,\omega),\ t \in \mathbf{R}\}$ as a stochastic flow on M with the property

$$x(t+s,\omega) = x(t,\theta_s\omega) \circ x(s,\omega), \quad \text{for all} \quad t,s \in \mathbf{R}.$$

2.3. ASSOCIATED CONTROL SYSTEMS, SUPPORT THEOREM

To the stochastic system (2.2) we associate a (deterministic) control system by formally replacing the Wiener process by control functions:

$$\dot{x}(t) = X_0(x(t)) + \sum_{i=1}^{m} u_i X_i(x(t)), \tag{2.4}$$

with the class of admissible control functions

$U = \{u = (u_1,\dots,u_m) : [0,\infty) \to \mathbf{R}^m$ piecewise constant$\}$. The trajectories $\phi(t,p,u)$, starting at $p \in M$ under the control action $u \in U$ are of the form

$$\Phi_n(t_n) \circ \dots \circ \Phi_1(t_1)p,$$

where the diffeomorphisms $\Phi_k(t_k)$ correspond to the solution of (2.4) with control value $u^k \in \mathbb{R}^m$ for $k = 1,\dots,m$. The time t_k of course has to be in the time interval $I_{\Phi_{k-1}(t_{k-1}) \circ \dots \circ \Phi_1(t_1)p}$ as described in the beginning of this section.

We denote by W_p the set of all trajectories of (2.4), i.e.

$$W_p = \{\psi(\cdot,p) : [0,\beta(p)) \to M, \ \psi \text{ solves (2.4) for some } u \in U, \ \psi(0) = p\}.$$

The relation between the solutions of (2.2) and (2.4) is expressed by the support theorem:

11. Theorem: supp $P_p = \overline{W_p}$ where the measure P_p is defined in Theorem 3 and the closure is taken in $\underset{\sim}{C}_p = \{\psi : [0,\infty) \to M,$ continuous, $\phi(0) = p\}$ w.r.t. the topology of uniform convergence on compact sets.

Theorem 11 is an extension of the Wong-Zakai approximation in $\mathbb{R}^1$ and was first proved by Stroock and Varadhan (1972), the above version is due to Kunita (1974). It suggests the study of the control properties of (2.4) in order to obtain information about the stochastic system (2.2).

It is left as an exercise to verify that with suitable switchings between control values the system (2.4) cannot only move in the infinitesimal directions given by $F = \{X_0 + \sum u_i X_i, \ (u_i) \in \mathbb{R}^m\}$, but also in directions described by the Lie bracket of vector fields in F. (The Lie bracket of two vector fields X and Y on M is defined by $[X,Y] = XY - YX$.) We therefore define the Lie algebras

$\underset{\sim}{L} = \underset{\sim}{L}(F) = \underset{\sim}{L}(X_0,\dots,X_m)$, generated by F

$\underset{\sim}{B} = \underset{\sim}{L}(X_1,\dots,X_m)$ generated by the control (or diffusion) vector fields

$\underset{\sim}{J}$ the ideal in $\underset{\sim}{L}$ generated by $\{X_1,\dots,X_m\}$.

Of course $\underset{\sim}{B} \subset \underset{\sim}{J} \subset \underset{\sim}{L}$.

A Lie algebra $\underset{\sim}{X}$ of vector fields on M defines a <u>distribution</u> $\Delta_{\underset{\sim}{X}}$ on M, i.e. a smooth map $\Delta_{\underset{\sim}{X}} : M \to TM$, the tangent bundle of M, such that $\Delta_{\underset{\sim}{X}}(p) = \{X_p \in T_pM, \ X \in \underset{\sim}{X}\} \subset T_pM$. A submanifold $N \subset M$ is an integral manifold of $\Delta_{\underset{\sim}{X}}$, if for all $p \in N$: $T_pN = \Delta_{\underset{\sim}{X}}(p)$, and $\Delta_{\underset{\sim}{X}}$ (or $\underset{\sim}{X}$) is integrable, if through any point $p \in M$ there passes a maximal integral manifold N_p. Examples of integrable distributions include constant dimensional distributions or analytic situations, i.e. M and $X_0, \ldots, X_m$ are real analytic. (Consult the tutorial in this volume for a more detailed discussion of these concepts.) From now on we will use

12. Assumption: All vector field in $\underset{\sim}{L}$ are complete.

Then it is possible to define the systems group and semi group of diffeomorphisms:

$$\underset{\sim}{G} = \{\Phi_n(t_n) \circ \ldots \circ \Phi_1(t_1), \quad t_k \in \mathbf{R}, \quad 1 \leq k \leq n \in \mathbf{N}\}$$

$$\underset{\sim}{S} = \{\Phi_n(t_n) \circ \ldots \circ \Phi_1(t_1), \quad t_k > 0, \quad 1 \leq k \leq n \in \mathbf{N}\}$$

where the diffeomorphisms $\Phi_k(t_k)$ correspond to some vector field in F.

The orbit and the positive orbit of (2.4) are given by

$$\underset{\sim}{O}(p) = \{g(p), \quad g \in \underset{\sim}{G}\}$$

$$\underset{\sim}{O}^+(p) = \{g(p), \quad g \in \underset{\sim}{S}\}$$

$$\underset{\sim}{O}_t^+(p) = \{g(p), \quad g \in \underset{\sim}{S}, \quad \textstyle\sum t_k = t \text{ in the representation of } g\}$$

Chow's Theorem now says, that if $\underset{\sim}{L}$ is integrable, then the orbit from p is the maximal integral manifold through p, i.e. $\underset{\sim}{O}(p) = N_p$ for all $p \in M$. We can thus restrict our control system in this case to the maximal integral manifolds, where we have $\dim \Delta_{\underset{\sim}{L}}(m) = \dim N_p$ for all $m \in N_p$. We therefore introduce the following conditions.

(A),(B),(C) $\quad \dim \Delta_{\underset{\sim}{L}}(p) = d$ (or $\dim \Delta_{\underset{\sim}{B}}(p) = d$, or $\dim \Delta_{\underset{\sim}{J}}(p) = d$) for all $p \in M$.

The following theorem lists some easy consequences of the support theorem:

13. Theorem:

(i) If $\underset{\sim}{L}$ is integrable, then the vector fields $X_0,\dots,X_m$ can be considered on the maximal integral manifolds N_p of $\underset{\sim}{L}$, and N_p is an invariant set for the solution $x(t,q)$ of (2.2) for all $q \in N_p$.

(ii) supp $P(t,q,\cdot) = \overline{O_t^+(q)}$ for $q \in N_p$, in the topology of N_p,

(iii) If (C) holds, then $\underset{\sim}{O}_t^+(p)$ has nonvoid interior for all $t > 0$, $p \in M$, the transition probabilities $P(t,p,\cdot) = p(t,p,m)dm$ have C^∞-densities w.r.t. the Riemannian volume dm on M, and are continuous in $p \in M$.

The existence of smooth densities follows from Hörmander's hypoellipticity theorem applied to the operator $G + \frac{\partial}{\partial t}$. In particuar we have $O_t^+(p) = M$ for all $p \in M$, $t > 0$ if e.g. (B) holds or (C) and $[\underset{\sim}{B},\underset{\sim}{J}] \subset \underset{\sim}{B}$.

So far we have concentrated on equation (2.2), which represents the white noise case (1.9). For the colored noise system (1.11) we can proceed as follows: We split the system into its dynamics and noise part and obtain the control systems

$$\dot{x} = X(x,\xi)$$

$$\dot{\xi} = X_0(\xi) + \sum_{i=1}^{r} u_i X_i(\xi) \qquad \text{on } M \times N, \text{ with control values in } \mathbf{R}^r$$

and

$$\dot{x} = X(x,v) \qquad \text{on } M, \text{ with control values in } N.$$

The connection between these systems is given by: Consider the Lie algebras $\underset{\sim}{L}_1 = \underset{\sim}{L}(X(\cdot,\xi),\ \xi \in N)$ on M and $\underset{\sim}{L} = \underset{\sim}{L}(X+X_0,X_1,\dots,X_r)$ on $M \times N$. If (B) holds for $\underset{\sim}{B} = \underset{\sim}{L}(X_1,\dots,X_r)$, then the following statements are equivalent if $\underset{\sim}{L}_1$ and $\underset{\sim}{L}$ are integrable:

- $\dim \Delta_{\underset{\sim}{L}_1}(x) = \dim M$ for all $x \in M$
- $\dim \Delta_{\underset{\sim}{L}}(x,p) = \dim M + \dim N$ for all $(x,p) \in M \times N$.

The support theorem now holds for the diffusion pair process $(x(t),\xi(t))$ described by equation (1.11). The remark above shows that, under Assumption (B), Condition (A) can be replaced by

$\dim \Delta_{\underset{\sim}{L}_1}(x) = \dim M$ for all $x \in M$ in this context, which only refers to the x-component of the system.

The results obtained in this paragraph allow the analysis of stochastic systems w.r.t. invariant sets, stationarity and ergodicity:

2.4. STATIONARY AND ERGODIC STOCHASTIC SYSTEMS

We first discuss the question, which subsets of M can carry invariant probabilities.

14. Definition: A set $C \subset M$ is called an <u>invariant control set</u> for the control system (2.4), if $\overline{\underset{\sim}{O}^+(x)} = \overline{C}$ for all $x \in C$.

Under Assumption (A) we have for invariant control sets $C \subset M$

- C is closed, $\text{int } C \neq \emptyset$, $C = \overline{\text{int } C}$, and C is connected
- C and $\text{int } C$ are invariant sets for (2.4).

The support theorem and Hörmander's hypoellipticity result imply:

15. Theorem: Assume (A). If μ is an extremal (i.e. indecomposable) invariant probability for (2.2), then there is an invariant control set $C \subset M$ such that $\text{supp } \mu = C$. Invariant probabilities on invariant control sets are unique and admit C^∞ densities. For all Borel sets $A \subset C$ we have for all $x \in C$:

$$(2.5) \qquad \lim_{t \to \infty} \frac{1}{t} \int_0^t P(t,x,A)dt = \mu(A).$$

Furthermore, if (C) holds, then $\lim_{t \to \infty} P(t,x,A) = \mu(A)$.

The stochastic system (2.2) is therefore uniquely ergodic, iff the control system (2.4) has exactly one invariant control set that carries an invariant probability. In this case the ergodic solution is attractive in the Cesari sense (2.5) for all $x \in C$. If (C) holds, the transition probabilities themselves converge and hence any solution $x(t,x)$ converges towards μ in distribution as $t \to \infty$.

If there is a compact invariant set $K \subset M$ for (2.4), then K always contains an invariant control set $C \subset K$ with an invariant probability, and if C is unique, the above results even hold for all $x \in K$. Furthermore, under (C), the transition densities $p(t,x,y) \to p(y)$, the density of μ, for all $x,y \in K$ as $t \to \infty$. (See Ichihara-Kunita (1974), Kliemann (1987), Arnold-Kliemann (1987).) In the noncompact case, existence of invariant probabilities can sometimes be found via stochastic Lyapunov functions, see e.g. Ethier-Kurtz (1986) or Hasminskii (1980).

For the colored noise system (1.11) we obtain under the assumptions mentioned above (i.e. $\underset{\sim}{L}_1$ and $\underset{\sim}{L}$ integrable, $\dim \Delta_{\underset{\sim}{L}_1}(x) = \dim M$ for all $x \in M$, and (B) holds): The invariant control sets of the pair system (x,ξ) on $M \times N$ are of the form $C \times N$, where C is an invariant control set of the system $\dot{x} = X(x,v)$ on M. Therefore if $\xi(t)$ is stationary and ergodic with invariant probability ν, then $(x(t),\xi(t))$ has a unique stationarily connected solution iff the control system

(2.6) $\qquad \dot{x} = X(x,u)$, u piecewise constant with values in N

has exactly one invariant control set C that carries an invariant probability. All other results from above remain true with the obvious modifications, and (C) is replaced by the condition that the ideal $\underset{\sim}{J}_1$ in $\underset{\sim}{L}$ generated by $X_1,\ldots,X_r$ has full rank on $M \times N$. Once a stationarily connected solution $(x^0(t),\xi(t))$ is found with invariant probability μ on $C \times N$, then $x^0(t)$ is itself stationary and the corresponding ergodic theorems hold. Of course the marginal of μ on N is ν.

Several examples for this analysis of stochastic systems will be demonstrated in Section 5.

3. LYAPUNOV EXPONENTS

Lyapunov's original intention for introducing the characteristic exponents of a dynamical system was his study of the stability behavior (w.r.t. the steady state $x \equiv 0$) of the nonlinear system $\dot{x} = A(t)x + f(t,x)$ via its linear part $\dot{x} = A(t)x$. As mentioned in the introduction, such a conclusion can actually be drawn, if the vector fields are time independent, and also in the periodic case. For the general time dependent situation additional reguarity is necessary, which in general is not easy to check.

If one wants to pursue a similar route for stochastic systems (with stationary noise $\xi(t)$) $\dot{x}(t) = X((t),\xi(t))$, i.e. study the stability of the stationary pair $(x^0(t),\xi(t))$ w.r.t. neighboring solutions $x(t,p)$, one has to consider the linearization $\dot{z}(t) = D_x X(x^0(t),\xi(t))z(t)$, where $D_x X(x^0(t),\xi(t))$ turns out to be a stationary process. The beauty of Oseledec' multiplicative ergodic theorem lies in the fact, that this situation is shown to be regular, i.e. stability and stable manifolds can be discussed via the linearization. In this section we give a brief outline of Oseledec theorem and some of its consequences.

3.1. COCYCLES AND LYAPUNOV EXPONENTS

Let $\{\Phi_t, t \in \mathbb{R}\}$ be a <u>measurable flow</u> on a measurable space (E,ζ), i.e.

- o $\Phi_t : E \to E$ is a bi-measurable bijection for all $t \in \mathbb{R}$
- o $\{\Phi_t, t \in \mathbb{R}\}$ is a group, i.e. $\Phi_{t+s} = \Phi_t \circ \Phi_s$
- o for all $f : E \to \mathbb{R}$ measurable ($\mathbb{R}$ carries the Borel σ-algebra) the map $f(\Phi_\bullet)(\cdot) : \mathbb{R} \times E \to \mathbb{R}$ is measurable.

We attach to each $e \in E$ a d-dimensional linear space F_e.

A <u>cocycle</u> $C(t,e)$ associated with Φ is a measurable function $C(\cdot,\cdot) : \mathbb{R} \times E \to L(F_e,F_{\Phi_t e}) \cong G\ell(d,R)$ into the bijective linear maps from F_e into $F_{\Phi_t e}$, i.e. the nonsingular $d \times d$ matrices, such that

$$C(t+s,e) = C(t,\Phi_s e) \cdot C(s,e) .$$

measurable flow with an associated cocycle defines a skew product flow

$$\tilde{\Phi}_t(e,v) \quad \text{on} \quad \bigcup_{e \in E} \{e\} \times F_e \quad \text{by}$$

$$\tilde{\Phi}_t(e,v) = (\Phi_t e, C(t,e)v).$$

For $\tilde{\Phi}$ we define the Lyapunov exponent of $C(t,e)$ at $e \in E$ in direction $v \in F_e$ as

(3.1) $$\lambda(e,v) = \overline{\lim_{t \to \infty}} \frac{1}{t} \log |C(t,e)v| ,$$

where $|\cdot|$ is a norm on the spaces $\{F_e, e \in E\}$.

1. Example: The linear system.

Consider $\dot{x} = Ax$ in $\mathbf{R}^d$. Its one parameter group of diffeomorphisms $\{e^{tA}, t \in \mathbf{R}\}$ constitutes a cocycle $C(t,0) = e^{tA} \in Gl(d,\mathbf{R})$. (The measurable space E here consists simply of the point $0 \in \mathbf{R}^d$, $\Phi_t 0 \equiv 0$, the trivial solution.) The Lyapunov exponents are then defined for $v \in \mathbf{R}^d$ as

$$\lambda(0,v) = \overline{\lim_{t \to \infty}} \frac{1}{t} \log |e^{tA} v| = \overline{\lim_{t \to \infty}} \frac{1}{t} \log |x(t,v)|.$$

2. Example: Nonlinear systems.

Let $\dot{x}(t) = X(x(t))$ be a smooth system on a Riemannian manifold M, and $\{\Phi_t, t \in \mathbf{R}\}$ its one parameter group of diffeomorphisms on $E = M$ (compare the discussion of (2.1) in Section 2). The cocycle now is given by the linearized flow $C(t,p) = D\Phi_t(p)$, and the skew product flow $\tilde{\Phi}_t = T\Phi_t$ lives on the tangent bundle TM: $\tilde{\Phi}_t(p,v) = (\Phi_t p, D\Phi_t(p)v) \in T_{\Phi_t p} M$. The Lyapunov exponents are given by

$$\lambda(p,v) = \overline{\lim_{t \to \infty}} \frac{1}{t} \log \| C(t,p)v \| = \overline{\lim_{t \to \infty}} \frac{1}{t} \log \| D\Phi_t(p)v \| ,$$

where $\|\cdot\|$ stands for the Riemannian metric, i.e. a smooth field of positive definite bilinear forms on M.

If $M = \mathbf{R}^d$ with the Euclidian metric, then $D\Phi_t(p)$ is simply given by the Jacobian at $p \in \mathbf{R}^d$, $TM = \mathbf{R}^d \times \mathbf{R}^d$, i.e. $T_pM = \mathbf{R}^d$ for all $p \in \mathbf{R}^d$, and $\|\cdot\|$ is the Euclidian norm. More precisely: The linearization of $\dot{x} = X(x)$ with solution $\Phi_t p$ for $x(0) = p \in \mathbf{R}^d$ obeys the equation

$$\dot{v}(t) = A(\Phi_t p)v(t) = \frac{\partial X}{\partial x}\Big|_{\Phi_t p} v(t), \quad v(0) = v \quad \text{in} \quad \mathbf{R}^d.$$

The fundamental matrix $\psi(t,p)$ of this linear equation satisfies

$$\psi(t,p)v = D\Phi_t(p)v = C(t,p)v ,$$

and hence

$$\lambda(p,v) = \overline{\lim_{t \to \infty}} \frac{1}{t} \log\|\psi(t,p)v\| .$$

3. Example: Nonlinear stochastic systems

Let $dx(t) = X_0(x(t)) + \sum_{i=1}^{m} X_i(x(t)) \circ dW^i$ be a smooth stochastic system on M. Let (Ω,F,P) be the canonical probability space of the m-dimensional Wiener process $(W^1,\ldots,W^m)$, with P the Wiener measure. On (Ω,F,P) the canonical flow of shift operators $\{\theta_t, t \in \mathbf{R}\}$ is defined by

$$\theta_t\omega(\cdot) = \omega(t+\cdot) - \omega(t) \quad \text{for} \quad \omega \in \Omega, \quad t \in \mathbf{R}.$$

The Wiener measure P is invariant and ergodic with respect to $\{\theta_t, t \in \mathbf{R}\}$. Assume that the system defines a stochastic flow of diffeomorphisms $\phi_t(\cdot,\omega) : M \to M$ (compare Section 2.2). On $E = \Omega \times M$ we are then given the combined flow Φ_t by

$$\Phi_t(\omega,p) = (\theta_t\omega, \phi_t(p,\omega)).$$

We now linearize the system: Let the vector field X_j for $j = 0,\ldots,m$ be given locally by

$$X_j = \sum_{k=1}^{d} \alpha_{kj}(x) \frac{\partial}{\partial x_k}$$

and denote the Jacobian of the coefficient functions by

$$A_j(x) = \left(\frac{\partial \alpha_{kj}(x)}{\partial x_\ell} \right) .$$

Then the linearized vector fields $TX_j = (X_j, DX_j)$ on the tangent bundle TM are locally given by

$$TX_j(x,v) = \big(\alpha_j(x), A_j(x)\big).$$

The linearized solution $(Tx)(t)$ on TM obeys the SDE

$$d(Tx)(t) = TX_0\big(Tx(t)\big)dt + \sum_{i=1}^{m} TX_i\big(Tx(t)\big) \circ dW^i,$$
$$Tx(0) = (p,v),$$

which actually is a coupled pair of SDE, given locally through

$$dx(t) = \alpha_0\big(x(t)\big)dt + \sum_{i=1}^{m} \alpha_i\big(x(t)\big) \circ dW^i, \quad x(0) = p \quad \text{on} \quad M$$

$$dv(t) = A_0\big(x(t)\big)v(t)dt + \sum_{i=1}^{m} A_i\big(x(t)\big)v(t) \circ dW^i, \quad v(0) = v,$$

on the tangent spaces. Denote by $\psi(t,p)v$ the solution of the linear equation for v, i.e. locally $\psi(t,p)$ is the fundamental matrix. Then we have $Tx(t)(\omega,p,v) = \big(x(t,\omega,p), \psi(t,\omega,p)v\big)$ and the cocycle is defined through $C(t,\omega,p) = D\phi_t(p,\omega) = \psi(t,\omega,p)$. The skew product flow $\tilde{\Phi}_t$ then is of course

$$\tilde{\Phi}_t(\omega,p,v) = \big(\Phi_t(\omega,p),\ C(t,\omega,p)v\big) = \big(\theta_t\omega,\ Tx(t)(\omega,p,v)\big)$$

on $\Omega \times TM$.

The Lyapunov exponents

$$\lambda(p,\omega,v) = \overline{\lim_{t \to \infty}} \frac{1}{t} \log \| D\phi_t(p,\omega)v \|$$

may now depend on $(\omega,p) \in \Omega \times M$. $\|\cdot\|$ is again a Riemannian metric.

3.2. THE MULTIPLICATIVE ERGODIC THEOREM

Oseledec Theorem describes the regularity of the Lyapunov exponents in the stationary case:

4. **Theorem:** Let $\tilde{\Phi}_t$ be a skew product flow and assume

(i) there is a proability measure ρ on (E,ζ), which is invariant under the flow Φ_t, i.e. $\Phi_t \rho = \rho$ for all $t \in \mathbf{R}$,

(ii) $\int_E \sup_{-1 \leq t \leq 1} \log^+ \| C^{\pm 1}(t,e) \| \, d\rho < \infty$.

Then there exists a set $\tilde{E} \subset E$ of ρ-measure 1, such that for all $e \in \tilde{E}$

- there are ℓ numbers $\lambda_e^\ell < \dots < \lambda_e^1$, $\ell(e) \leq d$ and subspaces $0 \subset W_e^\ell \subset \dots \subset W_e^1 = F_e$ of dimension $d_e^\ell < \dots < d_e^1 = d$ such that for all $i = 1,\dots,\ell$

  $v \in W_e^i \setminus W_e^{i+1}$ iff $\lim_{t \to \infty} \frac{1}{t} \log |C(t,e)v| = \lambda_e^i$.

- If furthermore ρ is ergodic for Φ_t (i.e. all Φ_t invariant sets have ρ-measure 0 or 1), then ℓ, λ^i, d^i are constants, but the spaces W_e^i may still depend on $e \in \tilde{E}$. $\{\lambda_\ell,\dots,\lambda_1\}$ is called the <u>Lyapunov spectrum</u> of the flow.

We apply this theorem to the above examples:

5. **Example: The linear system**

Here the conditions (i) and (ii) (and ergodicity) are trivially satisfied. The Lyapunov exponents $\lim_{t \to \infty} \frac{1}{t} \log|x(t,v)|$ are of course the real parts $\lambda_\ell < \dots < \lambda_1$ of the eigenvalues of A, and a solution $x(t,v) = e^{tA}v$ has exponent λ_i iff $v \neq 0$ is in the eigenspace V_i corresponding to λ_i. The subspaces W_i are therefore $W_i = \bigoplus_{k=1}^{\ell} E_k$.

In a similar way the periodic linear system $\dot{x} = A(t)x$ with period T can be treated: Define

$$E = [0,T], \text{ with } 0 \text{ and } T \text{ identified}$$

$$\rho = \frac{1}{T} m, \quad m \text{ the Lebesgue measure on } [0,T]$$

$$\Phi_t e = t + e \bmod T \quad \text{for } e \in [0,T].$$

Let $F_e = \mathbf{R}^d$ for all $e \in [0,T]$. Denote by $\psi(t,e) = P(t)e^{(t-e)R}P^{-1}(e)$ the fundamental matrix and define $C(t,e) = \psi(t,e)$ for $t \geq 0$. Now Oseledec' Theorem applies for a set $\tilde{E} \subset E$ of ρ-measure 1. In particular for $e = 0$: The λ_i's are the characteristic exponents of the Floquet representation and the W_i's are again the direct sums of (generalized) eigenspaces for the λ_i's, compare Example 1.2.

6. Example: Nonlinear systems

Assume there exists a measure ρ, Φ_t invariant and equivalent to the Riemannian volume volume dm on M and

$$\int_M \sup_{-1 \leq t \leq 1} \log^+ \| D\Phi_t(p)^{\pm 1} \| \, d\rho(p) < \infty ,$$

then Oseledec' theorem holds. The subspace filtration now decomposes each tangent space T_pM into the W_p^i, $i = 1,\ldots,\ell(p)$ according to the λ_p^i. Note that on compact manifolds the integrability condition is always satisfied.

Let us be more specific for systems $\dot{x} = X(x)$ in $\mathbf{R}^d$: Suppose first that $p_0 \in \mathbf{R}^d$ is a steady state, then (with the notations from Example 2)

$$\Phi_t p_0 = p_0 \quad \text{for } t \geq 0, \quad A(\Phi_t p_0) = A(p_0), \quad \dot{v}(t) = A(p_0)v(t).$$

We set $E = \{p_0\}$, $\rho = \delta_{p_0}$, the Dirac measure, then $\tilde{E} = \{p_0\}$. By Example 5, the Lyapunov exponents are the real parts of the eigenvalues of $A(p_0)$.

Now suppose that $p^0(t)$ is a periodic solution of $\dot{x} = X(t)$ with period T, then $\dot{v}(t) = A(p^0(t))v(t)$ and we define

$E_1 = [0,T]$, 0 and T identified

$\rho_1 = \frac{1}{T} m$, m the Lebesgue measure on $[0,T]$

$\theta_t e = t + e \bmod T$ for $e \in [0,T]$.

Furthermore we set

$$E = \{(e,p),\ e \in E_1,\ p = p^0(e)\} \subset E_1 \times \mathbb{R}^d, \quad \text{the graph of } p^0,$$

$$\rho(C \times D) = \rho_1\{e,\ e \in C \text{ and } p^0(e) \in D\} \quad \text{for Borel sets } C \times D \text{ in } [0,T] \times \mathbb{R}^d.$$

Then $\rho(E) = 1$ and Oseledec Theorem applies (again for the cocycle defined by $\dot{v} = A(p^0(t))v$ as in Example 5).

7. Example: Nonlinear stochastic systems

Assume that the system has a stationary solution, i.e. an invariant probability measure ν. Since the shift group θ_t on the Wiener space leaves the measure P invariant, $P \otimes \nu$ is invariant for the flow $\Phi_t(\omega,p)$. Assume that the integrability condition (ii) is satisfied, i.e.

$$\int_{\Omega\times M} \sup_{0 \leq t \leq 1} \log^+ \| Tx(t)(\omega,p)^{\pm 1} \| \, dP \otimes \nu < \infty .$$

Then for a subset $\tilde{E} \subset \Omega \times M$ of full $P \otimes \nu$ measure Oseledec theorem holds for the linearized stochastic flow $\tilde{\Phi} = (\theta, T\phi)$. The integrability condition (ii) is satisfied e.g. if $M = \mathbb{R}^d$ and the linearized vectorfields, i.e. the Jacobians $A_0,\dots,A_m$, are of the type

$$\int \|A_0(x)\| d\nu < \infty, \quad \int \|A_i(x)\|^2 d\nu < \infty \quad \text{for } i = 1,\dots,m.$$

In the compact case, the system always generates a stochastic flow of diffeomorphisms and conditions (i) and (ii) are automatically satisfied in this set-up, see Carverhill (1985). Furthermore the Lyapunov spectrum is independent of the choice of the Riemannian metric in this case.

In particular if the system is uniquely ergodic (see Theorem 2.15 and the following remarks), then the Lyapunov exponents are independent of (ω,p) w.p.1. We investigate this situation further in Sections 4 and 5.

Theorem 4 is almost an exact analogue of the situation for the linear, deterministic case, with the steady state $x \equiv 0$ replaced by a flow with invariant probability. The only difference being that Theorem 4 deals with the flag $W^{\ell} \subset \ldots \subset W^{1}$ of linear subspaces instead of the generalized eigenspaces themselves. This an be removed by considering forward and backward in time Lyapunov exponents. For the nonlinear stochastic system in Examples 3. and 7. this amounts to defining a more general notion of invariant measures on $E = \Omega \times M$, and thus stationary processes $(\theta_t\omega, x(t,\omega))$, such that $X(t,\omega)$ is in general not a solution of the SDE in the usual sense any more, but is well defined as a stochastic flow. For some aspects of stochastic bifurcation theory this generalized point of view has its merits, compare Arnold (1987).

The results in Section 2.4 on stationarity, ergodicity, invariant probabilities etc. depend only on the generator $G = X_0 + \frac{1}{2}\sum X_i^2$, hence on the transition probabilities $P(t,x,\cdot)$ or the Markovian semigroup P_t. The convergence results in Theorem 2.15 are therefore in the weak sense. Lyapunov exponents on the other hand are defined for flows, i.e. in the stochastic case for stochastic flows of diffeomorphisms that describe the multi point motion of a system. They do <u>not</u> only depend on the generator, as the following example shows:

8. Example: Let $M = S^1$, the one dimensional sphere, and define a Wiener process on S^1 in two different ways:

- as a projection of the Wiener process in $\mathbb{R}^2$ onto S^1, i.e. $dx(t) = -\sin x(t)\, dW^1 + \cos x(t)\, dW^2$, where $x(t)$ denotes the angle,
- as a one dimensional Wiener process on S^1, $dy(t) = dW$.

Both diffusion processes have generator $G = \frac{1}{2}\Delta$, Δ the Laplace-Beltrami operator on S^1. But $x(t)$ has Lyapunov exponent $-\frac{1}{2}$, while for $y(t)$ we obviously obtain 0, compare Carverhill et al. (1986). The trajectories of $x(t)$ converge exponentially, while those of $y(t)$ stay at fixed distances w.p.1. (For further examples in this direction see e.g. Baxendale (1986).)

So far we have established the existence of Lyapunov exponents for the linearized flows of stochastic systems. The local behavior of the nonlinear system itself is now described by the stable manifold theorem, which we give in its manifold version (compare Carverhill 1985):

9. Theorem: Consider the uniquely ergodic stochastic system $dx(t) = X_0(x) + \sum_{i=1}^{m} X_i(x) \circ dW^i$ on a compact Riemannian manifold M. Assume that the Lyapunov exponent λ_i is stricly negative. Define for the Riemannian distance $d(\cdot,\cdot)$ on M the set

$$\underset{\sim}{M}^i_{(p,\omega)} = \left\{ y \in M,\ \overline{\lim_{t\to\infty}} \frac{1}{t} \log d\left(x(t,p,\omega),x(t,y,\omega)\right) < \lambda_i \right\} ,$$

consisting of all starting points in M such that the trajectories converge exponentially fast with rate $< \lambda_i$ towards $x(\cdot,p,\omega)$.

Then $\underset{\sim}{M}^i_{(p,\omega)}$ is an (immersed) submanifold of M of dimension d_i, tangent to the subspace $W^i_{(p,\omega)} \subset T_pM$ at p, for all $(\omega,p) \in \tilde{E}$.

In particular, if $\lambda_1 < 0$, then $M^1_{(p,\omega)}$ has dimension $d = \dim M$ and thus contains an open neighborhood of p. Therefore locally around p the trajectories of $x(t,p,\omega)$ and $x(t,p,y)$ converge exponentially fast at least with rate λ_1, for P-almost all ω. (This holds for ρ-almost all $p \in M$.) Note that for ergodic systems λ_1 is a constant almost surely, while the domain of attraction $M^1(p,\omega)$ of a solution starting at $p \in M$ is random!

3.3. THE UPPER LYAPUNOV EXPONENT FOR LINEAR STOCHASTIC SYSTEMS

Linear stochastic systems, besides being of interest on their own, can be used to study the local behavior of nonlinear systems. See Theorem 9. To avoid too many technical details, we study the $\mathbf{R}^d$-version, but in a unified treatment, so that both the white and the colored noise case are covered. Consider

$$dx(t) = A_0\big(\xi(t)\big)x(t)dt + \sum_{i=1}^{m} A_i\big(\xi(t)\big)x(t) \circ dW^i \quad \text{in } \mathbf{R}^d \tag{3.2}$$

where

- $\xi(t)$ is stationary diffusion processes on a compact manifold N, given by

$$d\xi(t) = X_0\big(\xi(t)\big)dt + \sum_{j=1}^{r} X_j\big(\xi(t)\big) \circ dV^j ,$$

The Wiener processes W^i and V^j are independent and the Lie algebra $\underset{\sim}{L}(X_1,\ldots,X_r)$ has full rank, i.e. satisfies Condition (B) in Section 2.3. By Theorem 2.15 $\xi(t)$ is uniquely ergodic with invariant probability $\rho = p(m)dm$.

o $A_i : N \to g\ell(d,\mathbb{R})$ are C^∞ for $i = 1,\ldots,m$.

To apply Oseledec's Theorem consider the following set up: Let Ω_1 be the trajectory space of $\{\xi_t, t > 0\}$, i.e. $\Omega_1 = \underset{\sim}{C}(\mathbb{R}^+,N)$ with measure P_1 induced by the stationary, ergodic process ξ, see Theorem 2.3. On Ω_1 consider the flow $\theta_t^1\,\omega(\cdot) = \omega(t+\cdot)$, then P_1 is invariant and ergodic with respect to θ^1 and ξ defines a stochastic flow ϕ^1 on N. Let Ω_2 be again the canonical Wiener space of $W = (W^1,\ldots,W^m)$ with flow θ^2. Then the solution $x(t,\omega^1,\omega^2,\cdot) : M \to M$ of equation (3.2) is stochastic flow of diffeomorphisms with respect to (θ^1,θ^2). on $E = \Omega_1 \times \Omega_2$. The associated cocycle $C(t,e)$ is of course the fundamental matrix of the linear equation (3.2).

The probability measure $P_1 \otimes P$ is invariant under (θ^1,θ^2) and the integrability condition in Oseledec's Theorem reads (compare Example 7.)

$$\int \|A_0(\xi)\| d\rho < \infty\ , \qquad \int \|A_i(\xi)\|^2 d\rho < \infty \quad \text{for } i = 1,\ldots,m\ .$$

We now obtain the Lyapunov exponents

$$\lambda(\omega^2,x_0) = \lim_{t\to\infty} \frac{1}{t} \log|x(t,\omega^1,\omega^2,x_0)|$$

for each initial value $x_0 \in \mathbb{R}^d \setminus \{0\}$ of (3.2). These numbers are independent of ω^1 because of the ergodicity of ξ, while the corresponding subspaces $W_e^i \subset \mathbb{R}^d$ depend on $e \in \Omega_1 \times \Omega_2$.

We are interested in the stability behavior of (3.2), i.e. in its upper Lyapunov exponent $\lambda = \lambda_1$, and look for conditions that assure $\lambda(\omega,x_0) = \lambda$ a.s. for all $x_0 \neq 0$. This means that the smaller exponents, obtained in W_e^2, are not realized w.p.1 for fixed initial values x_0, and hence for all variables independent of the Wiener processes.

For the upper Lyapunov exponent it is appropriate to project the system (3.2) onto the sphere $S^{d-1} \subset \mathbb{R}^d$ (see the discussion of Furstenberg boundaries in Furstenberg (1963) or Bougerol-Lacroix (1985)), where because of periodicity, we can identify $s \in S$ with $-s \in S$, so that the projected system actually lives on the real projective space $\mathbf{P}^{d-1} \subset \mathbb{R}^d$. Its equations are

$$(3.3) \qquad ds(t) = h_0(\xi(t),s(t))dt + \sum_{i=1}^{m} h_i(\xi(t),s(t)) \circ dW^i ,$$

where $q_i(\xi,s) = s^* A_i(\xi)s$

$$h_i(\xi,s) = A_i(\xi)s - q_i(\xi,s), \quad \text{the projected vector fields of (3.2),}$$

with * denoting transpose. For future use we also define

$$Q(\xi,s) = q_0 + \frac{1}{2} \sum_{i=1}^{m} h_i(q_i) .$$

According to the discussion of Equations (1.9) and (1.11) the pair process $(\xi(t),s(t))$ is a diffusion on $N \times \mathbf{P}$ with generator $L = G + h_0 + \frac{1}{2} \sum_{i=1}^{m} h_i^2$, where $G = X_0 + \frac{1}{2} \sum_{j=1}^{r} X_j^2$ is the generator of $\xi(t)$. By Corollary 2.9 the system (3.3) induces a stochastic flow of diffeomorphisms on $N \times \mathbf{P}$.

On the other hand, according to Section 2.3, we can associate deterministic control systems to (3.3), e.g.

$$\begin{aligned} \dot{\phi} &= h_0(\psi,\phi) + \sum u_i h_i(\psi,\phi) \\ \dot{\psi} &= X_0(\psi) + \sum v_j X_j(\psi) \end{aligned} \qquad \text{on } N \times \mathbf{P}$$

with $u = (u_i,\ i = 1,\ldots,m)$ and $v = (v_j,\ j = 1,\ldots,r)$ piecewise constant in $\mathbb{R}^m$ and $\mathbb{R}^r$. Or

$$(3.4) \qquad \dot{\phi} = h_0(u_0,\phi) + \sum u_i h_i(u_0,\phi) \qquad \text{on } \mathbf{P}$$

with u_0 piecewise constant in M. The associated Lie algebras are

$$\tilde{\underset{\sim}{L}} = \underset{\sim}{L}(X_0+h_0, X_1,\ldots,X_r, h_1,\ldots,h_m) \qquad \text{on} \quad N \times \mathbf{P}$$

$$\underset{\sim}{L} = \underset{\sim}{L}(h_0(w),\ldots,h_m(w),\ w \in \mathbf{N}) \qquad \text{on} \quad \mathbf{P}$$

$$\underset{\sim}{B} = \underset{\sim}{L}(X_1,\ldots,X_r, h_1,\ldots,h_m)$$

$\underset{\sim}{J}$ the ideal in $\tilde{\underset{\sim}{L}}$ generated by $\underset{\sim}{B}$.

10. Assumption: $\dim \Delta_{\tilde{\underset{\sim}{L}}}(\xi,s) = \dim N + d-1$, i.e. $\tilde{\underset{\sim}{L}}$ has full rank.

This according to the discussion at the end of Section 2.3, is equivalent in the present set up to $\underset{\sim}{L}$ having full rank on $\mathbf{P}$, which we will use from now on.

To compute the systems group and semigroup for (3.4) denote

$$U = \{A_0(u_0) + \sum u_i A_i(u_0),\ u_0 \in N,\ u_1,\ldots,u_m \in \mathbf{R}\}$$

$$U_1 = \{\sum u_i A_i(u_0),\ u_0 \in N,\ u_1,\ldots,u_m \in \mathbf{R}\}\ .$$

We define

$$\underset{\sim}{G} = \{e^{t_n B_n} \cdot \ldots \cdot e^{t_1 B_1},\ t_i \in \mathbf{R},\ B_i \in U,\ i = 1,\ldots,m \in \mathbf{N}\}$$

$$\underset{\sim}{S} = \{e^{t_n B_n} \cdot \ldots \cdot e^{t_1 B_1},\ t_i > 0,\ B_i \in U,\ i = 1,\ldots,n \in \mathbf{N}\}$$

$$\underset{\sim}{H} = \{e^{t_n B_n} \cdot \ldots \cdot e^{t_1 B_1},\ t_i \quad \mathbf{R},\ B_i \in U_1,\ i = 1,\ldots,n \in \mathbf{N}\}$$

$\underset{\sim}{G}$ and $\underset{\sim}{S}$ are the group and semigroup of the linear control system associated with (3.2). $\underset{\sim}{H}$ corresponds to the control system given by $\underset{\sim}{B}$. If we define $g(s) = \frac{g \cdot s}{|g \cdot s|}$ for $g \in \underset{\sim}{G}$ and $s \in \mathbf{P}$, then these groups also describe (3.4), since h_i is the projected vector field of $A_i x$.

Now let us consider the upper Lyapunov exponent $\bar{\lambda}$ of (3.2). Itô's calculus yields for $x_0 \neq 0$, $s_0 = \frac{x_0}{|x_0|}$:

$$(3.5)\qquad |x(t,x_0)| = |x_0| \exp\{\int_0^t Q(\xi(t),s(\tau,s_0))\,d\tau + \sum_{i=1}^{m}\int_0^t q_i(\xi(t),s(\tau,s_0))\,dW^i\}$$

and hence

$$(3.6)\qquad \lim_{t\to\infty}\frac{1}{t}\log|x(t,x_0,\omega)| = \lim_{t\to\infty}\frac{1}{t}\int_0^t Q(\xi(\tau,\omega),s(\tau,s_0,\omega))\,d\tau ,$$

because the Wiener terms tend to 0, being regular martingales. (This is the reason, why we had to transform the Stratonovic integral to the Ito integral in (3.5) and obtained the correction terms in $Q(\xi,s)$.)

Equation (3.6) shows that the independence of λ of $x_0 \neq 0$ and ω is the question of (unique) ergodicity of the process $(\xi(t),s(t))$. According to Theorem 2.15 this is the problem of the uniqueness of the invariant control set of (3.4).

11. Theorem: Assume that $\underset{\sim}{\tilde{L}}$ has full rank on $N \times \mathbf{P}$. Then

- the control system (3.4) has a unique invariant control set $C \subset \mathbf{P}$,
- the diffusion process $(\xi(r),s(t))$ has a unique invariant probability μ on $N \times \mathbf{P}$ with $\operatorname{supp}\mu = N \times C$, μ has a C^∞-density,
- $\lambda(\omega,x_0) = \lambda$ a.s. for all $x_0 \neq 0$, $\lambda = \lambda_1$ the upper Lyapunov exponent from Oseledec' theorem.
- $\lambda = \int_{N\times\mathbf{P}} Q(\xi,s)\,d\mu = \lim_{t\to\infty}\frac{1}{t}\log\|\Phi(t)\|$,

where $\Phi(t)$ is the fundamental matrix of (3.2). This is called the Hasminskii formula for λ.

For a proof see Arnold et al (1986a), Arnold et al. (1986b). Actually the last two statements of the theorem would follow from the weaker condition that the group $\underset{\sim}{G}$ acts irreducibly on $\mathbf{P}^d$, see Furstenberg (1963) or Bougerol-Lacroix (1985). Our Lie algebra condition however is easier to verify and needed for the detailed study of the stability behavior in the following sections.

So far we have studied the upper Lyapunov exponent λ_1. For information on the whole Lyapunov spectrum $\{\lambda_\ell,\ldots,\lambda_1\}$ it is appropriate to project the system (3.2) onto the flag manifold

$\mathbb{F}$ over $\mathbb{R}^d$, as again suggested by Furstenbergs boundary theory. The approach outlined above goes through with the obvious modifications. In particular the average exponent $\bar{\lambda} = \sum_{i=1}^{\ell} d_i \lambda_i$ can be studied via the determinant: $\bar{\lambda} = \lim_{t \to \infty} \frac{1}{t} \log|\det \Phi(t)|$, where $\Phi(t)$ is the fundamental matrix of (3.2), see Arnold-Kliemann (1986) for more details.

4. APPLICATIONS TO THE ANALYSIS OF STOCHASTIC SYSTEMS: STABILITY AND LARGE DEVIATIONS

4.1. THE LINEAR SYSTEM

We use the set-up of Section 3.3 and analyze the stability behavior of

$$dx(t) = A_0(\xi(t))x(t)dt + \sum_{i=1}^{m} A_i(\xi(t))x(t) \circ dW^i . \tag{4.1}$$

The steady state solution $x \equiv 0$ is asymptotically stable in probability iff it is asymptotically stable w.p.1 iff $\lim_{t \to \infty} x(t,x_0) = 0$ w.p.1 for all $x_0 \neq 0$. Let $\lambda_\ell(\omega) < \dots < \lambda_1(\omega)$ be the Lyapunov exponents of (4.1), then the system is asymptotically exponentially stable w.p.1 iff $\lambda_1(\omega) < 0$ w.p.1, and unstable if $\lambda_1(\omega) > 0$ with positive probability. Theorem 3.11 gives sufficient conditions for $\lambda = \lambda_1$ being the same constant for all $x_0 \neq 0$. (Note that this theorem also provides a recipe for calculating $\lambda = \int Q \, d\mu$ via the law of large numbers, hence $\lambda_1 < 0$ can be checked at least numerically.)

In the engineering literature it is common to consider the stability of moments, in particular the first and second moment. We include these concepts by defining

$$g(p,x_0) = \overline{\lim_{t \to \infty}} \frac{1}{t} \log E|x(t,x_0)|^p, \quad p \in \mathbf{R}, \quad x_0 \neq 0,$$

the Lyapunov exponent for the p-th moments.

Oseledec' theorem and the Perron-Frobenius theory for positive semigroups imply (compare Arnold 1984):

1. Theorem: Assume the Lie algebra $\underset{\sim}{J}$ has full rank on $N \times \mathbf{P}$. Then

- o $g(p,x_0) = g(p) = \lim_{t \to \infty} \frac{1}{t} E|x(t,x_0)|^p \in \mathbf{R}$ for all $p \in \mathbf{R}$, $x_0 \neq 0$,
- o $g : \mathbf{R} \to \mathbf{R}$ is convex, analytic and $g(0) = 0$,
- o $g'(0) = \lambda$.

In particular: If the p-th moment is exponentially stable for some $p > 0$, then the system is a.s. exponentially stable and $\lambda < \frac{g(p)}{p}$. Vice versa: if the system is a.s. exponentially stable, then for small p the p-th moments are exponentially stable. Theorem 1 expresses a relation between the a.s. and the moment Lyapunov exponents for small p. This brings up the question, whether for large p it is possible that $\lambda < 0$ BUT $\frac{g(p)}{p} > 0$, i.e. the system is a.s. stable, but some moments are unstable.

We investigate this question further and introduce

$$\gamma(p) = \begin{cases} \lambda & \text{if } p = 0 \\ g(p)/p & \text{if } p \neq 0, \end{cases} \qquad \gamma_{\pm} = \lim_{p \to \pm\infty} \gamma(p).$$

$\gamma : \mathbf{R} \to \mathbf{R}$ is analytic and increasing.

Now: Either $\gamma(p) \equiv \lambda$ iff $g(p) = \lambda p$ iff $\gamma_- = \lambda = \gamma_+$ $(= g'(0))$
or $\gamma(p)$ is strictly increasing iff $g(p)$ is strictly convex iff $\gamma_- < \lambda = g'(0) < \gamma_+$.

Of course $\gamma_+ < 0$ means that all moments of (4.1) are exponentially stable.

To obtain estimates on $\gamma_-, \lambda, \gamma_+$ we use the groups $\underset{\sim}{G}$ and $\underset{\sim}{H}$ introduced in Section 3.3, and also: Define

$$A_i^0(\xi) = A_i(\xi) - \frac{1}{d} \operatorname{trace} A_i(\xi) \mathrm{Id}, \quad i = 0, \ldots, m,$$

$$U^0 = \{A_0^0(u_0) + \sum u_i A_i^0(u_0),\ u_0 \in N,\ u_1, \ldots, u_m \in \mathbf{R}\}$$

similarly for U_1^0. Introduce the groups $\underset{\sim}{G}^0$ and $\underset{\sim}{H}^0$, defined by U^0 and U_1^0 instead of U and U_1. Now $\underset{\sim}{H}^0 \subset \underset{\sim}{G}^0 \subset S\ell(d,\mathbf{R})$, the nonsingular matrices with determinant 1, thus the ^{0}-system amounts to neglecting the systematic drift in (4.1).

The following theorem gives criteria in terms of compactness of the groups $\underset{\sim}{G}, \underset{\sim}{H}, \ldots$. This rather "compact" way of formulating the results can be made computational by observing that $\underset{\sim}{G}$ is compact iff there is a transformation matrix $T \in G\ell(d,\mathbf{R})$ such that simultaneously $T\{A_i(N),\ i = 0, \ldots, m\}T^{-1} \subset so(d,\mathbf{R})$, the skew symmetric matrices, i.e. $\underset{\sim}{G}$ consists only of rotation matrices in some coordinate system of $\mathbf{R}^d$, similarly for $\underset{\sim}{G}^0, \underset{\sim}{H}, \underset{\sim}{H}^0$.

2. **Theorem:** Assume the Lie algebra $\underset{\sim}{J}$ has full rank on $N \times \mathbf{P}$. Then

1. If $\bar{\underset{\sim}{H}}$ is not compact, then $-\infty = \gamma_- < \lambda < \gamma_+ = +\infty$.

2. If $\bar{\underset{\sim}{H}}$ is compact (and hence $\underset{\sim}{H} = \underset{\sim}{H}^0$), then $-\infty < \gamma_- \leq \lambda \leq \gamma_+ < +\infty$. In particular:

2.1.1. If $\underset{\sim}{G}$ is not compact and $\underset{\sim}{G}^0$ is not compact, then $\gamma_- < \lambda < \gamma_+$.

2.1.2. If $\underset{\sim}{G}$ is not compact, but $\underset{\sim}{G}^0$ is compact, then

a) if $\frac{1}{d}$ trace $A_0(\xi) \equiv c$, then $\gamma_- = \lambda = \gamma_+ = c$

b) if $\frac{1}{d}$ trace $A_0(\xi) \not\equiv c$, then

$$\gamma_- = \frac{1}{d} \min_{\xi \in N} \text{trace } A_0(\xi) <$$

$$< \lambda = \frac{1}{d} \int_N \text{trace } A_0(\xi)\rho(d\xi) < \gamma_+ = \frac{1}{d} \max_{\xi \in N} \text{trace } A_0(\xi)$$

2.2. If $\underset{\sim}{G}$ is compact (and hence $\underset{\sim}{G} = \underset{\sim}{G}^0$), then $\gamma_- = \lambda = \gamma_+ = 0$.

For a proof see Arnold-Kliemann (1986) and note that $\underset{\sim}{G}$ and $\underset{\sim}{G}^0$ are always closed in our set-up. From Theorem 2 the dichotomy above now extends to

Either $\gamma(p) \equiv \lambda$ iff $\underset{\sim}{G}^0$ is compact and $\frac{1}{d}$ trace $A_0(\xi) \equiv c$

iff $g''(0) = 0$

or $\gamma(p)$ is strictly increasing

iff $\underset{\sim}{G}^0$ is not compact or $\frac{1}{d}$ trace $A_0(\xi) \not\equiv c$

iff $g''(0) > 0$.

The results above also allow certain estimates on λ: Denote by (4.1^0) the system (4.1) with A_1 replaced by A_1^0, then $\lambda = \lambda^0 + \frac{1}{d}$ trace $E\,A_0(\xi)$, where λ^0 corresponds to (4.1^0). Of course $\lambda^0 \geq 0$ and $= 0$ iff $\underset{\sim}{G}^0$ is compact. In general it is not possible to compute λ directly for a given system (4.1) with given noise $\xi(t)$. We therefore have to resort to numerical methods, compare Section 5.2.

Remark (for the probabilist) Theorem 3.11 is a law of large numbers for the exponential growth rate $\log|x(t,x_0)|$. The corresponding central limit theorem is

$$\frac{1}{\sqrt{t}}\{\log|x(t,x_0)| - \lambda t\} \Rightarrow \underset{\sim}{N}(0,g''(0)),$$

where $\Rightarrow$ denotes convergence in law and $\underset{\sim}{N}(0,g''(0))$ is a Gaussian distribution with mean 0 and variance $g''(0)$. The above dichotomy characterizes the case $g''(0) = 0$.

As far as stability of (4.1) is concerned, we have obtained so far:

- If the matrices in U^0 can simultaneously be made skew symmetric, i.e. there is a transformation matrix $T \in G\ell(d,\mathbb{R})$ such that $T \cdot \{A_i(u_0) - \frac{1}{d}\text{ trace } A_i(u_0)Id,\ u_0 \in N,\ i = 0,\dots,m\} T^{-1} \subset s_0(d,\mathbb{R})$, and if $\frac{1}{d}\text{ trace } A_0(u_0) \equiv c$, then the system is a.s. and p-th moment exponentially stable if $c < 0$, and unstable iff $c > 0$.
- Otherwise $g(p)$ is strictly convex, and the system is
  - a.s. exponentially stable iff $\lambda < 0$
  - p-th moment exponentially stable for all $p > 0$ if $\gamma_+ < 0$.

  $\lambda > 0$ implies that all p-th moments for $p > 0$ are exponentially unstable,
  $\lambda < 0$ implies that (at least) for small $p > 0$ the p-th moments are exponentially stable.

In Section 5.2 we will discuss an example, where $\lambda < 0$, but $\gamma_+ > 0$.

We now address the question: What is the behavior of the system, if $\lambda < 0$, but $\gamma_+ > 0$? The answer is given by the theory of large deviations.

A sequence Q_n of Borel measures on a complete, separable metric space M obeys a large deviations (LD) principle with rate function $I(r)$, if there is a sequence $a_n \to \infty$ of real numbers, such that

- $I(r)$ is lower semicontinuous and has compact level sets,
- $\overline{\lim_{n\to\infty}} \frac{1}{a_n} \log Q_n(F) \leq - \inf_{r \in F} I(r)$ for F closed in M,
- $\underline{\lim_{n\to\infty}} \frac{1}{a_n} \log Q_n(G) \geq - \inf_{r \in G} I(r)$ for G open in M.

The rate function $I(r)$, if it exists, is unique.

In our set up we are interested in a LD-principle for the sequence of measures $Q_t(\cdot) = P\{\frac{1}{t} \log \frac{|x(t,x_0)|}{|x_0|} \in \cdot\}$ on $\mathbf{R}$, describing the deviation from the growth rate as given by the upper Lyapunov exponent as $t \to \infty$. Inspired by the i.i.d case (where $I(r)$ is the convex conjugate of the log of the moment generating function) and a theorem of Ellis (1985), we set

$$I(r) = \sup_{p \in \mathbf{R}} \left(rp - g(p)\right) \quad \text{for} \quad r \in \mathbf{R},$$

the convex conjugate or Legendre-Fenchel transform of $g(p)$. Then Theorem II.6.1 in Ellis (1985) yields

3. Theorem: If the Lie algebra $\underset{\sim}{J}$ has full rank, then the sequence Q_t defined above satisfies a LD-principle with rate function $I(r)$.

In order to understand this principle better, we use some convex analysis to obtain:

- $I(r) > 0$, $I(r) = 0$ iff $r = \lambda$, $-I(0) = \inf_{p \in \mathbf{R}} g(p)$
- if $\gamma_- = \gamma_+$, then $I(r) = 0$ for $r = \lambda$, and $+\infty$ otherwise
- if $\gamma_- < \gamma_+$, then $I(r)$ is finite on (γ_-, γ_+), $+\infty$ outside $[\gamma_-, \gamma_+]$, $I(r)$ is strictly convex and analytic on (γ_-, γ_+), strictly decreasing on (γ_-, λ), strictly increasing on (λ, γ_+).

With this in mind we have the following

4. Corollary: Assume $\lambda < 0$ and $K > 0$ a constant, then

- If $\gamma_- = \lambda = \gamma_+$, then for each $x_0 \neq 0$

$$\lim_{t \to \infty} \frac{1}{t} \log P\{ \frac{|x(t,x_0)|}{|x_0|} > K\} = -\infty$$

- If $\gamma_- < \lambda < \lambda_+$, then for each $x_0 \neq 0$

$$\lim_{t \to \infty} \frac{1}{t} \log P\{ \frac{|x(t,x_0)|}{|x_0|} > K\} = -I(0).$$

Or in other words for the case $\gamma_- < \gamma_+$:

o If $\gamma_+ > 0$, hence $I(0) \in \mathbb{R}$, then with positive probability $\sim e^{-I(0)t}$ the system will cross any level $K > 0$.

o If $\gamma_+ < 0$, hence $I(0) \in \mathbb{R}$, then $P\{\frac{|x(t,x_0)|}{|x_0|} > K\}$ will settle down faster than any exponential function: There exists a constant $\hat{K} > 1$ such that $\sup_{t \geq 0} \frac{|x(t,x_0)|}{|x_0|} \leq \hat{K}$ with probability one.

While Corollary 4 gives an exponential estimate on $P\{\frac{|x(t,x_0)|}{|x_0|} > K\}$ as $t \to \infty$ for any level $K > 0$, the probability of crossing depends on K in the following way: If $\lambda < 0 < \gamma_+$, then there is a constant $c \geq 1$ such that for all $0 \leq |x_0| < K$

$$\frac{1}{c}\left(\frac{|x_0|}{K}\right)^a \leq P\{\sup_{t \geq 0} |x(t,x_0)| \geq K\} \leq c\left(\frac{|x_0|}{K}\right)^a$$

where a is the unique number > 0 with $g(a) = 0$. For further information on large deviations for the system (4.1) see Arnold-Kliemann (1986).

The remaining task is to compute γ_+. While λ can be obtained in a reliable way via numerics, using the strong law of large numbers in Theorem 3.11, the computation of high moments in the formula $\gamma_+ = \lim_{p \to \infty} \frac{g(p)}{p}$ does not provide a reasonable method. A way to avoid this is to formulate γ_+ as the solution of a (deterministic) optimal control problem: According to Theorem 2 we only have to consider the case where $\bar{\underset{\sim}{H}}$ is compact. Assume therefore that a coordinate system in $\mathbb{R}^d$ has been chosen such that all matrices $A_i(\xi)$, $i = 1,\dots,m$, $\xi \in N$ are simultaneously skew symmetric. Consider the control system

$$(4.2) \qquad \dot{\phi} = A_0(u_0)\phi + \sum_{i=1}^{m} u_i A_i(u_0)\phi$$

with admissible controls $\underset{\sim}{U} = \{u = (u_0,\dots,u_m) : \mathbb{R}^+ \to N \times \mathbb{R}^m$ piecewise constant$\}$. The Lyapunov exponents of the solutions of (4.2) are given by

$$\lambda(u,x_0) = \overline{\lim_{t\to\infty}} \frac{1}{t}\int_0^t q_0\big(u_0(\tau),s(\tau,s_0,u)\big)\,d\tau .$$

Under the conditions of Theorem 2 one can show:

$$\gamma_+ = \sup_{x_0 \neq 0} \sup_{u \in \underset{\sim}{U}} \lambda(u,x_0)$$

$$= \sup_{u \in \underset{\sim}{U}_p} \frac{1}{t}\int_0^t q_0\big(u_0(\tau),s(\tau,u)\big)\,d\tau ,$$

where $\underset{\sim}{U}_p$ is the set of periodic controls in $\underset{\sim}{U}$ such that $\big(u,s(\cdot,u)\big)$ is T-periodic for some $T > 0$ (see Colonius-Kliemann 1986, Arnold-Kliemann (1986)). This characterization sometimes allows the precise computation of γ_+, and reliable numerics in any case.

Once γ_+ is computed, one can find the LD constant $-I(0) = \inf_{p \in \mathbf{R}} g(p)$: If $\gamma_+ < 0$, then $-I(0) = -\infty$. If $\gamma_+ > 0$, then $g(p)$ has to be computed only for a bounded p-range. We will give examples in Section 5.2.

4.2. THE NONLINEAR SYSTEM IN $\mathbf{R}^d$

We consider white and colored noise nonlinear systems in $\mathbf{R}^d$ and obtain stability results via linearization around stationary solutions.

The white noise system (1.9) is

$$(4.3) \qquad dx(t) = X_0\big(x(t)\big)dt + \sum_{i=1}^{m} X_i\big(x(t)\big) \circ dW^i.$$

Assume the situation of Example 3.3 in $\mathbf{R}^d$, i.e. all representations are global, and of Example 3.7, i.e. there is a stationary solution $x^0(t)$ of (4.4). Then we can study the stability of $x^0(t)$ with respect to neighboring solutions via linearization:

$$(4.4) \qquad dv(t) = A_0\big(x^0(t)\big)v(t)dt + \sum_{i=1}^{m} A_i\big(x^0(t)\big)v(t) \circ dW^i.$$

The Lyapunov exponents of (4.4) were studied in Sections 3.3 and 4.1. Now the stable manifold theorem (compare Theorem 3.9) implies: If

$\lambda_1 < 0$, then, except for a set of $P \otimes \nu$-measure 0, for all $(\omega,x) \in \Omega \times \mathbf{R}^d$ the stable manifold $M_{(x,\omega)}$ is an open set around x and for all $y \in M_{(x,\omega)}$

$$\overline{\lim_{t \to \infty}} \frac{1}{d} \log|x(t,x,\omega) - x(t,y,\omega)| \leq \lambda_1 .$$

The problem therefore boils down to finding the random stable manifolds, for which even in the deterministic case, there is no general technique.

Two special cases for the above set up should be mentioned:

- If the control system associated with (4.3) has a unique compact, invariant control set $C \subset \mathbf{R}^d$, then there always is a staionary and ergodic solution $x^0(t)$ of (4.4) in C and the analysis outlined above can be carried out.
- If (4.3) has a steady state, i.e. a point $x^0 \in \mathbf{R}^d$ with $X_j(x^0) = 0$ for all $j = 0,\dots,m$, then $x(t) \equiv x^0$ is a stationary, ergodic solution. Equation (4.4) now has constant matrices $A_0(x^0), \dots, A_m(x^0)$.

For stability considerations, only the upper Lyapunov exponent is of interest, but in general it is difficult to compute. The average Lyapunov exponent

$$\bar{\lambda} = \sum_{i=1}^{\ell} d_i \lambda_i$$

describing the change of volume, has a simpler representation:

$$\bar{\lambda} = \int_{\mathbf{R}^d} \Big[\operatorname{div} X_0(x) + \frac{1}{2} \sum_{i=1}^{m} X_i\big(\operatorname{div} X_i(x)\big)\Big] d\nu . \tag{4.5}$$

For $d = 1$, $\bar{\lambda} = \lambda$ and (4.5) provides an explicit formula, compare Section 5.4.

For the colored noise system (1.11), given by

$$\dot{x}(t) = X\big(x(t),\xi(t)\big) \tag{4.6}$$

we assume that $\xi(t)$ satisfies the assumptions of Section 3.3. For Oseledec's Theorem we set Ω_1, P_1, θ^1 as in Section 3.3, then $x(t,\omega,\cdot) : \mathbf{R}^d \to \mathbf{R}^d$ is a stochastic flow. Assume that there is a stationarily connected ergodic solution $(x^0(t),\xi(t))$ of (4.5). The linearization reads

$$(4.7) \qquad \dot{v}(t) = A(x^0(t,\omega),\xi(t,\omega))v(t)$$

and under the familiar integrability condition, Oseledec's Theorem holds. The Lyapunov exponents of (4.7) are investigated in Sections 3.3 and 4.1, with $A_1 = \dots = A_m = 0$, and the corresponding stable manifold theorem describes the stability of (4.5) around $x^0(t)$.

An interesting special case again arises if x^0 is a steady state of Equation (4.6), i.e. $X(x^0,\xi) = 0$ for all $\xi \in N$. Then $(x^0,\xi(t))$ is stationarily connected and the theory developed above applies.

4.3. THE NONLINEAR SYSTEM ON A MANIFOLD

Oseledec' theorem is available for nonlinear stochastic flows on compact manifolds. We restrict ourselves to this case and treat the white and the colored noise system in a unified way:

$$(4.8) \qquad dx(t) = Y_0(\xi(t),x(t))dt + \sum_{i=1}^{m} Y_i(\xi(t),x(t)) \circ dW^i$$

on a compact manifold M, where $\xi(t)$ is given as in Section 3.3 on N. Assume that the pair $(x^0(t),\xi(t))$ is a uniquely ergodic solution on $N \times M$ with invariant probability ρ. The linearized flow of (4.8), denoted by $Tx(t)$ is given by

$$(4.9) \qquad dTx(t) = TY_0(\xi(t),Tx(t))dt + \sum_{i=1}^{m} TY_i(\xi(t),Tx(t)) \circ dW^i$$

and lives on the tangent bundle TM. (Here for a vector field $Y(\xi,x)$ on M $TY(\xi,Tx) = (Y(\xi,x),D_xY(\xi,x))$ on TM.) Combining Example 3.3, 3.7 and the discussion in Section 3.3, we can apply Oseledec' theorem to this set up.

In order to recover the results from Section 3.3 and 4.1 for nonlinear systems on manifolds, we proceed as follows: Note that $(\xi(t),Tx(t))$ is again a diffusion process. We project the linearized flow (4.9) onto the projective bundle $\mathbb{P}M$, i.e. onto the projective spaces $\mathbf{P}_m$ in the tangent spaces $T_mM \cong \mathbb{R}^d$ for each $m \in M$. Thus we obtain the projected system $\mathbb{P}x(t)$:

$$(4.10)\qquad d\mathbf{P}x(t) = \mathbf{P}Y_0(\xi(t),\mathbf{P}x(t))dt + \sum_{i=1}^{m} \mathbf{P}Y_i(\xi(t),\mathbf{P}x(t)) \circ dW^i,$$

where $\mathbf{P}Y$ is the projection of the vector field TY on TM onto $\mathbf{P}M$, i.e. $\mathbf{P}Y = (Y,h(DY))$, with h as defined in equation (3.3).

Associated with (4.8) is a control system

$$(4.11)\qquad \dot{x}(t) = Y_0(u_0,x(t)) + \sum_{i=1}^{m} u_i Y_i(u_0,x(t)) \quad \text{on } M$$

with controls $\underset{\sim}{U} = \{u = (u_0,\dots,u_m) : \mathbf{R}^+ \to N \times \mathbf{R}^m$ piecewise constant$\}$. Denote by $\underset{\sim}{G}$ the group of diffeomorphisms of (4.11) and by $\underset{\sim}{O}(x)$ the orbit from $x \in M$, compare Section 2.3. From $\underset{\sim}{G}$ we construct the groups $G(x) \subset G\ell(d,\mathbf{R})$: Let $\underset{\sim}{G}_x$ be the subgroup of $\underset{\sim}{G}$ that leaves x invariant, i.e. $\underset{\sim}{G}_x = \{\phi \in \underset{\sim}{G},\ \phi(x) = x\}$. Then the Jacobian $D\phi(x)$ of ϕ at x is a d×d-matrix (choose some coordinate system in $T_xM \cong \mathbf{R}^d$). Define $\underset{\sim}{G}(x) = \{D\phi(x),\ \phi \in \underset{\sim}{G}_x\}$. $\underset{\sim}{G}(x)$ acts on $\mathbf{P}$ in the way described in Section 3.3 (and replaces the group $\underset{\sim}{G}$ defined there). The assumption in Theorem 3.11 is now replaced by:

(a) (4.11) is controllable on $\underset{\sim}{O}(x)$,

(b) $\underset{\sim}{G}(x)$ is transitive on $\mathbf{P}$, i.e. $\{g(s),\ g \in \underset{\sim}{G}(s)\} = \mathbf{P}$ for all $s \in \mathbf{P}$.

5. Theorem: Fix $x \in M$ and assume the above conditions. Then

- the control system associated with (4.10) has exactly one invariant control set $C \subset \mathbf{P}M|_{\underset{\sim}{O}(x)}$,
- the diffusion $(\xi(t),\mathbf{P}x(t))$ has a unique invariant probability μ on $N \times \mathbf{P}M|_{\underset{\sim}{O}(x)}$ with $\operatorname{supp}\mu = N \times C$,
- the Lyapunov exponents $\lambda(\omega,v)$ for $v \in TM|_{\underset{\sim}{O}(x)} \setminus \{0\}$ satisfy $\lambda(\omega,v) = \lambda = \lim_{t\to\infty} \frac{1}{t} \log\|Dx(t,\omega)v\|$ w.p.1, and λ is the upper Lyapunov exponent from Oseledec' theorem,
- $\lambda = \int_{N\times C} Q(\xi,s)d\mu$, where Q involves the Riemannian covariant derivative and the Riemannian curvature tensor, see e.g. Baxendale (1986).

Two special cases are of particular importance for us:

- $\underset{\sim}{O}(x) = M$, i.e. the system is transitive on M. Then assumption (a) requires that (4.11) is controllable, which follows e.g. if the Lie algebra $\underset{\sim}{B} = \underset{\sim}{L}(Y_1,\ldots,Y_m)$ has full rank on M, or more general, if the ideal $\underset{\sim}{J}$ in $\underset{\sim}{L}(Y_0,\ldots,Y_m)$ generated by $\underset{\sim}{B}$ has full rank and $[\underset{\sim}{B},\underset{\sim}{J}] \subset \underset{\sim}{B}$, see Kunita (1976).
- $\underset{\sim}{O}(x) = x$ i.e. x is a steady state for (4.11). In this case (a) is always satisfied.

See San Martin-Arnold (1986) and San Martin (1986) for further details.

The linear system, discussed in Sections 3.3 and 4.1 fits into this framework: There $x \equiv 0$ is the unique steady state, so we consider the linear system in $T_0\mathbb{R}^d \cong \mathbb{R}^d$, where it takes the form (3.2), and Theorem 5 is just Theorem 3.11.

To obtain the results of Section 4.1 for the linearized system (4.9), we define the function $g(p,v) = \lim_{t\to\infty} \frac{1}{t} \log E\|Dx(t,\omega)v\|^p$ for $p \in \mathbb{R}$.

Again, if the transition probabilities of $(\xi(t),\mathbb{P}x(t))$ have smooth densities (which follows e.g. from the assumption that the ideal in $\underset{\sim}{L}(X_0+\mathbb{P}Y_0,X_1,\ldots,X_r,\mathbb{P}Y_1,\ldots,\mathbb{P}Y_m)$ generated by $\{X_1,\ldots,X_r,\mathbb{P}Y_1,\ldots,\mathbb{P}Y_m\}$ has full rank), then $g(p,v) = g(p)$ and the techniques developed in Section 4.1 go through.

The stable manifold theorem (Theorem 3.9) now describes the local behavior of (4.8).

5. EXAMPLES OF STOCHASTIC MECHANICAL SYSTEMS

5.1. STATIONARY AND ERGODICCITY OF NONLINEAR SYSTEMS

In this section we apply the theory, developed in Section 2.4 to the examples 1.3-1.5 and investigate the existence and uniqueness of stationary solutions.

1. Example: Van der Pol oscillator with colored noise

Consider the SDE

$$\ddot{y}(t) + \xi(t)[y(t)^2 - 1]\dot{y}(t) + y = 0 \tag{5.1}$$

where $\xi(t)$ is a stationary, ergodic diffusion process in the interval $N = (r_1, r_2)$, $0 < r_1 < r_2 < \infty$, i.e. for each fixed $\xi \in N$ the system has a stable limit cycle. It now can be shown that the corresponding control system

$$\dot{x} = \begin{pmatrix} x_2 \\ -x_1 - ux_1^2x_2 + ax_2 \end{pmatrix}, \quad u \text{ piecewise constant with values in } N,$$

has exactly one compact invariant control set $N \times C \subset N \times \mathbb{R}^2$ (compare Rümelin (1983)) and therefore (5.1) has a unique ergodic solution $(y^0, \dot{y}^0, \xi)$ stationarily connected with ξ. The process $(y^0(t), \dot{y}^0(t))$ is itself stationary, its invariant distribution μ can be obtained via numerics, see e.g. Arnold-Kliemann (1983), page 71. The support of μ of course is anulus-like around the point (0,0) and reflects the shape of the limit cycles.

2. Example: The Brusselator with colored noise

Here we discuss the stochastic system

$$\begin{pmatrix} \dot{x}_1 \\ \dot{x}_2 \end{pmatrix} = \begin{pmatrix} \xi_1 - (\xi_2+1)x_1 + x_1^2x_2 \\ \xi_2x_1 - x_1^2x_2 \end{pmatrix} \tag{5.2}$$

where $(\xi_1(t), \xi_2(t))$ is a stationary, ergodic diffusion process in $N_1 \times N_2 = N$, $N_i = (r_1^i, r_2^i)$, $0 < r_1^i < f_2^i < \infty$ for $i = 1,2$. Here the system for fixed $(\xi_1, \xi_2) \in N$ may have a unique stable steady state, or a unique stable limit cycle, depending on (ξ_1, ξ_2), see Example 1.5.

Nevertheless, control analysis shows that again there s a unique compact invariant control set for the system in $N \times \mathbb{R}^{+2}$. Hence (5.2) has a unique stationary and ergodic solution (see Ehrhardt (1983)), whose density is concentrated on the stable configurations, compare Ehrhardt (1983) for numerical studies.

3. Example: Lotka-Volterra model with colored noise

Consider the stochastic Lotka-Volterra model

$$(5.3) \qquad \begin{pmatrix} \dot{x}_1 \\ \dot{x}_2 \end{pmatrix} = \begin{pmatrix} \xi_1 x_1 - x_1 x_2 \\ x_1 x_2 - \xi_2 x_2 \end{pmatrix} \qquad \text{in } \mathbb{R}^+ \times \mathbb{R}^+,$$

with $(\xi_1(t), \xi_2(t))$ as in Example 2. In this case the system for fixed $(\xi_1, \xi_2) \in N$ is marginally stable, see Example 1.4. Control analysis shows that the only control set is $\mathbb{R}^+ \times \mathbb{R}^+$ (if either $\xi_1(t)$ or $\xi_2(t)$ take on at least two values). A precise stochastic analysis shows that Equation (5.3) has no stationary solution, see Kliemann (1983). If however (ξ_1, ξ_2) is not a Markov process, this resonance phenomenon can be avoided in some cases, see Arnold et al (1979).

These three examples suggest as a rule of thumb: If the "frozen" systems for fixed $\xi \in N$ are asymptotically stable with overlapping domain of attraction, then the stochastic system will have a stationary, ergodic solution. A more precise statement can be found in Colonius-Kliemann (1986).

5.2. STABILITY OF THE LINEAR STOCHASTIC OSCILLATOR

A paradigm for the study of linear systems--and of linearized nonlinear systems--is the oscillator

$$(5.4) \qquad \ddot{y} + 2b\dot{y} + \big(1 + \sigma f(\xi(t))\big)y = 0,$$

where $b \in \mathbb{R}$ is the damping constant and $\sigma \in \mathbb{R}^+$ the amplitude parameter of the noise. We assume as in Section 3.3 that ξ is an ergodic diffusion process on a compact manifold N, satisfying condition (B). (5.4) can be written as

$$(5.5) \qquad \dot{x} = \begin{pmatrix} 0 & 1 \\ -1-\sigma f(\xi) & -2b \end{pmatrix} x = A(\xi)x, \qquad \text{in } \mathbb{R}^2$$

compare Example 1.1. To fulfill the conditions of Sections 3.3 and 4.1, we assume that $f : N \to \mathbb{R}$ is analytic and assumes at least two values, compare Arnold et al. (1986a).

Let us first analyze the undamped oscillator

$$\dot{x} = \begin{pmatrix} 0 & 1 \\ -1-\sigma f(\xi) & 0 \end{pmatrix} x .$$

The projected system (compare Equation (3.3)) for the angle $\phi \in [0,2\pi) \sim S^1$ is

$$\dot{\phi} = -\sin^2\phi - \big(1 + \sigma f(\xi)\big)\cos^2\phi .$$

Theorem 3.11 implies: $\lambda^0 = \lambda^0(x_0,\omega)$ a.s. for $x_0 \in \mathbb{R}^2\setminus\{0\}$ and

$$\lambda^0 = \int_{N\times P} q(\xi,s)d\mu = \int_{N\times[0,2\pi)} -\frac{1}{2}\sigma f(\xi)\sin 2\phi \; d\mu$$

where μ is the unique invariant probability of $\big(\xi(t),\phi(t)\big)$ on $N \times [0,2\pi)$. Theorem 4.2 implies

$$-\infty < \gamma_-^0 < \lambda^0 < \gamma_+^0 < +\infty$$

because $\{A(\xi),\ \xi \in N\}$ cannot simultaneously be made skew symmetric. Furthermore trace $A(\xi) = 0$, hence $\underset{\sim}{G} = \underset{\sim}{G}^0$ and therefore

$$0 < \lambda^0 < \gamma_+^0 .$$

The a.s. exponential growth λ^0 depends on the dynamics, i.e. the vectorfields of the noise $\xi(t)$, while γ_+^0 depends only on the noise interval $\sigma f[N]$, compare the discussion of Equation (4.2).

The dependence of λ^0 on the noise amplitude σ is given by the expansion formula (for small σ)

(5.6) $$\lambda_\sigma^0 = \frac{\sigma^2}{8} \int_0^\infty \cos(2t)\, C(t)\, dt + O(\sigma^3),$$

where $C(t)$ is the covariance function of $f(\xi(t))$, compare Arnold et al. (1986c).

Consider now the damped system (5.5): Use the transformation $\bar{y} = ye^{bt}$ to obtain

(5.7) $$\dot{\bar{x}} = \begin{pmatrix} 0 & 1 \\ -1 + b^2 - \sigma f(\xi) & 0 \end{pmatrix} \bar{x}\,.$$

We therefore obtain for the stability parameters λ, γ_-, γ_+ of (5.5):

$$\lambda = \bar{\lambda} - b$$

$$\gamma_\pm = \bar{\gamma}_\pm - b$$

where the $^-$-quantities correspond to the system (5.7). From the discussion above we know that $0 < \bar{\lambda} < \bar{\gamma}_+$, and hence

$$\lambda < 0 \quad \text{iff} \quad \bar{\lambda} < b$$

$$\gamma_+ < 0 \quad \text{iff} \quad \bar{\gamma}_+ < b$$

and therefore for each $\sigma > 0$ there is a parameter area in b such that

$$\lambda < 0 \quad \text{but} \quad \gamma_+ > 0,$$

which implies the existence of large deviations in the system.

Furthermore the expansion formula (5.6) now reads for $b^2 < 1$:

(5.8) $$\lambda_\sigma = -b + \frac{\sigma^2}{8(1-b^2)} \int_0^\infty \cos\left(2\sqrt{1-b^2}\, t\right) C(t)dt + O(\sigma^3)$$

and for $b^2 > 1$:

(5.9) $$\lambda_\sigma = -b + \sqrt{b^2-1} + \frac{\sigma^2}{4(1-b^2)} \int_0^\infty e^{-2\sqrt{b^2-1}\, t}\, C(t)dt + O(\sigma^3).$$

We compare these findings with the deterministic case $\sigma = 0$:

for $b^2 < 1$, the exponential growth rate is $-b$, while
for $b^2 > 1$, we obtain $-b + \sqrt{b^2-1}$.

Therefore for small noise amplitudes σ:

if $b^2 < 1$, then λ_σ increases with σ, and hence the noise has a destabilizing effect,

if $b^2 > 1$, then λ_σ decreases for small σ, hence in this case the noise has a stabilizing effect.

Finally we present some numerical results:

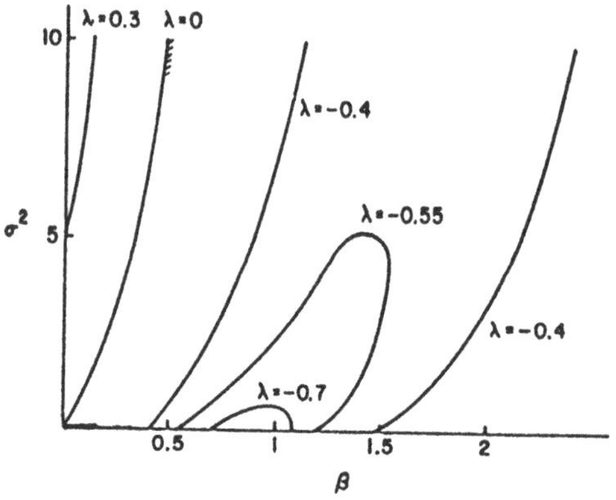

Figure 1: Lyapunov exponents of (5.4). Shown are the level curves $\lambda(b,\sigma)$
$f(\xi)$ is an Ornstein-Uhlenbeck process with variance 1, compare Arnold-Kliemann (1983).

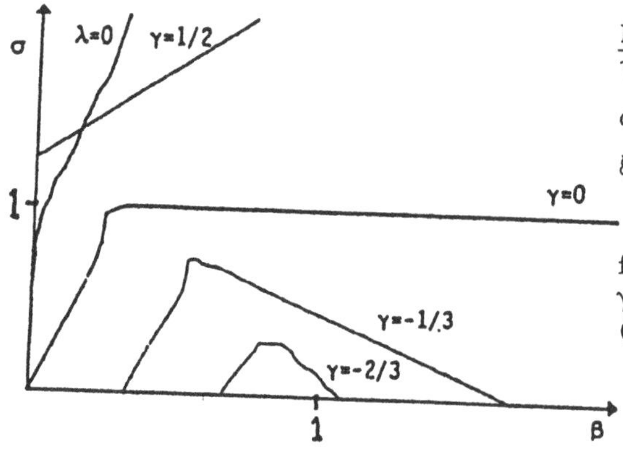

Figure 2: Large deviation constant γ_+ of (5.4). Shown are the level curves $\gamma_+(b,\sigma)$
ξ is standard Brownian motion on S^1, $f(\xi) = \cos \xi$, hence $f[S^1] = [-1,1]$ determines the γ-curves, compare Arnold-Kliemann (1986).

5.3. STABILITY OF THE RANDOMLY DISTURBED HELICOPTER BLADE MOTION

In this section we investigate the motion of helicopter blades parametrically disturbed by white oise, see Example 1.2 for the deterministic system. We will follow closely the exposition in Pardoux-Pignol (1984).

The assumption of white noise disturbances leads to the Statonovic SDE

$$(5.10)\qquad dx(t) = [A(t)x(t) + F(t)]dt + \sum_{i=1}^{m} [B_i(t)x(t) + G_i(t)] \circ dW^i$$

where $A(t)$, $B(t)$, $F(t)$, $G_i(t)$ are periodic with period T. Just as in the deterministic case, we first analyze the homogeneous part of Equation (5.10):

$$(5.11)\qquad dy(t) = A(t)y(t)dt + \sum_{i=1}^{m} B_i(t)y(t) \circ dW^i .$$

Denote by $\Phi(t)$ the fundamental matrix of (5.11). Now: if there exists $\eta < 0$ such that for the upper Lyapunov exponent with probability 1

$$\overline{\lim_{t\to\infty}} \frac{1}{t} \log\|\Phi(t)\| < \eta ,$$

then (5.10) has a T-periodic solution $x^p(t)$ (compare the definitions at the end of Section 1.) and all Markovian solutions $x(t,x_0)$ of (5.10) tend to $x^p(t)$ as $t \to \infty$, i.e. $x^p(t) - x(t,x_0) \to 0$ almost surely. The stability of (5.10) with respect to periodic solutions is thus reduced to the problem of computing the upper Lyapunov exponent of (5.11).

Again, under the corresponding rank condition for the Lie algebra generated by $\{A(t)x, B_1(t)x, \ldots, B_m(t)x\}$, the upper Lyapunov exponent λ is attained w.p.1 for all $x_0 \in \mathbb{R}^d\setminus\{0\}$ and Hasminskii type formula (compare Theorem 3.11) holds.

In the periodic case, discretization in time with respect to the period T leads to a suitable numerical sceme for the calculation of λ: Define again for any $d \times d$ matrix $C : C^0 = C - \frac{1}{d}$ trace C and consider the reduced system

$$d\bar{y}(t) = A^0(t)\bar{y}(t) + \sum_{i=1}^{m} B_i^0(t)\bar{y}(t) \circ dW^i.$$

Under the corresponding rank conditions, one obtains for $y_0 \in \mathbb{R}^d \setminus \{0\}$

$$\lambda^0 = \lim_{n \to \infty} \frac{1}{nT} \log |\bar{y}(nT, y_0)| > 0$$

and the upper Lyapunov exponent of (5.11) is

$$\lambda = \lambda^0 + \frac{1}{dT} \int_0^t \text{trace } A(s)\, ds .$$

Further results on the approximte evaluation of λ are given in Pardoux-Pignol (1984).

5.4. NONLINEAR SYSTEMS WITH WHITE NOISE: STABILITY AND STOCHASTIC BIFURCATIONS

For the discussion of the effect of white noise on the behavior of nonlinear systems we restrict ourselves to one-dimensional systems, i.e. $M \subset \mathbb{R}^1$, because the invariant probabilities and the Lyapunov exponents can be computed explicitly via the Fokker-Planck equation.

Consider the system

$$\text{(5.12)} \qquad dx(t) = f(x(t))dt + g(x(t)) \circ dW \qquad \text{in } \mathbb{R}^1$$

where $g(x) > 0$. If (5.12) has a stationary solution, then the density of the invariant probability $p^0(x)$ is given by

$$\text{(5.13)} \qquad p^0(x) = c \cdot \frac{1}{f(x)} \exp\{ 2 \int_{x_0}^{x} \frac{f(u)}{g(u)^2} du\}$$

where c is a normalization constant, compare e.g. Gihman-Skorohod (1979). Denote by $X^0(t)$ the stationary solution associated with $p^0(x)$, then its Lyapunov exponent is given by (compare (4.5))

$$\text{(5.14)} \qquad \lambda = \int [f'(x) + \frac{1}{2} g'(x) g''(x)] p^0(x) dx.$$

As an example we consider the Verhulst equation

$$\text{(5.15)} \qquad dx(t) = [ax(t) - x(t)^2]dt + \sigma x(t) \circ dW \quad \text{in } \mathbb{R}^+ = [0, \infty) .$$

The corresponding deterministic equation

$$\dot{x} = ax - x^2 \tag{5.16}$$

has the following behavior for $x \in \mathbf{R}^+$: $x \equiv 0$ and $x \equiv a$ (for $a > 0$) are steady states, 0 is stable for $a < 0$, a is stable for $a > 0$. At the point $a_0 = 0$ the system therefore has a transcritical bifurcation.

For the stochastic system the density (5.13) is integrable on $(0,\infty)$ iff $a > 0$, i.e. for $a \leq 0$ the only stationary density is the Dirac measure at the steady state 0. Formula (5.14) yields: the Lyapunov exponent $\lambda_0(a)$ of the steady state solution is $\lambda_0(a) = a$, hence 0 is stable for $a < 0$ and the Lyapunov exponent changes from negative to positive at $a_0 = 0$: a_0 is stochastic bifurcation point.

For $a > 0$ there exists a second stationary density $p_a^0(x)$ on $(0,\infty)$, hence a second stationary solution. By plugging $p_a^0(x)$ into formula (5.14), we see that the Lyapunov exponent $\lambda_1(a)$ for this stationary solution is always negative, hence no further bifurcations occur.

Summing up we see that for the white noise Verhulst model (5.15) the stochastic bifurcation point is the same as for the deterministic model (5.16). This need not always be the case, as Equation (5.14) shows: Consider the SDE in $\mathbf{R}^1$

$$dx(t) = f\big(a,x(t)\big)dt + g\big(x(t)\big) \circ dW \tag{5.17}$$

such that $x^0(t) \equiv x_0$ is a steady state of (5.17) for all a. Then the invariant density for this stationary solution is δ_{x_0}, hence

$$\lambda_0(a) = f'(a,x_0) + \frac{1}{2} g'(x_0)g''(x_0) .$$

The bifurcation behavior of the deterministic system $\dot{x} = f\big(a,x(t)\big)$ is described through the linearized equation

$$\dot{v}(t) = f'(a,x_0)v(t) ,$$

hence determined by the sign of $f'(a,x_0)$. For the stochastic system (5.17) we obtain

- if $g'(x_0)g''(x_0) = 0$, then the deterministic and the stochastic bifurcation point coincide,
- if $g'(x_0)g''(x_0) > 0$ (< 0), then the stochastic bifurcation occurs earlier (or later) than the deterministic one. Hence the noise has destabilizing (or stabilizing) effect on the steady state x_0.

A simple example is

$$dx(t) = \left[ax(t) - x(t)^2\right]dt + \left[\exp\{bx(t)\} - 1\right] \circ dW,$$

where $\lambda_0(a) = a + \frac{1}{2} b^3$ and the shift of the bifurcation point depends on the sign of b.

Model (5.15) has another interesting feature concerning the nontrivial stationary density $p_a^0(x)$ for $a > 0$: If $0 < a < \frac{1}{2}\sigma^2$, then $p_a^0(x)$ has a pole at $x = 0$, while for $a > \frac{1}{2}\sigma^2$ the density $p_a^0(x)$ has a unique maximum, see Horsthemke-Lefever (1984). These authors also discuss situations, where invariant densities change the number of maxima etc. It should be pointed out however that these phenomena are not always accompanied by a change of sign of the Lyapunov exponent, i.e. are not stochastic bifurcations in the way the term is used here.

5.5. NONLINEAR SYSTEMS WITH COLORED NOISE: STABILITY AND STOCHASTIC BIFURCATIONS OF A RANDOM VAN DER POL OSCILLATOR

We apply the results of section 4.2 to a Van der Pol oscillator with colored noise of the form

$$\text{(5.18)} \qquad \ddot{y}(t) - 2b\left(y(t)^2 - 1\right)\dot{y}(t) + \left(1 + \sigma f(\xi)\right)y(t) = 0$$

where ξ and f satisfy the conditions of Section 5.2. Rewrite (5.18) as

$$\text{(5.19)} \qquad \begin{pmatrix} \dot{x}_1 \\ \dot{x}_2 \end{pmatrix} = \begin{pmatrix} x_2 \\ 2b(x_1^2-1)x_2 - \left(1 + \sigma f(\xi)x_1\right) \end{pmatrix} = X(x_1,x_2) .$$

The origin $(0,0)$ is a steady of (5.19) and we linearize around this point:

$$\text{(5.20)} \qquad \dot{v}(t) = \begin{pmatrix} 0 & 1 \\ -1 - \sigma f(\xi) & -2b \end{pmatrix} v(t) = D_x X(0)v(t)$$

The resulting system (5.20) was discussed in Section 5.2. The stability diagram in the b-σ-plane shows the (local) stochastic stability behavior of the Van der Pol Oscilator (5.12) around its steady state.

We now consider the bifurcation behavior of the system (5.12) via its Lyapunov exponents. In this sense a stochastic system admits a bifurcation, if its upper Lyapunov exponents wanders through 0.

The deterministic analogue of (5.12) was discussed in Example 1.3 with $-2b = a$. Taking b as the bifurcation parameter, we see that the deterministic system exhibits a Hopf bifurcation at $b_0 = 0$ when the sign of b changes from positive to negative.

For the stochastic system assume that f takes on at least 2 values and the amplitude $\sigma > 0$. At the parameter value $b_0 = 0$ we have seen in Section 5.2 that $\lambda(b_0) > 0$, while for b large enough one has $\lambda(b) < 0$. Thus the system (5.12) exhibits a stochastic bifurcation, for $\sigma > 0$ fixed, at a point $b_\sigma > 0$. The randomness therefore leads to a shift of the bifurcation point: the stochastic bifurcation takes place earlier, i.e. at a parameter value where for the deterministic system the origin is still exponentially stable.

The expansion formulas (5.8) and (5.9) can be used to obtain expansions for the shift of the bifurcation point as a function of the noise amplitude σ. This approach, further examples and more mathematical fundations are work in progress.

REFERENCES

Arnold, L. (1984) A formula connecting sample and moment stability of linear stochastic systems, SIAM J. Appl. Math. 44, 793-802.

Arnold, L. (1987) Lyapunov exponents of linear stochastic systems, to appear in: G. I. Schueller and F. Ziegler (eds.) Nonlinear Stochastic Dynamic Engineering Systems, Springer Verlag: Berlin.

Arnold, L. Horsthemke, W. and J. Stucki (1979) The influence of external real and white noise on the Lotka-Volterra model, Biometrical J. 21, 451-471.

Arnold, L. and W. Kliemann (1986) Large deviations of linear stochastic differential equations, to appear in Proceedings of the Conference in Eisenach, GDR", Springer Lecture Notes in Control and Information Sciences.

Arnold, L. and W. Kliemann (1987) On unique ergodicity for degenerate diffusions, to appear in Stochastics.

Arnold, L., W. Kliemann and E. Oeljeklaus (1986a) Lyapunov exponents of linear stochastic systems, in Arnold-Wihstutz (1986), 85-125.

Arnold, L., E. Oeljeklaus and E. Pardoux (1986b) Almost sure and moment stability for linear Itô equations, in Arnold-Wihstutz (1986), 129-159.

Arnold, L., Papanicolaou, G. and V. Wihstutz (1986c) Asymptotic analysis of the Lyapunov exponent and rotation number of the random oscillator, SIAM J. Appl. Math. 46, 427-450.

Arnold, L. and V. Wihstutz (eds.) (1986) Lyapunov Exponents, Springer Lecture Notes in Mathematics 1186, Springer Verlag: Berlin.

Baxendale, P. (1980) Wiener processes on manifolds of maps, Proc. Royal Soc. Edinburgh 87A, 127-152.

Baxendale, P. (1986) The Lyapunov spectrum of a stochastic flow of diffeomorphisms, in Arnold-Wihstutz (1986), 322-337.

Baxendale, P. (1986b) Asymptotic behavior of stochastic flows of diffeomorphisms, Proceedings of 15th SPA, Nagoya 1985, Lecture Notes in Mathematics, Vol. 1203, Springer Verlag: Berlin.

Bougerol, P. and J. Lacroix (1985) Products of Random Matrices with Applications to the Schrödinger Operator, Birkhäuser: Boston.

Carr, J. (1986) Applications of Center Manifold Theory, Springer Verlag: Berlin.

Carverhill, A.P. (1985) Flows of stochastic dynamical systems: Ergodic theory, Stochastics 14, 273-318.

Carverhill, A.P., M.J. Chappell, and K.D. Elworthy (1986) Characteristic exponents for stochastic flows, in Springer Lecture Notes in Mathematics 1158, 52-80.

Chappell, M. J. (1986) Bounds for average Lyapunov exponents of gradient stochastic systems, in: Arnold-Wihstutz (1986), 308-321.

Colonius, F. and W. Kliemann (1986) Infinite time optimal control and periodicity, IMA preprint no. 240, University of Minnesota, Minneapolis, submitted to Appl. Math. Opt.

Ehrhardt, M. (1983) Invariant probabilities for systems in a random environment. Bull. Math. Biol. 45, 579-590.

Ehrhardt, M. and W. Kliemann (1872) Controllability of linear stochastic systems, Systems and Control Letters 2, 145-153.

Ellis, R.S. (1985) Entropy, Large Deviations and Statistical Mechanics, Springer: New York.

Elworthy, K.D. (1978) Stochastic dynamical systems and their flows, in: A. Friedman and M. Pinsky (eds.) Stochastic Analysis, Academic Press: New York, 79-95.

Ethier, S.N. and T.G. Kurtz (1986) Markov Processes, Wiley: New York.

Furstenberg, H. (1963) Noncommuting random products, Trans. AMS 108, 377-428.

Gihman, I.I. and A.V. Skorohod (1979) The Theory of Stochastic Processes Vol. III, Springer: New York.

Hasminskii, R.Z. (1980) Stochastic Stability of Differential Equations, Sijthoff and Nordhoff: Alphen.

Horsthemke, W. and R. Lefever (1984) Noise-Induced Transitions, Springer Verlag: Berlin.

Ichihara, K. and H. Kunita (1974) A classification of the second order degenerate elliptic operators and its probabilistic characterization. Z. Wtheorie verw. Gebiete 30, 235-254 and 39(1977), 81-84.

Ikeda, N. and S. Watanabe (1981) Stochastic Differential Equations and Diffusion Processes, North Holland: Amsterdam.

Kliemann, W. (1987) Recurrence and invariant measures for degenerate diffusions, to appear Ann. Prob.

Kunita, H. (1974) Diffusion processes and control systems, Lecture Notes, University of paris VI.

Kunita, H. (1978) Supports of diffusion processes and controllability problems. In: Proc. Intern. Symp. Stochastic Diff. Equ., Kyoto, Wiley: New York, 163-185.

Kunita, H. (1982) Stochastic differential equations and stochastic flows of diffeomorphisms. Ecole d' Ete de Probabilité de Saint-Flour XII.

Malliavin, P. (1978) Stochastic calculus of variations and hypoelliptic operators. In: Proc. Intern. Symp. Stochastic Diff. Equa., Kyoto. Wiley: New York, 195-263.

Pardoux, E. and M. Pignol (1984) Etude de la stabilité de la solution d'une EDS bilinéaire à coefficients periodiques. In: Lecture Notes in Control and Information Sciences, 63, 92-103.

Ruelle, D. (1979) Ergodic theory of differentiable dynamical systems. Publ. IHES 50, 275-305.

Rümelin, W. (1983) Stationary solutions of nonlinear systems disturbed by Markovian noise, IEEE Trans. AC 28, 244-246.

San Martin, L. and L. Arnold (1986) A control problem related to the Lyapunov spectrum of stochastic flows, Matematica Aplicada e Computacional 4, 21-64.

San Martin, L. (1986) Invariant control sets on fibre bundles, report no. 165 University of Bremen.

Stroock, D.W. and S.R.S. Varadhan (1972) On the support of diffusion processes with applications to the strong maximum principle, Proc. 6th Berkeley Symp. Math. Stat. Probab. 3, 333-359.

LECTURES ON LINEAR AND NONLINEAR FILTERING

M. Hazewinkel
CWI, Amsterdam, The Netherlands

1. Introduction

Quite generally the *filtering problem* can be described as follows. Given a stochastic process $x(t)$, $t \in I \subset \mathbf{R}$, i.e. a sequence of random variables, and a (more or less related) second process $y(t)$, $t \in I$, it is desired to find the *best estimate* of x at time t, i.e. the best estimate of $x(t)$, given the (past observations) $y(s)$, $0 \leqslant s \leqslant t$. Usually $I = \mathbf{Z}$ (discrete time) or $I = \mathbf{R}$ (continuous time).

Much related problems are *prediction:* calculate the best estimate of $x(t)$ given $y(s)$, $0 \leqslant s \leqslant t - r$, and *smoothing:* calculate the best estimate of $x(t-r)$ given $y(s)$, $0 \leqslant s \leqslant t$. In all these it may of course be the case that $y(t)$ and $x(t)$ are the same stochastic process.

In these lectures we shall be concerned with the (model) case that the continuous time processes $x(t)$ and $y(t)$ are related as follows

$$dx(t) = f(x(t))dt + G(x(t))dw(t), \quad x(t) \in \mathbf{R}^n, \ w(t) \in \mathbf{R}^m, \tag{1.1}$$

$$dy(t) = h(x(t))dt + dv(t), \ y(t) \in \mathbf{R}^p, \ v(t) \in \mathbf{R}^p \tag{1.2}$$

with initial conditions $x(0) \in \mathbf{R}^n$, $y(0) \in \mathbf{R}^p$. Here $w(t)$ and $v(t)$ are supposed to be independent Wiener noise processes also independent of the initial random variable $x(0)$, and $f(x)$, $G(x)$ and $h(x)$ are known vector and matrix valued functions. Thus $\dot{w}(t)$ and $\dot{v}(t)$ are white noise and (1.1) can be looked at as a dynamical system

$$\dot{x}(t) = f(x(t)) \tag{1.3}$$

subject to continuous random shocks whose direction and intensity is further modified (apart from being random) by $G(x(t))$. And equation (1.2), the observation equation, says that the observations at time t

$$y(t) = y(0) + \int_0^t h(x(s))ds, \quad \dot{y}(t) = h(x(t)) \tag{1.4}$$

are corrupted by further (measurement) noise $v(t)$. Technically speaking, equations (1.1), (1.2) are to be regarded as Ito stochastic differential equations; cf section 5 below for more remarks.

The phrase 'best estimate' of $x(t)$, or, more generally, of an interesting function $\phi(x(t))$, is to be understood in the mathematical sense of conditional expectation $\hat{x}(t) = E[x(t) | y(s), 0 \leqslant s \leqslant t]$, or, in the more general case, $E[\phi(x(t)) | y(s), 0 \leqslant s \leqslant t]$. This is a mathematically well defined object.

Unfortunately the (mathematical) proof of this statement contains nothing in the way of methods of calculating these conditional expectations (effectively).

There are many techniques and approaches to filtering. It is definitely not the idea of these lectures to give a general survey of the field. Instead I shall try to give an account of one particular approach pioneered by Roger Brockett, Martin Clark and Sanjoy Mitter, [6,7,8,9,53,54] which is variously known as the Lie-algebra approach, the reference probability approach, or the unnormalized density approach. This is a rather recent set of ideas, which has several merits. First, it takes geometrical aspects of the situation into account. Second, it explains convincingly why it is easy to find exact recursive filters for linear dynamical systems while it is very hard to filter something like the cubic sensor - for over 20 years a notoriously hard case to handle. The notion of a recursive exact filter will be discussed below in section 2. Thus excitement about this approach was high in the very first years of the 1980's. The book [34] well reflects this. Since then interested and excitement have waned perceptibly. There are also several connected reasons for this. First the method itself indicates clearly - through this remains to be proved in one sense or another - that one can not expect many cases (beyond the case of linear systems) where finite dimensional exact recursive filters exist. 'Generally', it seems, such filters will not exist and though there remains the tantalizing possibility of whole new classes of useful models for which they do exist, there are at the moment no clear ideas as to how and where to look for them. All the same a number of new filters, both 'model cases' and filters of importance in practice, have been discovered using these Lie-algebra ideas [4,13,18,19,47-51,56]. Since exact finite dimensional filters can not exist in many cases it is natural to look for approximate ones. Here it is not immediately apparent how to proceed on the basis of the Lie-algebra approach, and little has been done.

Still there are a number of very promising (heuristic) ideas, which definitely work in some cases. It is the second purpose of these lectures to examine some of these ideas for obtaining approximate recursive filters. All seem to lead to far from trivial unsolved, and possibly quite difficult, mathematical questions, which invite major research efforts.

2. Recursive filters

The basic quantities we have available at time t are observations up to and including time t, i.e. the $y(s)$, $0 \leq s \leq t$. A priori an algorithm to calculate $\hat{x}(t)$, say, could involve all the $y(s)$. Now if the observations come in at a high rate and the algorithm really needs all the $y(s)$ each time an estimate is calculated, one is likely to run into (i) long computation times, and (ii) storage (memory) problems. In such a situation it would be much more practical and much nicer if it were possible to calculate $\hat{x}(t+dt)$ on the basic of $\hat{x}(t)$ and the new information $y(t+dt)$ which has come in. (It is easier to think of this situation in discrete time with $dt=1$.)

This turns out to be too optimistic. Such filters almost cannot exist (in nontrivial situations). The next best thing would be the existence of some other quantity $\xi(t)$ which does have the recursive update property "$\xi(t+dt)=\alpha(\xi(t),y(t+dt))$" and from which the desired quantity can be directly calculated. Of course then $\xi(t)$ must be a reasonable quantity and not some hard to handle infinite dimensional object like the time history function $\{y(s):0\leq s\leq t\}$ itself. E.g. $\xi(t)$ could be a finite dimensional quantity, or something in an infinite dimensional space describable by a finite number of parameters or well approximatable in terms of a finite numbers of parameters.

Such filters do exist sometimes. E.g. in the case of linear time invariant systems

$$\begin{aligned} dx(t) &= Ax(t)dt + Bdw(t), \quad x \in \mathbf{R}^n,\ w \in \mathbf{R}^m \\ dy(t) &= Cx(dt + dv(t), \quad y \in \mathbf{R}^p,\ v \in \mathbf{R}^p \end{aligned} \tag{2.1}$$

where A,B,C are constant real matrices of dimensions $n\times n$, $n\times m$, $p\times n$ respectively, and where $w(t)$ and $v(t)$ are Wiener noise processes, independent of each other and also independent of the initial random vector $x(0)\in\mathbf{R}$. (One sets $y(0)=0$). In this case one has the well-known Kalman-Bucy filter for the conditional state

$$dP(t) = (AP(t)+P(t)A^T+BB^T-P(t)C^TCP(t))dt \tag{2.2}$$
$$dm(t) = Am(t)dt+P(t)C^T(dy(t)-Cm(t)dt)$$

Thus in this case the quantity $\xi(t)=(P(t),m(t))$ has the desired "recursive updating property", and the quantity we want to filter for, i.e. the quantity $\hat{x}(t)$ in this case, is obtained by a simple projection $\hat{x}(t)=m(t)$.

All this leads to the following initial definition of a *finite dimensional exact recursive filter* for a quantity (statistic) $\widehat{\phi(x(t))}$. By definition such a filter is a finite dimensional dynamical system of the form

$$d\xi(t) = \alpha(\xi(t))dt+\sum_{j=1}^{p}\beta_j(\xi(t))dy_j(t)\,,\quad \xi(t)\in M \tag{2.3}$$

where $y_j(t)$ is the j-th component of $y(t)$, together with an output map

$$\widehat{\phi(x(t))} = \gamma(\xi(t)) \tag{2.4}$$

Here M is supposed to be a finite dimensional manifold and the α and β_j are smooth vectorfields on M. (One can usually think of M as an $\mathbf{R}^n$ so that (2.3) becomes an ordinary stochastic differential equation).

Of course, more generally, one could let the α and β_j in (2.3) depend on the y_j as well. This does not bring very much more because we can, so to speak, add the $y_1,\dots,y_p$ to the state variables $x_1,\dots,x_n$. However, certainly more general potential filters could be considered then (2.3); in particular one can allow the output map γ at time t to depend explicitly on $y_1(t),\dots,y_p(t)$, and we shall have occasion to use this. Again this can be taken care of by extension. This time by extending the filter state vector $\xi(t)$ to the filter state vector $(\xi(t),y(t))$.

3. Robustness

The $y(t)$ are stochastic processes. As it stands (2.3) is a stochastic (partial) differential equation and as such its solutions are only defined apart from a set of measure zero. On the other hand the possible observations paths are piece wise differentiable and these constitute a set of measure zero (in the space of all paths under Wiener measure or therewith mutually continuous measures). Thus solutions of (2.3) may, so to speak, well be undefined precisely on the possible observation paths, [14].

More importantly - in my view - in actual situations we do not have available the stochastic process $y(t)$ but just one possible realization of it. Thus it would be nice if (2.3) made sense pathwise and if it could be replaced with something involving just functions of the $y_1(t),\dots,y_p(t)$, say polynomials, and not their derivatives.

For the filter (2.2) this can be done. The transformation $\bar{m}=m-PC^Ty$ yields

$$d\bar{m} = dm-PC^Tdy-dPC^Ty=Amdt+PC^Tdy-PC^{T^T}Cmdt-PC^Tdy-dPC^Ty$$
$$= (A-PC^TC)\bar{m}dt-(PA^TC^T+BB^TC^T)ydt$$

or

$$\frac{d}{dt}\bar{m} = (A-PC^TC)\bar{m}-(PA^TC^T+BB^TC^T)y$$
$$\frac{d}{dt}P = AP+PA^T+BB^T-PC^TCP \tag{3.1}$$
$$\hat{x}(t) = \bar{m}(t)+P(t)C^Ty(t)$$

a set of equations which makes perfect sense for an arbitrary single continuous but possibly almost everywhere non-differentiable observation path $y(t)$.

Such filters are called robust.

4. The unnormalized density approach

One obvious approach to try to find a filter for, say, the conditional state $\hat{x}(t)$ is to try to derive a differential equation for it. This can actually be done [44-46] and yields

$$d\hat{x} = \hat{f} - (\widehat{(xh_T)} - \hat{x}\hat{h}^T)\hat{h}dt + (\widehat{(xh_T)} - \hat{x}\hat{h}^T)dy \tag{4.1}$$

where f and h stand for $f(x(t))$ and $h(x(t))$ and a $\hat{}$ over a symbol means taking the conditional expectation. The trouble with this equation is that it involves the expectations $\hat{f}$, $\hat{h}$ and $\widehat{(xh_T)}$ (and these are for nonlinear f,h of course not equal to $f(\hat{x})$, $h(\hat{x})$ and $\hat{x}h^T(\hat{x})$). One can also write down equations for the conditional quantities $\hat{f}$, $\hat{h}$, $\widehat{xh^T}$, but these then involve conditional expectations of still more complicated expressions etc, etc. As a rule this process will not stop and there results an infinite system of equations.

The conditional density $p(x,t)$, that is the density of the stochastic variable $\hat{x}$ at time t, satisfies a nicer looking equation

$$dp = \bar{\pounds}pdt + (h - \hat{h})^T(dy - \hat{h}dt)p \tag{4.2}$$

Given $p(t,x)$, $\hat{h}$ (at time t) is calculated by $\hat{h} = \int(hx)p(x,t)dx$ and inserting this gives in any case an integro partial differential equation of recursive type.

Still nicer is the equation satisfied by a certain unnormalized version $\rho(x,t)$ of the conditional density. And it is this equation, the so called Duncan-Mortensen-Zakai equation, or DMZ-equation, which is at the basis of the Lie algebraic approach. Here unnormalized means that $\rho(x,t) = r(t)p(x,t)$ for some (unknown) function $r(t)$ depending only on time (and not on x).

The DMZ equation for $\rho(x,t)$ reads ([20,33,34,55,65])

$$d\rho(x,t) = \bar{\pounds}\rho(x,t)dt + \sum_{j=1}^{p} h_j(x)\rho(x,t)dy_j(t) \tag{4.3}$$

where $\bar{\pounds}$ is the second-order semi-elliptic operator (in the $x_1,\ldots,x_n$) defined by

$$(\bar{\pounds}\phi)(x) = \frac{1}{2}\sum_{j,k}\frac{\partial_2}{\partial x_j \partial x_k}((G(x)G(x)^T)_{jk}\phi) - \sum_{i=1}^{n}\frac{\partial}{\partial x_i}(f_i(x)\phi). \tag{4.4}$$

Here $(G(x)G^T(x))_{jk}$ is the (j,k)-entry of the $n \times n$ matrix GG^T and f_i is the i-th component of f.

Note that (4.2) is recursive, but being a *partial* stochastic differential equation, it is of course infinite dimensional. Note also that for the calculation of conditional expectations the unknown factor $r(t)$ does not matter much. Indeed $r(t) = \int\rho(x,t)dx$ and correspondingly, if $\phi(x)$ is some interesting function of the state, one has that

$$\widehat{\phi(x(t))} = \int_{\mathbf{R}^n}\phi(x)\rho(x,t)dx \Big/ \int_{\mathbf{R}^n}\rho(x,t)dx \tag{4.5}$$

5. Ito and Fisk-Stratonovič stochastič differential equations

Most of the equations written down so far, e.g. (1.1), (1.2), (2.1), (2.2), (2.3), (4.1), (4.2), (4.3) are *stochastic* differential equations. It is definitely not my intention to give extensive explanations of what this means, but a few words seem in order. The meaning of (1.1) e.g. is that there is a well defined notion of stochastic integral such that

$$x(t) = x(0) + \int_0^t f(x(t))dt + \int_0^t G(x(t))dw(t) \tag{5.1}$$

There are in fact several possible definitions. Two of these are the Ito integral and the Fisk-Stratonovič integral [1]. And they are definitely different in the sense that different stochastic processes result depending on whether the second integral (5.1) is interpreted in the Ito or Fisk-

Stratonovič sense. Both definitions have their advantages and disadvantages. The Fisk-Stratonovič integral has the major advantage that the usual rules of the differential-integral calculus still hold. This makes it the preferred interpretation on manifolds. Thus equation (2.3) is intended to be read as a Fisk-Stratonovič equation.

The original system (1.1), (1.2) however is a set of Ito equations and the DMZ equation (4.3) in also an Ito equation.

There is a fairly simple conversion rule from Ito equations to Fisk-Stratonovič equations and vice versa as follows.

Let

$$dx = f(x)dt + G(x)dw(t)\,, \quad x\in\mathbf{R}^n,\ w\in\mathbf{R}^m \tag{5.2}$$

be an Ito stochastic differential equation. Then the equivalent Stratonovič equation is

$$dx = f(x)dt - \frac{1}{2}\sum_{j=1}^{m}\sum_{i=1}^{n}\left(\frac{\partial G}{\partial x_i}\right)_j G_{ij}dt + G(x)dw(t) \tag{5.3}$$

Here $\left(\frac{\partial G}{\partial x_i}\right)_j$ is the j-th column of the $n\times n$ matrix $\frac{\partial G}{\partial x_i}$, i.e. $\left(\frac{\partial G}{\partial x_i}\right)_j = \left(\frac{\partial G_{1j}}{\partial x_i},\ldots,\frac{\partial G_{nj}}{\partial x_i}\right)^T$, and G_{ij} is the (i,j)-entry of the $n\times m$ matrix G. Here equivalent means that the same stochastic processes $x(t)$ occur as solutions of the Ito equation (5.2) and the Fisk-Stratonovič equation (5.3).

Note in particular that for an Ito equation of bilinear type

$$dx = Axdt + \sum_{k=1}^{m} B_k x dw_k(t) \tag{5.4}$$

the equivalent Fisk-Stratonovič equation is

$$dx = Axdt - \frac{1}{2}\sum_{k=1}^{m}(B_k^2 x)dw_k(t) + \sum_{k=1}^{m} B_k x dw_k(t) \tag{5.5}$$

The equivalent Fisk-Stratonovič equation to the DMZ filter equation (4.3) is

$$d\rho = \pounds\rho dt + \sum_{j=1}^{p} h_j \rho dy_j(t) \tag{5.6}$$

where now $\pounds$ is the operator given by

$$\pounds(\phi) = \frac{1}{2}\sum\frac{\partial^2}{\partial x_i\partial x_j}((GG^T)_{ij}\phi) - \sum_i\frac{\partial}{\partial x_i}(f_i\phi) - \frac{1}{2}\sum_j h_j(x)^2\phi. \tag{5.7}$$

Though I have not given a discussion of the Ito-Stratonovič equivalence for partial stochastic differential equations this is easily understood by analogy from (5.5) if nothing else. Note that in (4.3) the unknown is ρ and that the $h_j(x)$ are (commuting) diagonal linear operators $\rho\mapsto h_j(x)\rho$.

As is turns out the stochastic aspects of the filtering problem in this approach largely disappear. This happens because there is an equivalent version of the Fisk-Stratonovič type DMZ equation (5.6) which is robust and can be interpreted pathwise, i.e. as a family of deterministic partial differential equations indexed by the possible observation paths, say, by the continuous functions $\mathbf{R}(\geqslant 0)\to\mathbf{R}^p$; cf below.

If from now on a stochastic differential equation appears then unless the contrary is explicitly stated, it will always be a Fisk-Stratonovič one.

6. THE ROBUST VERSION OF THE DMZ EQUATION

Consider the (Fisk-Stratonovič) equation (5.6) for an unnormalized conditional density $\rho(x,t)$. This involves the $dy_j(t)$. Now, as also mentioned in section 3 above, what we have available in terms of observations is one realization, one possible path, of a stochastic process $y(t)$. Hence, apart from smoothing effects introduced by the measurements process an almost surely nowhere differentiable function, which makes it more difficult to handle the integrals involved and to find numerical approximations.

Consider the time-dependent (gauge) transformation

$$\tilde{\rho}(x,t) = e^{-h_1(x)y_1(t)-\ldots-h_p(x)y_p(t)}\rho(x,t) \tag{6.1}$$

As we are dealing with a Fisk-Stratonovič equation the ordinary rules of calculus apply and equation (5.6) transforms into an equation

$$\frac{\partial\tilde{\rho}}{\partial t} = £\tilde{\rho} - \sum_{i=1}^{p} y_i(t)£_i\tilde{\rho} + \sum_{i,j=1}^{p} y_i(t)y_j(t)£_{ij}\tilde{\rho} \tag{6.2}$$

where the differential operators $£_i$ and $£_{ij}$ are given by

$$£_i = [h_i,£] = h_i£ - £h_i\ , \quad £_{ij} = \frac{1}{2}[h_i,[h_j,£]]. \tag{6.3}$$

Cf. below for a derivation of (6.2). The terms $dy_j(t)$ cancel after this transformation and we are left with a family of partial differential equation indexed by the possible observations paths, i.e. with one equation from this family in a given filtering situation.

If $\tilde{\rho}(x,t)$ (for a particular path $y(t)$) has been found from (6.2), then $\rho(x,t)$ is given, as a function of x and t, by formula 6.1. Note in particular that the $h_i(x)$ in (6.2) should not be read as functions of the stochastic process $x(t)$; instead (6.1) is simply the exponential of the known multiplication operator $\rho \mapsto (\Sigma_{i=1}^{p} h_i(x)y_i(t))\rho$ on densities.

To obtain (6.2) observe that substituting (6.1) into the DMZ equation (5.6) gives

$$\frac{\partial\tilde{\rho}}{\partial t} = e^{-\sum h_j(x)y_j(t)}£e^{\sum h_j(x)y_j(t)}\tilde{\rho} \tag{6.4}$$

Thus writing A for the operators of multiplication with $\sum h_j(x)y_j(t)$, we have to calculate $e^{-A}£e^{A}$. By the adjoint action formula (cf the short tutorial on Lie algebras in this volume, [28]) this is equal to

$$e^{-A}£e^{A} = £ - [A,£] + \frac{A,[A,£]]}{2!} - \frac{[A,[A,[A,£]]]}{3!} + \ldots \tag{6.5}$$

In our case $£$ is a second order differential operator and A is multiplication with a function. Hence $[A,£]$ is a first order differential operator, $[A,[A,£]]$ is a zero-th order differential operator, i.e. (multication with) a function, and $[A,\ldots,[A,£]\ldots]=0$ if three or more A's occur. Writing out $[A,£]$ and $[A,[A,£]]$ yields (6.2)-(6.3).

Even though now we can work with nonstochastic partial differential equations (6.2) the numerics of the situation are daunting, cf however, also [57]. Typically x is a large dimensional vector of, say, dimension 27, in certain practical problems involving helicopters. So we have a second order semi-elliptic PDE in 27 space dimensions and one time dimension. This rules out standard approximation schemes. Also of course we need a solution method which deals in principle not with one instance of equation (6.2) but with the whole family (6.2). I.e. the parameters $y_1(t),\ldots,y_p(t)$ must enter into the calculation algorithm in a reasonable way. These remarks constitute some of the motivation for the approach via Wei-Norman equations discussed in the sections below.

7. Wei-Norman theory [64]

It is important to note that the filtering equation (6.2) (or (5.6)) is of the general form

$$\dot{x} = (A_1x)u_1 + \ldots + (A_kx)u_k \tag{7.1}$$

where the A_i are linear operators and the u_i known functions of time. Of course in (6.2) the role of x is played by $\tilde{p}$, an infinite dimensional object. Here, for the moment, lets consider (7.1) as a finite dimensional object. Let us also assume that the $A_1,\ldots,A_k$ who are now, say, $n \times n$ matrices, form the basis of a Lie algebra. (By adding a few more terms with corresponding u_i equal to zero this can of course always be assured.) Let us look for solutions of the form (an Ansatz)

$$x(t) = e^{g_1A_1}e^{g_2A_2}\ldots e^{g_kA_k}x(0) \tag{7.2}$$

where the $g_i(t)$ are still to be determined functions of t. Differentiating (7.2) gives

$$\dot{x} = \dot{g}_1A_1e^{g_1A_1}e^{g_2A_2}\ldots e^{g_kA_k}x(0) + e^{g_1A_1}\dot{g}_2A_2e^{g_2A_2}\ldots e^{g_kA_k}x(0) + \ldots \tag{7.3}$$

and inserting

$$e^{-g_{i-1}A_{i-1}}\ldots e^{-g_1A_1}e^{g_1A_1}\ldots e^{g_{i-1}A_{i-1}}$$

just after $\dot{g}_iA_i$ in the i-the term, equation (7.3) can be rewritten

$$\begin{aligned}\dot{x} &= \sum_{i=1}^{k}\dot{g}_i(A_i + \sum_{\substack{j_1,\ldots,j_{i-1} \\ j_1+\ldots+j_i-1>0}} \frac{g_1^{j_1}\ldots g_{i-1}^{j_{i-1}}}{j_1!\ldots j_i-1!} ad_{A_1}^{j_1}\ldots ad_{A_{i-1}}^{j_{i-1}}(A_i))x \\ &= \sum_{i=1}^{k}\dot{g}_i(A_i + h_{ij}(g_1,\ldots,g_k)A_j)\end{aligned} \tag{7.4}$$

with $h_{ij}(0,\ldots,0)=0$, where, again, the adjoint action formula (6.5) has been used. Here $ad_A(B)=[A,B]$, $ad_A^i(B)=ad_A(ad_A^{i-1}(B))$. Thus it remains to solve (equating the coefficients of the basis elements A_i in (7.4) and (7.1))

$$\begin{aligned}\dot{g}_1 + \dot{g}_1h_{11}(g_1,\ldots,g_k) + \dot{g}_2h_{21}(g_1,\ldots,g_k) + \ldots + \dot{g}_kh_{k1}(g_1,\ldots,g_k) &= u_1 \\ \dot{g}_2 + \dot{g}_1h_{12}(g_1,\ldots,g_k) + \dot{g}_2h_{22}(g_1,\ldots,g_k) + \ldots + \dot{g}_kh_{k2}((g_1,\ldots,g_k) &= u_2 \\ &\ldots \\ \dot{g}_k + \dot{g}_1h_{1k}(g_1,\ldots,g_k) + \dot{g}_2h_{2k}(g_1,\ldots,g_k) + \ldots + \dot{g}_kh_{kk}(g_1,\ldots,g_k) &= u_4\end{aligned} \tag{7.5}$$

which can be done for small t and $g_1(0)=\ldots=g_k=(0)=0$ because $h_{ij}(0,\ldots,0)=0$. These equations are called the Wei-Norman equations of (7.1). In general a representation (7.2) for the solution is only possible for small t. However things change if the Lie-algebra in question is solvable [64], then there is such a representation for all t. More precisely there is a suitable basis such that there is such a representation for all t. How this comes about is easy to see in the case that the Lie algebra L is nilpotent:

$$L \underset{\neq}{\supset} L^{(1)} = [L,L] \underset{\neq}{\supset} L^{(2)} = [L,L^{(1)}] \underset{\neq}{\supset} \ldots \underset{\neq}{\supset} L^{(m)} = [L,L^{(m-1)}] = 0 \tag{7.6}$$

Indeed let $A_1,\ldots,A_{k_1}$, $A_{k_1+1},\ldots$, $A_{k_2},\ldots,A_{k_{m-1}}+1,\ldots,A_{k_m} = A_k$, $k_1<k_2<\ldots<k_m$ be a basis such that $A_{k_i+1},\ldots,A_{k_m}$ is a basis for $L^{(i)}$, $i=0,\ldots,m-1 (k_0=1, k_m=k)$. Then it immediately follows from (7.4) that $h_{ij}=0$ for $j \leqslant i$ and the set of equations (7.5) gets a nice triangular structure. Moreover $h_{ij}(g_1,\ldots,g_k)$ involves only $g_1,\ldots,g_{i-1}$ (this is always the case, cf (7.4), so the h_{1j} in (7.5) are always all zero) and the resulting equations (7.5) for the nilpotent case are therefore of the form

$$\begin{aligned}\dot{g}_1 &= u_1, \quad \ldots \quad ,\dot{q}_{k_1} = u_{k_1} \\ \dot{g}_{k_1+1} &= u_{k_1+1} + \alpha_{k_1+1}(u_1,\ldots,u_{k_1};g_1,\ldots,g_{k_1}),\ldots,\dot{g}_{k_2} = u_{k_2} + \alpha_{k_2}(u_1,\ldots,u_{k_1};g_1,\ldots,g_{k_1}) \\ \dot{g}_{k_2+1} &= u_{k_2+1} + \alpha_{k_2+1}(u_1,\ldots,u_{k_2};g_1,\ldots,g_{k_2}),\ldots,\dot{g}_{k_3} = u_{k_3} + \alpha_{k_3}(u_1,\ldots,u_{k_2};g_1,\ldots,g_{k_2}) \\ &\ldots\end{aligned} \tag{7.7}$$

where the α_j are known (universal) functions of the u's and g's. The initial conditions are $g_i(0)=0$, $i=1,...,k$.

It is quite important (for applications of Wei Norman theory) to note that the equations (7.5), i.e. the functions h_{ij} are universal and depend only on the abstract structure of the Lie algebra and the chosen basis. This means the following. That the matrices $A_1,...,A_n$ form the basis of a Lie algebra L of $n \times n$ matrices means that

$$[A_i,A_j] = A_iA_j - A_jA_i = \sum_r \gamma_{ij}^r A_r \tag{7.8}$$

for certain real numbers γ_{ij}^r, the structure constants of L with respect to this basis. Now let L' be a second Lie algebra, say of $m \times m$ matrices. Suppose that L and L' are isomorphic under $\phi: L \to L'$ and let $B_i = \phi(A_i)$. Then the $B_1,...,B_k$ are a basis for L' and because ϕ is an isomorphism

$$[B_i,B_j] = \sum_r \gamma_{ij}^r Br \tag{7.9}$$

with precisely the same γ_{ij}^r. As a result the Wei-Norman equations for the bilinear system of equations

$$\dot{y} = (B_1y)u_1 + ... + (B_ky)u_k \quad , y \in \mathbb{R}^m \tag{7.10}$$

are exactly the same.

This idea also extends to the case that we have a set $£_1,...,£_k$ of operators on some function space which form the basis of a Lie algebra L (so that for all i,j $[£_i,£_j] = \Sigma\gamma_{ij}^r £_r$ for certain γ_{ij}^r). Then again the Wei-Norman equations are identical to the ones of any finite dimensional copy L' of L (and by Ado's theorem, cf the short tutorial in Lie algebras in this volume, such a finite dimensional copy always exists). Of course in such a case of operators we still need to be able to calculate the $e^{g_i(t)L_i}$ for the individual operators L_i. Thus Wei-Norman theory can be seen as a method to integrate (solve) an equation of the form (7.1) in terms of the more elementary equations

$$\dot{x} = A_i x \tag{7.11}$$

Let me illustrate all this by means of an explicit example. Consider the four differential operators in one variable

$$£ = \frac{1}{2}\frac{d^2}{dx^2} - \frac{1}{2}x^2 \ , \ x, \ \frac{d}{dx}, \ 1 \tag{7.12}$$

We then have

$$[£,x]\phi = £(x\phi) - x£\phi = \frac{1}{2}\frac{d^2}{dx^2}(x\phi) - \frac{1}{2}x^3\phi - \frac{1}{2}x\frac{d^2}{dx^2}\phi + \frac{1}{2}x^3\phi = \frac{1}{2}\frac{d}{dx}(\phi + x\frac{d\phi}{dx}) - \frac{1}{2}x\frac{d^2\phi}{dx^2}$$

$$= \frac{1}{2}\frac{d\phi}{dx} + \frac{1}{2}\frac{d\phi}{dx} + \frac{1}{2}x\frac{d^2\phi}{dx^2} - \frac{1}{2}x\frac{d^2\phi}{dx^2} = \frac{d\phi}{dx} = \frac{d}{dx}(\phi).$$

Thus

$$[£,x] = \frac{d}{dx} \tag{7.13}$$

and similarly one finds

$$[£,\frac{d}{dx}] = x, \ [\frac{d}{dx},x] = 1, \ [£,1] = [x,1] = [\frac{d}{dx},1] = 0 \tag{7.14}$$

Thus the four operators (7.12) span a four dimensional Lie algebra. It is called the oscillator Lie algebra.

A finite dimensional copy in terms of 4×4 matrices of this algebra is given by the assignment

$$£ \mapsto \begin{pmatrix} 0 & 1 & 0 & 0 \\ 1 & 0 & 0 & 0 \\ 0 & 0 & 0 & 0 \\ 0 & 0 & 0 & 0 \end{pmatrix}, \ x \mapsto \begin{pmatrix} 0 & 0 & 1 & 0 \\ 0 & 0 & 0 & 0 \\ 0 & 0 & 0 & 0 \\ 0 & 1 & 0 & 0 \end{pmatrix}, \ \frac{\partial}{\partial x} \mapsto \begin{pmatrix} 0 & 0 & 0 & 0 \\ 0 & 0 & 1 & 0 \\ 0 & 0 & 0 & 0 \\ -1 & 0 & 0 & 0 \end{pmatrix}, \ 1 \mapsto \begin{pmatrix} 0 & 0 & 0 & 0 \\ 0 & 0 & 0 & 0 \\ 0 & 0 & 0 & 0 \\ 0 & 0 & -2 & 0 \end{pmatrix}$$

(Exercise: check this; NB this does not contradict the Stone-von Neumann theorem that it is impossible to represent the communication relation $[\frac{\partial}{\partial x},x]=1$ in terms of finite dimensional operators *in such a way that 1 is represented by the unit operator.)*

Now, by way of example, let us explicitly calculate the Wei-Norman equations for the Lie algebra (7.12).

So the equation we want to solve is

$$\rho_t = £\rho u_1 + x\rho u_2 + \frac{\partial}{\partial x}\rho u_3 + \rho u_4, \quad \rho(x,0)=\pi(x) \tag{7.15}$$

The 'Ansatz' is a solution of the form

$$\rho = e^{g_1(t)£} e^{g_2(t)x} e^{g_3(t)\frac{d}{dx}} e^{g_4(t)} \pi(x) \tag{7.16}$$

Differentiating (7.16) gives

$$\begin{aligned}
\frac{d}{dt}\rho &= £\dot g_1 e^{g_1£} e^{g_2x} e^{g_3\frac{d}{dx}} e^{g_4}\pi(x) \\
&+ e^{g_1£}\dot g_2 x e^{g_2x} e^{g_3\frac{d}{dx}} e^{g_4}\pi(x) \\
&+ e^{g_1£} e^{g_2x}\dot g_3 \frac{d}{dx} e^{g_3\frac{d}{dx}} e^{g_4}\pi(x) \\
&+ e^{g_1£} e^{g_2x} e^{g_3\frac{d}{dx}} \dot g_4 e^{g_4}\pi(x) \\
&= £\dot g_1\rho + e^{g_1£}\dot g_2 x e^{-g_1£}\rho + e^{g_1£} e^{g_2x}\dot g_3\frac{d}{dx} e^{-g_2x} e^{-g_1£}\rho \\
&+ e^{g_1£} e^{g_2x} e^{g_3\frac{d}{dx}} \dot g_4 e^{-g_3\frac{d}{dx}} e^{-g_2x} e^{-£g_1}\rho
\end{aligned}$$

Now $\dot g_4$ commutes with the operators $£$, x, $\frac{\partial}{\partial x}$ (cf (7.14))). So the last term above simply gives $\dot g_4\rho$. To calculate the two middle terms we use the adjoint action formula again

$$\begin{aligned}
e^{g_1£}xe^{-g_1£} &= x + g_1[£,x] + \frac{g_1^2}{2!}[£,[£,x]] + \frac{g_1^3}{3!}[[£,[£,[£,x]]] + \ldots \\
&= x + g_1\frac{d}{dx} + \frac{g_1^2}{2!}x + \frac{g_1^3}{3!}\frac{d}{dx} + \ldots \\
&= \cosh(g_1)x + \sinh(g_1)\frac{d}{dx} \\
e^{g_2x}\frac{d}{dx}e^{-g_2x} &= \frac{d}{dx} + g_2[x,\frac{d}{dx}] + \frac{g_2^2}{2!}[x,[x,\frac{d}{dx}]] + \ldots \\
&= \frac{d}{dx} - g_2 + 0 + 0 + 0 + \ldots \\
e^{g_1£}e^{g_2x}\frac{d}{dx}e^{-g_2x}e^{g_1£} &= e^{g_1£}(\frac{d}{dx} - g_2)e^{-g_1£} = \\
&= \frac{d}{dx} - g_2 + g_1[£,\frac{d}{dx}] + \frac{g_1^2}{2!}[£,[£,\frac{d}{dx}]] + \ldots
\end{aligned}$$

$$= \frac{d}{dx} - g_2 + g_1 x + \frac{g_1^2}{2!}\frac{d}{dx} + \frac{g_1^3}{3!} x + \ldots$$

$$= -g_2 + \sinh(g_1) x \cosh(g_1)\frac{d}{dx}$$

Thus we find

$$\frac{\partial \rho}{\partial t} = \dot{g}_1 £_\rho + \dot{g}_2 \cosh(g_1) x\rho + \dot{g}_2 \sinh(g_1)\frac{d}{dx}\rho + \dot{g}_3 \sinh(g_1) x\rho + \dot{g}_3 \cosh(g_1)\frac{d}{dx}\rho - \dot{g}_3 g_2 \rho + \dot{g}_4 \rho$$

Comparing this with the original equation (7.15) we find the following ordinary differential equations for the $g_1,\ldots,g_4$

$$\begin{aligned} &\dot{g}_1 = u_1 &&, g_1(0)=0 \\ &\cosh(g_1)\dot{g}_2 + \sinh(g_1)\dot{g}_3 = u_2 &&, g_2(0)=0 \\ &\sinh(g_1)\dot{g}_2 + \cosh(g_2)\dot{g}_3 = u_3 &&, g_3(0)=0 \\ &\dot{g}_4 - \dot{g}_3 g_2 = u_4 &&, g_4(0)=0 \end{aligned} \tag{7.17}$$

or

$$\begin{aligned} &\dot{g}_1 = u_1 &&, g_1(0)=0 \\ &\dot{g}_2 = \cosh(g_1)u_2 - \sinh(g_1)(u_3) &&, g_2(0)=0 \\ &\dot{g}_3 = -\sinh(g_1)u_2 + \cosh(g_1)u_3 &&, g_3(0)=0 \\ &\dot{g}_4 = u_4 + g_2\dot{g}_3 &&, g_4(0)=0 \end{aligned} \tag{7.18}$$

which are of course trivial to solve. In order to find $\rho(t,x)$ itself it now remains to calculate the $e^{g_1 £}, e^{g_2 x}, e^{g_3 \frac{e}{dx}}, e^{g_4}$; or, in other words to solve the simpler initial value problems

$$\frac{\partial \sigma}{\partial t} = \dot{g}_1 £\sigma,\ \frac{\partial \sigma}{\partial t} = \dot{g}_2 x\sigma,\ \frac{\partial \sigma}{\partial t} = \dot{g}_3 \frac{\partial}{\partial x}\sigma,\ \frac{\partial \sigma}{\partial t} = \dot{g}_4 \sigma$$

The last three of these are trivial and the first one is the harmonic oscillator. Some more remarks on solving 'harmonic oscillator type' equations occur below in section 9.8.

The oscillators Lie algebra (7.12) is solvable, and not nilpotent. Hence the occurrence of the coupled equations block consisting of the second and third equation of (7.17). This is typical for the case of a solvable a Lie algebra. In the nilpotent case the equations can always be solved by quadratures only.

8. The estimation Lie - algebra

In section 4, 5 and 7 above two things have become clear. Firstly that the DMZ equation (5.6) or (6.2) is of bilinear type, i.e. of the general form

$$\frac{\partial \rho}{\partial t} = (£_1 \rho)u_1(t) + \ldots + (£_n \rho)(u_n(t)) \tag{8.1}$$

where the $£_i$ are linear (differential) operators on some suitable space of unnormalized densities (functions), and secondly that for bilinear type equations the Lie algebra generated by the operators $£_1,\ldots,£_n$ is important. If this Lie algebra is finite dimensional we have at least small time solutions and if it is finite dimensional and solvable we have explicit methods to solve the initial value provided one can do the same for the simpler equations

$$\frac{\partial \rho}{\partial t} = v_i(t) £_i \rho \qquad i = 1,\ldots,n \tag{8.2}$$

Incidentally the phrase 'bilinear' for equations of type (8.1) comes out of system and control theory. Analysts would call this simply a system of linear equations. In control theory however, a linear dynamical system is one of the form

$$\frac{\partial \rho}{\partial t} = A\rho + Bu(t) \tag{8.3}$$

where A is a linear operator on the space of ρ's, the state space, and B is a linear operator from some space of imputs to state space. The term bilinear is used to denote control systems of the form

$$\frac{\partial \rho}{\partial t} = A\rho + \sum_{j=1}^{m} B_i \rho u_i(t) \tag{8.4}$$

with A and $B_1,\ldots,B_m$ operators on state space. In both cases the $u_i(t),\ldots,u_m(t)$ are thought of as inputs or controls.

Thus it is clear that the Lie-algebra generated by the operators occurring in the DMZ equations (5.6) is important and has much to say about how difficult the filtering problem is. Indeed, as will be explained, it can serve to formulate a necessary criterion (the *BC* principle) for the existence of a recursive finite dimensional filter and this can be used to prove that for certain system nontrivial exact finite dimensional recursive filters cannot exist. (E.g. for the cubic sensor, cf section 12 below).

By definition the *estimation Lie algebra* of a system (1.1) -(1.2) is the Lie algebra spanned by the operators occurring in the DMZ equation (5.6); i.e. it is the Lie algebra generated by the second order differential operator £ given by (5.7) and the multiplication operators $h_1,\ldots,h_p$. Notation: $ELie(\Sigma)$ if Σ denotes the system (1.1)-(1.2).

If one works with the robust version of the DMZ equation the natural object to study is the Lie-algebra:

$$E^s Lie(\Sigma) = \text{ Lie algebra generated by the operators } £,\ [£,h_i],\ [[£,h_i],h_j].$$

This is in any case a subalgebra; it is often equal to $ELie(\Sigma)$, and is in any case very similar to $ELie(\Sigma)$ as will be shown now. Indeed $E^sLie(\Sigma)$ is an ideal in $ELie(\Sigma)$. To see this it suffices to check that the generator of $ELie(\Sigma)$ when bracketed with the generators of $E^sLie(\Sigma)$ yield elements of $E^sLie(\Sigma)$. For the generators $£\in ELie(\Sigma)$ this is trivial because £ is also in $E^sLie(\Sigma)$ and for the generators h_k of $ELie(\Sigma)$ we have that $[h_k,£]$ and $[h_k,[£,h_i]]$ are in $E^sLie(\Sigma)$ by definition, and that $[h_k,[[£,h_i],h_j]]=0$. This is the case because £ is second order and the h's are functions; thus $[£,h_i]$ is first order, $[[£,h_i],h_j]$ is a function and hence $[[h_k,[[£,h_i],h_j]]=0$. Now consider the quotient of $ELie(\Sigma)$ by $E^sLie(\Sigma)$

$$0 \rightarrow E^sLie(\Sigma) \rightarrow ELie(\Sigma) \rightarrow Q \rightarrow 0$$

Q is generated by the images of the commuting operators $h_1,\ldots,h_p$ so that Q is abelian (= commutative; i.e. $[a,b]=0$ for all $a,b,\in Q$) of dimension $\leqslant p$. It follows that in particular that $ELie(\Sigma)$ is finite dimensional (resp. solvable) if and only if $E^sLie(\Sigma)$ is finite dimensional (resp. solvable). It also follows that doing Wei-Norman theory for $ELie(\Sigma)$ is practically the same as doing it for $E^sLie(\Sigma)$, the only difference being a number of initial quadratures, cf section 13 below.

9. Examples of Estimation Lie algebras

Let us look at some examples to see what kind of Lie algebras can arise as estimation Lie algebras.

9.1 Example *Wiener noise linearly observed.* This is the simplest non-zero linear system

$$\begin{aligned} dx(t) &= w(t), & x,w\in\mathbf{R} \\ dy(t) &= x(t)dt + dv(t), & y,v\in\mathbf{R} \end{aligned} \tag{9.2}$$

In this case $£=\frac{1}{2}\frac{d^2}{dx^2}-\frac{1}{2}x^2$, $h=x$. The Lie-algebra generated by this is the four-dimensional Lie algebra with basis £, x, $\frac{d}{dx}$, 1. Cf. section 7 above for some of the calculations.

This also means that starting from an arbitrary initial density $\phi(x)$ for x at time 0 we can solve the corresponding DMZ equation

$$\frac{\partial\rho}{\partial t} = (\frac{1}{2}\frac{\partial^2}{\partial x^2}-\frac{1}{2}x^2)\rho+x\rho\dot{y}(t) \tag{9.3}$$

by means of Wei-Norman theory. The explicit equation for g_1,g_2,g_3,g_4 occurring in the Ansatz

$$\rho(x,t) = e^{g_1(t)\pounds}e^{g_2(t)x}e^{g_3(t)\frac{d}{dx}}e^{g_4(t)}\phi(x) \tag{9.4}$$

are given by (7.18) above. They are

$$\dot{g}_1 = 1,\ \dot{g}_2=\cosh(t)\dot{y}(t),\ \dot{g}_3=-\sinh(t)\dot{y}(t),\ \dot{g}_4=-g_2\sinh(t)\dot{y}(t) \tag{9.5}$$

Not surprisingly the Kalman-Bucy filter for (9.2), given by

$$\begin{aligned}\dot{P} &= 1-P^2\\ \dot{m} &= P(\dot{y}-m)\end{aligned} \tag{9.6}$$

can be easily derived from (9.3). Indeed let us try for a solution of the form

$$\rho(x,t) = r(t)e^{-\frac{(x-m)^2}{2P}} \tag{9.7}$$

i.e. an unnormalized Gaussian density, where m and P are yet to be determined functions of t. One finds

$$\frac{\partial^2}{\partial x^2}\rho(x,t) = (\frac{(x-m)^2}{P^2}-\frac{1}{P})\rho(x,t)$$

$$\frac{\partial}{\partial t}\rho(x,t) = (\frac{(x-m)\dot{m}}{P}+\frac{(x-m)^2}{2P^2}\dot{P}+\dot{r})\rho(x,t)$$

Now substitute this in (9.3) and divide by $\rho(x,t)$. There results an expression of the form $ax^2+bx+c=0$ with a,b,c dependant on time alone. For this to hold we must have $a=0$, $b=0$, $c=0$. This gives

$$a=0:\ \frac{1}{2P^2}-\frac{1}{2}\frac{\dot{P}}{2P^2}=0, \qquad \text{i.e. } \dot{P}=1-P^2$$

$$b=0:\ \frac{2m}{2P^2}+\dot{y}(t)=\frac{\dot{m}}{P}-\frac{m\dot{P}}{P^2}, \ \text{ i.e., using } \dot{P}=1-P^2, \text{ on finds } \dot{m}=-Pm+P\dot{y}$$

Finally $c=0$ gives some (complicated) expression for $\dot{r}$. This shows that the solutions are in fact of the form (9.7) provided the initial density is also an unnormalized Gaussian density. The precise result for $r(t)$ (and hence the precise expression for $\dot{r}$) is largely irrelevant because of formula (4.4) for the conditional expectation $E[\phi(x)|y(s),0\leq s\leq t]=\widehat{\phi(x(t))}$ of a function $\phi(x)$ of the state.

9.8. Example. Linear systems.
Now let us consider general linear systems

$$\begin{aligned}dx &= Axdt+Bdw \quad x\in\mathbf{R}^n, w\in\mathbf{R}^m\\ dy &= Cxdt+dv \quad\ \ y\in\mathbf{R}^p, v\in\mathbf{R}^p\end{aligned} \tag{9.9}$$

The system is said to be completely reachable if the $n\times(n+1)m$ matrix

$$R(A,B) = (B\ AB\ A^2B...A^nB)$$

Consisting the $(n+1)$ $n\times m$ blocks A^iB, $i=0,...,n$, has rank n. This means for the associated control system $\dot{x}=Ax+Bu$ every element $x(1)\in\mathbf{R}^n$ can be reached from $x(0)=0$ by a suitable choice of input functions $u_1(t),\ldots,u_m(t)$, whence the terminology. Dually, the system (9.9) is completely observable if the $(n+1)p\times n$ matrix $Q(A,C)$

$$Q(A,C)^T = (C^T \ (CA)^T \ \ldots \ (CA^n)^T)$$

consisting of the blocks CA^i, $i=0,\ldots,n$ has rank n. This means that (with zero inputs) one can see from $y(t)$ whether two initial states $x(0)$ and $x'(0)$ were different or not. Whence the name of the concept.

Let us assume in (9.9) that (A,B) and (A,C) are completely reachable and completely observable pairs of matrices. Then it is not difficult to show that the estimation Lie algebra L of (9.9) is the $2n+2$ dimensional Lie algebra with basis

$$£,\ \frac{\partial}{\partial x_1},\ \ldots,\ \frac{\partial}{\partial x_n},\ x_1,\ldots,x_n,\ 1$$

and £ equal to

$$£ = \tfrac{1}{2}\sum_{ij}(BB^T)\frac{\partial^2}{\partial x_i\partial x_j} - \sum_{i,j}A_{ij}x_j\frac{\partial}{\partial x_i} - Tr(A) - \tfrac{1}{2}\sum_{i,j}(C^TC)_{ij}x_ix_j \tag{9.10}$$

The $(2n+1)$-dimensional Heisenberg algebra $\mathfrak{h}_n$ with basis

$$\frac{\partial}{\partial x_1},\ldots,\frac{\partial}{\partial x_n},x_1,\ldots,x_n,\ 1$$

is an ideal in L, i.e. $[£,\mathfrak{h}_n]\subset\mathfrak{h}_n$, cf the tutorial in Lie algebras [28] in this volume. Hence L is solvable and using Wei-Norman theory the DMZ equation can be integrated for arbitrary initial densities. Of course this requires that we be able to integrate the simpler equation

$$\frac{\partial\rho}{\partial t} = £\rho \quad , \quad \rho(0) = \pi(x) \tag{9.11}$$

All the others, i.e. $\partial\rho/\partial t = x_i\rho$ and $\partial\rho/\partial t = \partial\rho/\partial x_i$ are trivial, but £, cf (9.10), is itself a fairly complicated operator. One natural thing to try would be to try to do Wei-Norman theory once more, which leads tot the study of the Lie algebra generated by the various terms occurring in £, i.e. the $\frac{\partial^2}{\partial x_i\partial x_j}$, $x_j\frac{\partial}{\partial x_i}$, x_ix_j, 1. The constant $Tr(A)$ does not matter (as it commutes with everything). It turns out to be slightly more convenient to consider instead the operators

$$\begin{aligned}
&\frac{\partial^2}{\partial x_i\partial x_j} && , \ i,j=1,\ldots,n;\ i\leqslant j \\
&x_ix_j && , \ i,j=1,\ldots,n;\ i\leqslant j \\
&x_i\frac{\partial}{\partial x_j} && , \ i\neq j;\ i,j=1,\ldots,n \\
&x_i\frac{\partial}{\partial x_i}+\tfrac{1}{2} && , \ i=1,\ldots,n
\end{aligned} \tag{9.12}$$

It is a straightforward exercise to check these form in fact the basis of a $(2n^2+n)$ dimensional Lie algebra (of differential operators). As a matter of fact this Lie algebra is ismorphic to the Lie algebra of all $2n\times 2n$ sympletic matrices. i.e. the Lie algebra $sp_n(\mathbf{R})$

$$sp_n(\mathbf{R}) = \{ \begin{pmatrix} A & B \\ C & D\end{pmatrix} \in\mathbf{R}^{2n\times 2n}:\ A,B,C,D\in\mathbf{R}^{n\times n}, B=B^T, C=C^T, A=-D^T\} \tag{9.13}$$

This Lie algebra is simple and (thus) Wie-Norman theory only works for small time intervals. And indeed there are operators $M\in sp_n(\mathbf{R})$ such that $\frac{dx}{dt}=Mx$, $x\in\mathbf{R}^{2n}$, or, equivalently $\frac{\partial\rho}{\partial t}=\tilde{M}p$, where

$\tilde{M}$ is the differential operators corresponding to the $2n\times 2n$ matrix M, has finite escape time phenomena (for suitable initial conditions). However, it does turn out that this isomorphism of Lie algebras can be used effectively to integrate equations like (9.11). I shall not discuss this aspect further here, refering to [30] for all details. In the case of example 9.1. equation (9.11) is of course that of the harmonic oscillator which is well studied.

Again, there is of course a link of the DMZ equation with the Kalman-Bucy filter, and the latter can be deduced from the former [24]. As a matter of fact all Kalman-Bucy filters sort of fit together to define one large representation of the Lie algebra ls_n with basis

$$1;\ x_1,\dots,x_n;\ \frac{\partial}{\partial x_1},\dots,\frac{\partial}{\partial x_n};\ \frac{\partial^2}{\partial x_i \partial x_j}, i,\ j=1,\dots,n, i\leqslant j; \tag{9.14}$$

$$x_i x_j,\ i,j=1,\dots,n, i\leqslant j;\ x_i\frac{\partial}{\partial x_j} i,j=1,\dots,n$$

of dimension $2n^2+3n+1$, into the Lie-algebra of all vectorfields on $\mathbf{R}^N, N=\frac{1}{2}n^2+\frac{3}{2}n+1$. Cf. [24,29], for more details on this and the link with the so called oscillator or Segal-Shale-Weil representation of $sp_n(\mathbf{R})$. This is in fact the representation given by the operators (9.12) and it is precisely the fact that we know how to integrate this representation together with the availability of the matrix copy (9.13) of (9.12) which enables one to integrate equation like 9.11 [24].

9.15. Example. The cubic sensor
This is the system

$$\begin{aligned} dx &= dw \qquad x,w\in\mathbf{R} \\ dy &= x^3dt+dv \quad y,v\in\mathbf{R} \end{aligned}$$

In a certain sense this is the simplest nonlinear system. (The quadratic sensor where the observation equation is $dy=x^2dt+dv$ instead is perhaps still simpler; on the other hand the noninjectivity of $x\mapsto x^2$ seems to be asking for additional trouble; as it turns out both are 'equally hard'). This example has a substantial literature devoted to it and has a reputation of being quite hard to handle [11]. A first indication of why this might be the case is the structure of its estimation Lie algebra.

Let $W_1=\mathbf{R}<x,\frac{d}{dx}>$, i.e. the associative algebra generated by the symbols x and $\frac{d}{dx}$ subject to the relations suggested by the notation used, viz. $[\frac{d}{dx},x]=(\frac{d}{dx})x-x(\frac{d}{dx})=1$. Consider W_1 as a Lie algebra under the commutator bracket $[A,B]=AB-BA$. In other words W_1 is the Lie algebra of all differential operators in x (any order) with polynomial coefficients:

$$W_1 = \{\sum_{i,j}^{<\infty} c_{ij}x^i\frac{d^j}{dxj} : c_{i,j}\in\mathbf{R}\} \tag{9.16}$$

The estimation Lie algebra of (9.15) is the Lie-algebra generated by the two operators $\frac{1}{2}\frac{d^2}{dx^2}-\frac{1}{2}x^6,\ x^3$. It turns out that this is everything. I.e.

$$ELie(\text{cubic sensor}) = W_1 \tag{9.17}$$

This a very large infinite dimensional algebra and, as it turns out, cf below, a rather nasty one from the point of view of exact finite-dimensional filtering. For a proof of (9.17), cf [25].

9.18. Example. The quadratic sensor

$$dx = dw \qquad x, w \in \mathbf{R}$$
$$dy = x^2 dt + dv \quad y, v \in \mathbf{R}$$

Let W'_1 be the subalgebra of W_1 consisting of all differential operators of even total degree in x and $\frac{d}{dx}$ together:

$$W'_1 = \{ \sum_{i,j}^{<\infty} c_{ij} x^i \frac{d^j}{dx^j} \in W_1 : c_{ij} = 0 \text{ if } i \neq j \text{ mod} 2\} \tag{9.19}$$

The estimation Lie algebra of the quadratic sensor is generated by $\frac{d^2}{dx^2} - x^4$ and x^2. It turns out that

$$ELie(\text{quadratic sensor}) = W'_1 \tag{9.20}$$

9.12. Example. The weak cubic sensor

$$dx = dw \qquad x, \in \mathbf{R}$$
$$dy = (x + \epsilon x^3)dt + dv, \quad y, v \in \mathbf{R}$$

This time the generators of the estimation Lie algebra are $\frac{d^2}{dx^2} - (x + \epsilon x^3)^2$, $x + \epsilon x^3$. If $\epsilon = 0$ we have example (9.1) back. If $\epsilon \neq 0$ we have again [23], [25].

$$ELie(\text{weak cubic sensor}) = W_1 \ , \text{ if } \epsilon \neq 0. \tag{9.22}$$

9.23. Example.

$$dx_1 = dw_1, \ dx_2 = x_1^2 dt \qquad , x_1, x_2, w \in \mathbf{R}$$
$$dy_1 = x_1 dt + dv_1, dy_2 = x_2 dt + dv_2 \quad , y_i, v_i \in \mathbf{R}$$

Generalizing W_1, let W_n be the Lie algebra of all differential operators (any order) in n variables with polynomials coefficients, i.e.

$$W_n = \mathbf{R} < x_1, \ldots, x_n; \frac{\partial}{\partial x_1}, \ldots, \frac{\partial}{\partial x_n} > = \{ \sum_{\alpha,\beta}^{<\infty} c_{\alpha,\beta} x^\alpha \frac{\partial^\beta}{\partial x^\beta} : c_{\alpha,\beta} \in \mathbf{R}\}$$

where $\alpha = (a_1, \ldots, a_n)$ and β are multiindices, $a_i \in \mathbf{N} \cup \{0\}$ and where x^α and $\frac{\partial^{|\beta|}}{\partial x^\beta}$ are short for $x_1^{a_1} \ldots x_n^{a_n}$ and $\frac{\partial^{b_1}}{\partial x_1^{b_1}} \ldots \frac{\partial^{b_n}}{\partial x_n^{b_n}}$ respectively. The generators of the estimation Lie algebra are in this case

$$\frac{1}{2} \frac{\partial^2}{\partial x_1^2} - x_1^2 \frac{\partial}{\partial x_2} \ , \ x_1, \ x_2$$

and we find (again) [25]

$$ELie(\text{example 9.23}) = W_2 \tag{9.24}$$

9.25. Example.

$$dx = x^3dt + dw, \quad x,w \in \mathbf{R}$$
$$dy = xdt + dv, \quad y,v \in \mathbf{R}$$

This time the generators are $\frac{1}{2}\frac{d^2}{dx^2} - \frac{7}{2}x^2 - x^3\frac{d}{dx}$ and x and (again)

$$ELie(\text{example 9.25}) = W_1 \tag{9.26}$$

9.27. Example (mixed linear - bilinear type).

$$dx_1 = dw_1 \quad x_1,w_1 \in \mathbf{R}$$
$$dx_2 = x_1 + x_1dt + x_1dw_2 \quad x_2,w_2 \in \mathbf{R}$$
$$dy = x_2dt + dv \quad y,v \in \mathbf{R}$$

The generators are $\frac{1}{2}\frac{\partial^2}{\partial x_2^2} + \frac{1}{2}x_1^2\frac{\partial^2}{\partial_2^2} - x_1\frac{\partial}{\partial x_2} - \frac{1}{2}x_2^2$ and x_2, and (again) [25]

$$ELie(\text{example 9.27}) = W_2 \tag{9.28}$$

Thus it would appear that the infinite dimensional Lie algebras W_n have a habit of appearing very often. This seems indeed to be the case. In fact I have conjectured

9.29. Conjecture.
Consider stochastic systems (1.1)-(1.2) with polynomial f,G and h. Then for almost all f,G,h the estimation Lie algebra of (1.1)-(1.2) will be equal to W_n.

Here 'almost all' means an open dense set in the space of all triples (f,G,h) of vector and matrix valued functions of the right dimensions topologized by means of the natural topology on their sequences of coefficients. No proof of this conjecture appears to be in sight.

As we shall see the occurrence of W_n as the estimation Lie algebra of a stochastic system is bad news from the point of view of existence of exact recursive finite dimensional filters. So at least a few examples (besides the linear ones) where something else turns up would be welcome. One large class of such examples are the systems

$$\begin{aligned} dx &= f(x)dt + G(x)dw, \quad x \in \mathbf{R}^n, w \in \mathbf{R}^m \\ dy &= h(x)dt + dv, \quad y \in \mathbf{R}^p, v \in \mathbf{R}^p \end{aligned} \tag{9.30}$$

with the extra conditions that f,G and h are real analytic and that $f(0)=0, G(0)=0$, cf [25]. Another example is

9.31. Example [25]

$$dx_1 = dw, \quad x_1,w \in \mathbf{R}$$
$$dx_2 = x_1^2dt, \quad x_2 \in \mathbf{R}$$
$$dy = x_1dt + dv, \quad y,v \in \mathbf{R}$$

Here of course filtering for x_1, i.e. calculating $\hat{x}_1$, is straightforward by means of the Kalman filter. Finding $\hat{x}_2$ is a totally different matter. (NB: by the Ito formula $d(\frac{1}{3}x_1^3) = x_1^2dx_1 + x_1dt = x_1^2dw + x_1dt$ which does not have much to do with the equation for x_2). The generators in this case are

$$\pounds = \frac{1}{2}\frac{\partial^2}{\partial x_1^2} - x_1^2\frac{\partial}{\partial x_2} - \frac{1}{2}x_1^2 \ , \ x_1$$

and the Lie algebra generated by these two operators is the infinite dimensional Lie algebra with basis

$$£,\ x_1\frac{\partial^i}{\partial x_2^i},\ \frac{\partial}{\partial x_1}\frac{\partial^i}{\partial x_2^i},\ \frac{\partial^i}{\partial x_2^i},\ i=0,1,2,... \tag{9.32}$$

This is an infinite dimensional Lie algebra but still a solvable one in the appropriate sense, cf below.

9.33. Example. 'Polynomials' in n independent Brownian motions

The word 'polynomial' is here used in a very loose sense. What I mean are certain stochastic systems directly inspired by polynomials. E.g. the system corresponding to the 'polynomial' w^n where w is a one dimensional Wiener process is

$$\begin{aligned} dx_1 &= dw,\ dx_2=x_1dw,\ dx_3=x_2dw,...,\ dx_n=x_{n-1}dw \\ dy &= x_1dt+dv \end{aligned} \tag{9.34}$$

Here (9.34) are intended as Ito equations. Thus x_n does *not* really correspond to w^n. Indeed a system which has w^n as a second state variable, x_2, is

$$\begin{aligned} dx_1 &= dw \\ dx_2 &= nx_1^{n-1}dw+n(n-1)x_1^{n-2}dt \\ dy &= x_1dt+dv \end{aligned} \tag{9.35}$$

This one, obtained by adding the state variable, $x_2=x_1^n$, which is a function of the state variables already present, has an estimation Lie algebra which is isomorphic to the original system without the extra state variable. This is a general fact. Cf. section 10 below. It is a curious and somewhat remarkable fact that the estimation Lie algebra of (9.34) is also isomorphic to the oscillator Lie algebra. For this it is really necessary to 'unravel' w^n as in (9.34) by means of the intermediate states $x_2,...,x_{n-1}$: the system

$$\begin{aligned} dx_1 &= dw \\ dx_n &= nx_1^{n-1}dw \end{aligned}$$

does not have its estimation Lie algebra isomorphic to the oscillator Lie algebra. Instead, for $n \geqslant 4$, its estimation Lie algebra is a much more complicated infinite dimensional affair. (For $n=3$ the estimation Lie algebra is 5 dimensional with basis $£,\ x_1,\ \frac{\partial}{\partial x_1}+3x_1^2\frac{\partial}{\partial x_2},\ 1,\ \frac{\partial}{\partial x_2}$.)

More generally, consider systems 'corresponding to a polynomial'

$$P = \sum c_\alpha w^\alpha \tag{9.36}$$

where $\alpha=(a_1,...,a_n)$ is a multiindex and w^α is short for $w_1^{a_1}...w_n^{a_n}$. These systems are defined as follows. Let $\mu=(m_1,...,m_n)$ be such that $c_\alpha=0$ unless $a_k \leqslant m_k$ for $k=1,...,n$. Consider now the system with state variables p; $x_{i,r}$, $i=1,...,n$; $r=1,...,m_i$ given by

$$\begin{array}{ccccc} dx_{1,1}=dw_1 & & dx_{2,1}=dw_2 & & dx_{n,1}=dw_n \\ dx_{1,2}=x_{1,1}dw_1 & & dx_{2,2}=x_{2,1}dw_2 & & dx_{n,2}=x_{n,1}dw_n \\ \cdot & & \cdot & \cdots & \cdot \\ \cdot & & \cdot & & \cdot \\ \cdot & & \cdot & & \cdot \\ dx_{1,m_1}=x_{1,m_1-1}dw_1 & & dx_{2,m_2}=x_{2,m_2-1}dw_2 & & dx_{n,m_n}=x_{n,m_n-1}dw_n \end{array} \tag{9.37}$$

$$dp=\sum_\alpha\sum_{i=1}^n c_\alpha(x_{1,a_1}\cdots x_{n,a_n})x_{i,a_i}^{-1}dw_i$$

with the observation equations

$$dy_1 = x_1 dt + dv_1, \ldots, dy_n = x_n dt + dv_n \tag{9.38}$$

Then the Estimation Lie algebra of (9.37)-(9.38) is generated by the 2nd order operator £ and $x_1, \ldots, x_n$. It is finite dimensional with as basis

$$£,\ £_1 = [£, x_1], \ldots, £_n = [£, x_n],\ x_1, \ldots, x_n,\ 1. \tag{9.39}$$

Besides the $£_i = [£, x_i]$ the nonzero brackets are $[£, £_i] = x_i$ and $[£_i, x_i] = 1$. For a (rather computational) proof cf [31]. There could well be, indeed should be, a conceptual proof of this, but so far the arguments I have in this direction are unconvincing. This is a solvable Lie algebra and so Wei-Norman theory is applicable. Indeed the estimation Lie algebra is the same one as the one of n-independent completely observed Wiener processes, so the Wei-Norman equations are the same as those in that case. The individual operators £ and $£_i$ though are very different and quite complicated, cf the particular example 9.41 below. So it remains to deal with the equations

$$\frac{\partial \rho}{\partial t} = £\rho \text{ and } \frac{\partial \rho}{\partial t} = £_i \rho \tag{9.40}$$

It turns out that for both £ and $£_i$ the individual terms making up these operators themselves generate a solvable, albeit, as a rule, infinite dimensional Lie algebra. Here is a particular example

9.41. Example. System associated to the Brownian polynomial $w_1^2 + w_1 w_2 + w_2^3$

$dx_1 = dw_1,\ dx_2 = dw_2,\ dx_{2,2} = x_2 dw_2,\ dp = x_1 dw_1 + x_1 dw_2 + x_2 dw_1 + x_{22} dw_2$

$dy_1 = x_1 dt + dv_1,\ dy_2 = x_2 dt + dv_2$

In this (still quite simple) case the operators are equal to

$$£ = \frac{1}{2}\frac{\partial^2}{\partial x_1^2} + \frac{1}{2}\frac{\partial^2}{\partial x_2^2} + \frac{1}{2}x_2^2\frac{\partial^2}{\partial x_{22}^2} + \frac{1}{2}\{(x_1 + x_{22})^2 + (x_1 + x_2)^2\}\frac{\partial^2}{\partial p^2}$$

$$+ (x_1 + x_2)\frac{\partial}{\partial x_1}\frac{\partial}{\partial p} + (x_1 + x_{22})\frac{\partial}{\partial x_2}\frac{\partial}{\partial p} + x_2\frac{\partial^2}{\partial x_2 \partial x_{22}} + (x_1 x_2 + x_2 x_{22})\frac{\partial^2}{\partial x_{22}\partial p}$$

$$+ x_2\frac{\partial}{\partial p} + \frac{\partial}{\partial x_{22}} + \frac{\partial}{\partial p} - \frac{1}{2}x_1^2 - \frac{1}{2}x_2^2$$

$$£_1 = [£, x_1] = \frac{\partial}{\partial x_1} + (x_1 + x_2)\frac{\partial}{\partial p}$$

$$£_2 = [£, x_2] = \frac{\partial}{\partial x_2} + (x_1 + x_{22})\frac{\partial}{\partial p} + x_2\frac{\partial}{\partial x_{22}}$$

Besides the cases where the W_n occur as estimation Lie algebras we thus have several classes of systems which yield infinite dimensional but solvable Lie algebras, viz. real analytic systems (9.30) such that $f(0)=0$, $G(0)=0$, and systems like example 9.31. A further class is furnished by the systems which arise when a identification problem for linear systems is considered as a filtering problem (cf section 15 below). This filtering problem is then nonlinear, reflecting the essential nonlinearity of identification, but its estimation Lie algebra is again solvable but infinite dimensional. Finally in tackling the equations (9.40) which form part of the filtering of Brownian polynomials again infinite dimensional solvable Lie algebras arise.

This then is ample motivation for investigating whether something like infinite dimensional Wei-Norman theory exists. This is a topic which we will take up below in section 13.

9.42. Example. Beneš systems [4].

$$dx = f(x)dt + dw$$
$$dy = xdt + dv$$
$$f_x + f^2 = ax^2+bx+c$$

I.e. we have Wiener noise with an extra nonlinear drift term given by $f(x)dt$; this drift term is required to be such that $\frac{df}{dx}+f^2$ is a quadratic polynomial in x. In this case also the estimation Lie algebra is the oscillator Lie algebra.

9.43. Some open questions.
All in all very little is known about estimation Lie algebras. It seems very difficult to find other (non-trivial and interesting) examples of finite dimensional estimation Lie algebras. Besides the linear and Beneš case and the (new) case of 'Brownian polynomials' (9.37) very few examples are known (Wing Wong, Marcus, Lie, Ocone, . . .). In particular it is unknown whether finite dimensional simple Lie algebras can ever arise as estimation Lie algebras.

10. Invariance properties of the estimation Lie algebra.

This section discusses some questions much related to the subjects discussed so far and what will still come. But these questions are not essential for the remainder of this paper, and this section is somewhat more abstract than the remainder of this paper. It can be skipped if desired.

The estimation Lie algebra $ELie(\Sigma)$ is clearly an invariant of a system Σ: (1.1)-(1.2) in the following sense. If Σ and Σ' are two system of the type (1.1)-(1.2) on $\mathbf{R}^n$ respectively, and $\phi:\mathbf{R}^n\to\mathbf{R}^n$ is an isomorphism of Σ and Σ' (a transformation of variables), then the estimation Lie algebras of Σ and Σ' are isomorphic. Here, as we are dealing with Ito differential equations, isomorphism means that under the change of variables $x'=\phi(x)$ the equation

$$dx = f(x)dt + G(x)dw \tag{10.1}$$

transforms into the equation

$$dx' = f'(x')dt + G'(x')dw \tag{10.2}$$

under the Ito formula (transformation rule) which says that $\phi(x)=x'$ satisfies the differential equations

$$dx'^k = d\phi^k(x)\sum = (\sum_i \frac{\partial\phi_k}{\partial x_i} f_i(x) + \frac{1}{2}\sum_{i,j}\frac{\partial^2\phi}{\partial x_i \partial x_j}(GG^T)_{ij})dt + \sum_i \frac{\partial\phi_k}{\partial x_i} G_i(x)dw \tag{10.3}$$

where $f_i(x)$ is the i-th component of $f(x)$ and $G_i(x)$ is the i-th row of $G(x)$. Substituting $x = \phi^{-1}(x')$ in (10.3) must then yield the equations (10.2). On a general manifold M the transformation rule (10.3) has no real meaning and then to talk about equivalent stochastic systems it is better to start with systems in Fisk-Stratonovič form.

In addition the DMZ equation

$$\frac{\partial\rho}{\partial t} = \pounds\rho + \sum_{i=1}^{p} h_i(x)\rho\dot{y}_i \tag{10.4}$$

can be gauge transformed, $\tilde{\rho} = e^{\phi(x)}\rho$ to give an equation

$$\frac{\partial\tilde{\rho}}{\partial t} = \tilde{\pounds}\tilde{\rho} + \sum_{i=1}^{p} h_i(x)\tilde{\rho}\dot{y}_i \tag{10.5}$$

with $\tilde{\pounds} = e^{\phi(x)}\pounds e^{-\phi(x)}$ Correspondingly there is an isomorphism of the Lie algebra L generated by $\pounds,h_1,\ldots,h_p$ and the Lie algebra generated bij $\tilde{\pounds}$ and $h_1,\ldots,h_p$. This isomorphism is given by

$A \in L \mapsto e^{\phi(x)} A \phi^{-\phi(x)}$ in $\tilde{L}$. Sometimes non-isomorphic dynamical systems are gauge equivalent in this sense. This happens e.g. for the Beneš systems 9.42 and corresponding 1-dimensional linear systems. Cf. [2] for material on 'invariants' in this context.

More generally (than the case of, isomorphims, i.e. changes of variables), if $\Sigma \rightarrow \Sigma'$ is a morphism of stochastic dynamical systems then there is a corresponding homomorphism of their estimation Lie algebras. In particular consider a system (10.1) and let us add an additional state variable p which is a function of the original state: $p = \phi(x_1, ..., x_n)$. The resulting Ito differential equation for p is

$$dp = \sum_k \phi_{(k)} f_k dt + \sum_{k,l,j} \frac{1}{2} \phi_{(k,l)} G_{k,j} G_{l,j} dt + \sum_{k,l} \phi_{(k)} G_{k,l} dw_l \tag{10.6}$$

where

$$\phi_{(k)} = \frac{\partial \phi}{\partial x_k}, \; \phi_{(k,l)} = \frac{\partial^2 \phi}{\partial x_k \partial x_l}$$

Let L be the estimation Lie algebra of (10.1) with observations $dy_j = h_j(x)dt + dv_j$ and let $\tilde{L}$ be the estimation Lie algebra of (10.1) complemented with the p-equation (10.6) above and the same observations. Then the isomorphism $L \rightarrow \tilde{L}$ is induced by $\frac{\partial}{\partial x_i} \rightarrow \frac{\partial}{\partial x_i} + \phi_{(i)} \frac{\partial}{\partial p}$, $x_i \mapsto x_i$ and the inverse isomorphism $\tilde{L} \rightarrow L$ is induced by $x_i \mapsto x_i$, $\frac{\partial}{\partial x_i} \mapsto \frac{\partial}{\partial x_i}$, $\frac{\partial}{\partial p} \mapsto 0$. (These are, as is easily checked, homomorphisms of associative algebras $\mathbf{R}<x_1, ..., x_n; \frac{\partial}{\partial x_1}, ..., \frac{\partial}{\partial x_1}, \frac{\partial}{\partial p}> \;\longrightarrow\; \mathbf{R}<x_1, ..., x_n, \frac{\partial}{\partial x_1}, ..., \frac{\partial}{\partial x_n}>$.

Now, if

$$\pounds = \frac{1}{2} \sum_{i,j,k} \frac{\partial^2}{\partial x_i \partial x_j} G_{ik} G_{jk} - \sum_i \frac{\partial}{\partial x_i} f_i - \frac{1}{2} \sum_j h_j^2 \tag{10.7}$$

then the corresponding operator for the system extended with (10.6) is

$$\tilde{\pounds} = \pounds + \frac{1}{2} \sum_l \left(\sum_k \phi_{(k)} G_{k,l}\right)^2 + \sum_{k,l,i} \frac{\partial^2}{\partial x_i \partial p} G_{il} \phi_{(k)} G_{k,l} \tag{10.8}$$

$$- \frac{1}{2} \frac{\partial}{\partial p} \sum_{k,l,j} \phi_{(k,l)} G_{k,j} G_{l,j} - \sum_k \frac{\partial}{\partial p} \phi_{(k)} f_k$$

Replacing $\frac{\partial}{\partial x_i}$ with $\frac{\partial}{\partial x_i} + \phi_{(i)} \frac{\partial}{\partial p}$ in (10.7) yields the extra terms

$$\frac{1}{2} \sum_{i,j,k} \phi_{(i)} \frac{\partial}{\partial p} \frac{\partial}{\partial x_j} G_{ik} G_{jk} = \frac{1}{2} \sum_{i,j,k} \frac{\partial^2}{\partial x_j \partial p} \phi_{(i)} G_{ik} G_{jk} - \frac{1}{2} \sum_{i,j,k} \phi_{(i,j)} G_{ik} G_{jk} \frac{\partial}{\partial p} \tag{10.9}$$

$$\frac{1}{2} \sum_{i,j,k} \frac{\partial}{\partial x_i} \phi_{(j)} \frac{\partial}{\partial p} G_{ik} G_{jk} \tag{10.10}$$

$$\frac{1}{2} \sum_{i,j,k} \phi_{(i)} \phi_{(j)} G_{ik} G_{jk} \frac{\partial^2}{\partial p^2} \tag{10.11}$$

$$- \sum_i \phi_{(i)} \frac{\partial}{\partial p} f_i \tag{10.12}$$

Now the first term of the RHS of (10.9) and (10.10) combine to give the third term of the RHS of (10.8). The second term of the RHS of (10.9) is the fourth term of the RHS of (10.8); expression (10.11) is equal to the second term of the RHS of 10.8 and finally (10.12) is the last term of the RHS of 10.8. Thus $\frac{\partial}{\partial x_i} \mapsto \frac{\partial}{\partial x_i} + \phi_{(i)} \frac{\partial}{\partial p}$, $x_i \mapsto x_i$ does indeed take $\pounds$ into $\tilde{\pounds}$.

Inversely one can ask to what extent the estimation Lie algebra $ELie(\Sigma)$ determines the system Σ. Certainly nonisomorphic systems can have isomorphic estimation Lie algebras; e.g. the Beneš systems and one dimensional systems or the Brownian polynomial systems and the systems $dx_1 = dw_1, \ldots, dx_n, = dw_n,\ dy_1 = x_1 dt + dv_1, \ldots, dy_n = dt + dv_n$. But of course $ELie(\Sigma)$ is not just an abstract Lie algebra; it comes together with a natural (linear infinite dimensional) representation (on a suitable space of unnormalized densities). The more sensible question is therefore whether the pair ($ELie(\Sigma)$, corresponding representation) determines Σ up to isomorphism.

This is also not true as shown by the gauge equivalence of the Beneš systems with the corresponding linear ones. It would be nice to know more about just how much information is contained in the pair ($ELie(\Sigma)$, representation).

The question is akin to the following one for control systems of the form

$$\dot{x} = f(x)dt + \sum_{i=1}^{m} g_i(x)u_i, \quad x \in M, u_i \in \mathbf{R} \tag{10.13}$$
$$y = h(x), y \in \mathbf{R}^p$$

Associated to (10.13) we have the Lie algebra generated by f and the g_i; denote this one by $Lie(\Sigma)$. It also comes together with a natural representation. Indeed f and the g_i are vectorfields and hence are first order differential operators acting on functions on M, in particular the functions $h_1, \ldots, h_p$. Let $V(\Sigma)$ be the smallest subspace of $\mathfrak{F}(M)$ containing $h_1, \ldots, h_p$ and stable under D_f, D_{g_i}. Then $V(\Sigma)$ carries a linear representation of $Lie(\Sigma)$ and the question is to what extent the pair $(Lie(\Sigma), V(\Sigma))$ characterizes Σ up to isomorphism. A first problem here is to recover the manifold M from $(Lie(\Sigma), V(\Sigma))$. This is strongly related to the following question which has been studied in [59]. Given an n-dimensional manifold M let $V(M)$ be the Lie algebra of all vectorfields on M. Can one recover M from $V(M)$?

The reason I bring up these questions is the following. As we shall see in section 12 existence of an exact finite dimensional recursive filter implies the extence of a homomorphism of Lie algebras $ELie(\Sigma) \rightarrow V(M)$ where $V(M)$ is the Lie algebra of vectorfields on the manifold on which the filter for $\phi(x)$ exists (this filter is assumed to be of minimal dimension among all filters for $\phi(x)$.)

The questions briefly raised above relate to the inverse problem: given a homomorphism $ELie(\Sigma) \rightarrow V(M)$ for some M, plus suitable supplementary structure, does there exist a corresponding filter. Without additional hypotheses this is certainly not true cf e.g. the contributions by Krishnaprasad-Marcus and Hazewinkel-Marcus in [34].

11. The BC principle

We have already seen one set of reasons why $ELie(\Sigma)$ is important for filtering questions: If it is finite dimensional and solvable we can apply Wei-Norman theory; if it is at least finite dimensional we have in any case Wei-Norman theory for small time. If it is infinite dimensional but still solvable there are potential approximation schemes, cf below. Let me now describe a second reason why the estimation Lie algebra $ELie(\Sigma)$ of a system Σ is important for filtering problems. I like to call it the *BC principle,* not because it is very old, though it could have been maybe, nor is it named after Johny Hart's cartoon character; the BC stand for Brockett and Clark who first enunciated it, [9].

Suppose we have a filter (2.3)-(2.4) on a finite dimensional manifold M for a statistic $\widehat{\phi(x_t)}$. We may as well assume that it is minimal, i.e. has minimal $\dim(M)$. The α and $\beta_1, \ldots, \beta_p$ in (2.3) are vectorfields on M. Let $V(M)$ denote the Lie algebra of smooth vectorfields on M. Then the BC principle states the following:

11.1. BC Principle.
If (2.3)-(2.4) is a minimal filter for a statistic $\widehat{\phi(x_t)}$ of a system Σ then $\pounds \mapsto \beta_1, \ldots, h_p \mapsto \beta_p$ defines an antihomomorphism of Lie algebras from $ELie(\Sigma)$ into $V(M)$.

Here "anti" means the following: if $\phi: L_1 \to L_2$ is a map of vectorspaces from the Lie-algebra L_1 to the Lie-algebra L_2, it is called an antihomomorphism of Lie-algebras if $\phi([A,B]) = -[\phi(A), \phi(B)]$ for all $A, B \in L_1$.

11.2. Example.
Consider again the simplest nonzero linear system (9.2). It is linear so there is the Kalman-Bucy filter for the conditional state $\hat{x}$. This filter is (cf (9.6) and (2.2))

$$dP = (1-P^2)dt, \ dm = P(dy - m\,dt).$$

So the two vectorfields α and β of the filter are respectively

$$\alpha = (1-P^2)\frac{\partial}{\partial P} - Pm\frac{\partial}{\partial m}, \ \beta = P\frac{\partial}{\partial m}$$

where we have used the '$\frac{\partial}{\partial x_i}$ notation'; cf the tutorial on differentiable manifolds [27] in this volume. A simple calculation shows $[\alpha, \beta] = \frac{\partial}{\partial m}$, and it is now indeed a simple exercise to show that $\frac{1}{2}\frac{d^2}{dx^2} - \frac{1}{2}x^2 \mapsto \alpha$, $x \mapsto \beta$, induces an antimorphism of Lie-algebras. (It also induces a homomorphism, but that is an accident which happens for linear systems if the drift term Ax is absent).

A feeling of why the BC principle should be true can be generated as follows. Think for the moment of two automata with given initial state and with outputs (Moore automata), which, when fed the same string of input data, produce exactly the same string of output data. Suppose the second automaton is minimal. Then it is well known (and easy to prove by constructing the minimal automaton from the input-output data) that there is a homomorphism of the subautomaton of the first consisting of the states reachable from the initial state to the second automaton; this homomorphism so to speak makes visible that the two machines do the same job. A similar theorem holds for initialized finite dimensional systems [63], in particular for systems of the form

$$\dot{x} = \alpha(x) + \sum_{i=1}^{p} \beta_i(x)u_i, \ y = \gamma(x)$$

Here the picture produced by the theorem is the following commutative diagram

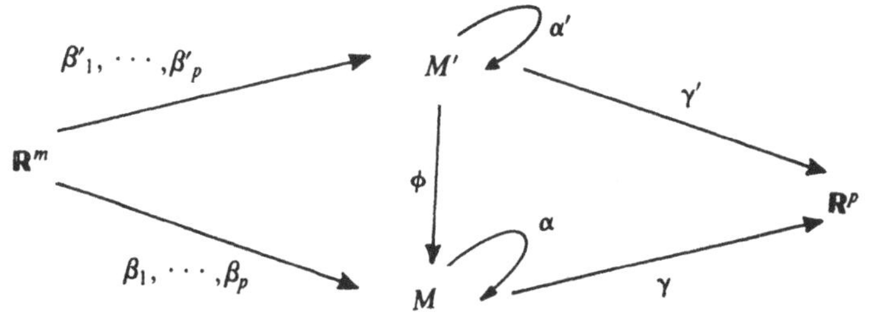

(The theorem assert the existence of a differentiable map ϕ defined on the reachable from x'_0 subset of M' which makes the diagram commutative. This in particular implies that $d\phi$ takes the vectorfields $\alpha', \beta'_1, \ldots, \beta'_m$ into $\alpha, \beta_1, \ldots, \beta_m$ respectively, and, ϕ being a differentiable map, $d\phi$ induces a homomorphism from the Lie algebra generated by $\alpha', \beta'_1, \ldots, \beta'_m$ to $V(M)$, cf [27].

In the case of the BC principle we also have two "machines" which do the same job: one is the postulated minimal filter, the other is the infinite dimensional machine given by the DMZ-equation (5.7) and the output map (4.4). So we are in a similar situation as above but with M' infinite dimensional. A proof in this case follows from considerations in [39].

The fact that in the case of the BC-principle we get an antihomomorphism instead of a homomorphism arises from the following. Given a linear space V and an operator A on it we can define a (linear) vectorfield on V by assigning to $v \in V$ the tangent vector Av. (So we are considering the equation $\dot{v}=Av$.) This defines an anti-isomorphism of the Lie algebra of operators on V to the Lie algebra of linear vectorfields on V.

What about a converse to the BC principle? I.e. suppose that we have given an antihomorphism of Lie-algebras $ELie(\Sigma) \to V(M)$ into the vectorfields of some finite dimensional manifold. Does there correspond a filter for some statistic of Σ. Just having the homomorphism is clearly insufficient. There are also explicit counterexamples [34]. This is understandable, for in any case we completely ignored the output aspect when making the BC-principle plausible. This is not trivial contrary to what the diagram above may suggest. It is not true that given ϕ and any γ one can take $\gamma' = \gamma \circ \phi$. The problem is that γ' as a function on M' = space of unnormalized densities is of a very specific type, cf (4.4).

Even apart from that things are not guaranteed. What we need of course is a ϕ making the left half of the diagram above commutative. Then, if $m' \in M'$ is going to the mapped on $m \in M$, obviously the isotropy subalgebra of $ELie(\Sigma)$ at m' will go into the isotropy subalgebra of $V(M)$ at m. The isotropy subalgebra $V(M)_{(x)}$ of $V(M)$ at $x \in M$ consists of all vectorfields which are zero at x. The isotropy subalgebra of $ELie(\Sigma)$ at $x \in M$ is $ELie(\Sigma) \cap V(M)_{(x)}$. For the case of finite dynamical systems there are positive results, [41], stating that in such a case this extra condition is also sufficient to guarantee the existence of ϕ locally.

The whole clearly relates to seeing to what extend a manifold can be recovered from its Lie algebra of vectorfields (via its maximal subalgebras of finite codimension) and whether differentiable maps can be recovered from the map between Lie-algebras they induce. This question has been examined in [59].

A more representation theoretic way of looking at things, already touched upon in section 10 above, is as follows. Both $ELie(\Sigma)$ and $V(M)$ come with a natural representation on the space of functionals on M' and the space of functions on M respectively. If there were a ϕ as in the diagram above ϕ would also induce a map between these representation spaces compatible with the homomorphism of Lie algebras. That therefore is clearly a necessary condition. This way of looking at things contains the isotropy subalgebra condition and also contains output function aspects. Thus the total picture regarding a converse to the BC-principle is not unpromising but nothing is established.

12. The cubic sensor

We have seen that the Weyl-Heisenberg algebra $W_n = \mathbf{R}<x_1,\ldots,x_n;\ \partial/\partial x_1,\ldots,\partial/\partial x_n>$ of all differential operators with polynomial coefficients often occurs in filtering problems, i.e. as an Estimation Lie algebra. Given the BC-principle it is therefore of interest to know something about its relations with another class of infinite dimensional Lie algebras, viz the Lie algebras $V(M)$ of smooth vectorfields on a finite dimensional manifold. The algebra W_n has a one-dimensional centre $\mathbf{R}.1$ consisting of the scalar multiples of the identity operator. Apart from that it is simple; i.e. the quotient algebra $W_n/\mathbf{R}.1$ is simple.

12.1. Theorem ([25]).
Let $\alpha: W_n \to V(M)$ or $W_n/\mathbf{R}.1 \to V(M)$ be a homomorphism or antihomomorphism of Lie algebras, where M is a finite dimensional manifold. Then $\alpha = 0$.

The original 12 page proof of this result, [25], was long and computational. Another much shorter proof has more recently been given by Toby Stafford. Perhaps inevitably this more conceptual proof is based on the Stone- von Neuman result on the impossibility of representing the 3-dimensional Heisenberg Lie algebra $\mathfrak{h}_1$ with basis x, $\frac{d}{dx}$, 1, $[\frac{d}{dx},x]=1$ by means of finite dimensional matrices in

such a way that 1 is represented by the unit matrix.
Now consider again the cubic sensor, i.e. the one-dimensional system

$$dx = dw, \; dy = x^3 dt = dv \tag{12.2}$$

consisting of Wiener noise, cubically observed with further independent noise corrupting the observations. As noted before (example 9.14)

$$ELie(\text{cubic sensor}) = W_1. \tag{12.3}$$

Now suppose that we have a finite dimensional filter for some conditional statistic $\widehat{\phi(x_t)}$ of the cubic sensor. By the BC-principle 11.1 if follows that there is an antihomomorphism of Lie-algebras $W_1 = ELie$ (cubic sensor) $\xrightarrow{\alpha} V(M)$. By theorem 12.1 it follows that $\alpha = 0$ and from this it is not hard to see that the only statistics of the cubic sensor for which there exists a finite dimensional exact recursive filter are the constants.

A direct proof of this, which sort of proves the BC-principle in this particular case along the way, is contained in [26].

A similar statement holds for all other systems whose estimation Lie algebras are isomorphic to a W_n or $W_n/\mathbf{R}$, and in fact also for the quadratic sensor whose estimation Lie algebra is 'the even subalgebra' W'_1 of W_1. As we have seen W_n occurs often as an estimation algebra so often exact finite dimensional recursive filters will not exist. This makes approximate recursive filters doubly important, a point to which we will return several times below.

13. Infinite dimensional Wei-Norman theory

We have already seen a number of cases where estimation Lie algebras were infinite dimensional and were claimed to be solvable in a suitable infinite dimensional sense. The precise definition of this is as follows.

13.1. Definition

Let L be a (finite or infinite dimensional) Lie algebra (over a field k; take $\mathbf{R}$ for convience). Then L is solvable if there exists a sequence of ideals $I_1, I_2, \ldots, I_n, \ldots$ such that $\bigcap_n I_n = 0$ and such that each quotient algebra L/I_n is finite dimensional and solvable (as a finite dimensional Lie algebra).

This is a good concept in the context of Wei-Norman theory because as we shall see in a few moments the Wei-Norman equations are well behaved with respect to quotients (and not at all well with respect to subalgebras).

13.2. Example

Consider again the estimation Lie algebra L of example 9.31. Recall that it had a basis consisting of the operators

$$\pounds = \frac{1}{2}\frac{\partial^2}{\partial x_1^2} - x_1^2\frac{\partial}{\partial x_2} - \frac{1}{2}x_1^2, \; x_1\frac{\partial^i}{\partial x_2^i}, \; \frac{\partial^i}{\partial x_2^i}, \; \frac{\partial}{\partial x_1}\frac{\partial^i}{\partial x_2^i}, i = 0,1,2,\ldots$$

Let $\mathfrak{h}_i$ be the subspace spanned by the operators

$$x_1\frac{\partial^j}{\partial x_2^j}, \; \frac{\partial^j}{\partial x_2^j}, \; \frac{\partial}{\partial x_1}\frac{\partial^j}{\partial x_2^j}, \quad j \geq i$$

It is easy to check that the $\mathfrak{h}_i$ are ideals, and not difficult to show that the quotients $L/\mathfrak{h}_i$ are finite dimensional and solvable.

13.3. Wei-Norman theory revisited

Now let us consider Wei-Norman theory again in the setting of a Lie-algebra L and an ideal $\mathfrak{a}$ in it. Let $\dim L = n$, $\dim \mathfrak{a} = n-k$, $0<k<n$. Choose a basis $A_1,\dots,A_k,A_{k+1},\dots,A_n$ for L in such a way that $A_{k+1},\dots,A_n$ are all in $\mathfrak{a}$ (i.e. they form a basis for the subspace $\mathfrak{a}$).

Recall from section 7, cf formulas (7.4) and 7.5), that the Wei-Norman equations are of the form

$$\dot{g}_r + \dot{g}_1 h_{1r}(g) + \dots + \dot{g}_n h_{nr}(g) = u_r \tag{13.4}$$

where g is short for $(g_1,\dots,g_n)$ and where the h_{jr} are such that

$$\sum_{k_1,\dots,k_{j-1}} \text{function of}(g_1,\dots,g_{r-1})\ ad_{A_1}^{k_1}\dots ad_{A_{j-1}}^{k_{j-1}}(A_j) = \sum h_{jr}(g)A_r \tag{13.5}$$

It follows first of all from (13.5) that

$$h_{jr}(g) \text{ only depends on } \ g_1,\dots,g_{r-1} \tag{13.6}$$

Second, let $j>k$, so that $A_j \in \mathfrak{a}$. Then $ad_{A_1}^{k_1}\dots ad_{A_{j-1}}^{k_{j-1}}(A_j) \in A$ and it follows that

$$\text{if } j>k,\ h_{jr}(g)=0 \text{ for } r \leqslant k \tag{13.7}$$

Now take $r \leqslant k$ in (13.6). Taking account of (13.6) and (13.7) we see that (13.4) is of the form

$$\dot{g}_r + \dot{g}_1 h_{1r}(g_1,\dots,g_{r-1}) + \dots + \dot{g}_k h_{kr}(g_1,\dots,g_{r-1}) = u_r \tag{13.4}$$

Thus in the situation under consideration of an ideal $\mathfrak{a}$ in L and a basis adapted to this situation the Wei-Norman equations for $g_1,\dots,g_k$ only involve the $g_1,\dots,g_k$, and can thus be written down, analysed and solved without any regard for the remainder of the Lie algebra.

As is now readily seen from what has been said, the Wei-Norman equations for $g_1,\dots,g_k$ are in fact the Wei-Norman equations belonging to the quotient algebra $L/\mathfrak{a}$ with respect to the basis $A_1+\mathfrak{a}$, $A_2+\mathfrak{a},\dots,A_k+\mathfrak{a}$ of this quotient.

We are now ready to consider the infinite dimensional case. So suppose that L has ideals $I_1,I_2,\dots,I_n,\dots$ such that $\cap I_n = 0$ and each L/I_n is dimensional and solvable. There is than a basis of L of the form

$$A_1,\dots,A_{k_1},\ A_{k_1+1},\dots,A_{k_2},\qquad k_1<k_2<\dots \tag{13.5}$$

such that

$$A_{k_n+1},\ A_{k_n+2},\dots\ \cdots \tag{13.6}$$

is a basis for I_n. We are, as usual, interested in solving an equation

$$\frac{\partial \rho}{\partial t} = \sum A_i \rho u_i(t) \tag{13.7}$$

where, in our case at least, the sum on the right is a finite one and where we can assume that the A_i are part of the basis 13.5 (otherwise write out the operators in (13.7) in terms of that basis.) We can in effect assume that, say, the sum runs from $i=1$ to $i=k_n$. This does not mean, however, that the solution can be written in terms of the $e^{g_j(t)A_j}$, $1 \leqslant j \leqslant k_n$; the higher A's will also tend to occur via higher brackets.

The idea now is to try an infinite product

$$e^{g_1A_1}\dots e^{g_{k_1}A_{k_1}} e^{g_{k_1+1}A_{k_1+1}}\dots e^{g_{k_2}A_{k_2}}\dots e^{g_{k_n}A_{k_n}}\dots \pi(x) \tag{13.8}$$

as Ansatz. By the remarks made about quotients above, the infinite system of Wei-Norman for the g_i is such that

$$\dot{g}_1,\dots,\dot{g}_{k_1} \quad \text{only involve } \ g_1,\dots,g_{k_1}\ ;\ u_1,\dots,u_{k_1}$$

$\dot{g}_1,...,\dot{g}_{k_2}$ only involve $g_1,...,g_{k_2}$: $u_1,...,u_{k_2}$

...

So that in any case the infinite system of Wei-Norman equations makes sense. We can now calculate a sequence of densities

$$e^{g_1A_1}...e^{g_{k_1}A_{k_1}}\pi(x)$$

...

$$e^{g_1A_1}...e^{g_{k_1}A_{k_1}}...e^{g_{k_n}A_{k_n}}\pi(x)$$

...

The question remains whether this sequence of densities converges in one sense or another. This is a largely uninvestigated mather. Scattered through the remainder of this article there are a number of comments on this.

There is more to be said about Wei-Norman type theory in infinite dimensional contexts. A number of estimation Lie algebras occur as solvable subalgebras of a Lie algebra of the form $\mathbf{R}[z_1,...,z_r]\otimes L$ where L is a finite dimensional Lie algebra of differential operators in $x_1,...,x_n$. The meaning of this symbolism is as follows. Let $A_1,...,A_s$ be a basis of L. Then a basis for $\mathbf{R}[z_1,...,z_r]\otimes L$ is formed by the differential operators

$$z^\alpha A_i,\quad \alpha=(a_1,...,a_r),\ a_j\in\mathbf{N}\cup\{0\} \text{ a multi index, } i=1,...,s$$

And the bracket between these basis elements is given by

$$[z^\alpha A_i,\ z^\beta A_j]=z^{\alpha+\beta}\ [A_i,A_j]$$

These are called current algebras and have been investigated in both the mathematics and the physics literature to a considerable extent [36-38,40,58]. The point here is that though $\mathbf{R}[z]\otimes L$ is infinite dimensional over $\mathbf{R}$ it is finite dimensional over the ring of functions $\mathbf{R}[z]$. Thus the natural object in which solutions of equations

$$\frac{\partial\rho}{\partial t}=\sum_{\alpha,i}u_{\alpha,i}(t)z^\alpha A_i$$

will live in something like the group of functions in $z_1,...,z_n$ to G, where G is the Lie group of L. In slightly more concrete terms this means that the Ansatz now becomes

$$\rho(t,x)=e^{g_1A_1}...e^{g_sA_s}\pi(x,z)$$

where now the g_i are supposed to be functions of both t and $z_1,...,z_r$, polynomial in z in this particular context.

The estimation Lie algebra of a linear system identification problem is of the 'subalgebra of current algebra' type, cf below in section 15. In [42] there are some more details on Wei-Norman theory and identification from this particular point of view.

14. The weak cubic sensor

Recall that this is the one dimensional system

$$dx=dw \qquad (14.1)$$
$$dy=(x+\epsilon x^3)dt+dv$$

with $\epsilon\neq0$. Recall also that its estimation Lie algebra is equal to W_1 for $\epsilon\neq0$ (and for $\epsilon=0$ it reduces of course to the oscillator Lie algebra). Thus by the arguments of section 12 above it follows that there are no exact finite dimensional recursive filters for any statistic of the weak cubic sensor. On the other hand it is intuitively hard to believe that the filter for $\epsilon=0$ will not give something of an approximation. Also when one actually calculates *ELie*(weak cubic sensor) one notices that the higher order differential operators and higher order polynomials and products of these appear with high powers of ϵ in front of them, suggesting that neglecting these will (i) not matter too much, and (ii) give us something finite dimensional to work with. This can be made precise as follows. Consider ϵ in (14.1) as a parameter. Calculate *ELie* (weak cubic sensor) in the usual way but with ϵ as a polynomial variable, i.e. calculate *ELie* as a subalgebra of $\mathbf{R}<\epsilon, x, \frac{d}{dx}>$, $[\epsilon, \frac{d}{dx}]=0$, i.e. treat ϵ as a second variable (besides x). Now introduce the extra rule $\epsilon^m=0$ for $m \geqslant n$. Then the resulting algebra is finite dimensional and solvable. Let us call this the estimation Lie algebra modulo ϵ^n, $ELie\, mod(\epsilon^n)$. Technically speaking we are considering $ELie \otimes_{\mathbf{R}[\epsilon]} \mathbf{R}[\epsilon]/\epsilon^n$. That $ELie\, mod(\epsilon^n)$ is finite dimensional and solvable in this case is an instance of a much more general phenomenon.

14.2. Theorem ([23,29]).
Let Σ_ϵ be a stochastic system of the form

$$dx = (Ax+\epsilon P_A(x)dt+(B+\epsilon P_B(x)dw \quad x\in\mathbf{R}^n,\ w\in\mathbf{R}^m$$
$$dy = (Cx+\epsilon P_C(x)dt+dv \qquad g\in\mathbf{R}^p,\ v\in\mathbf{R}^p$$

where $P_A(x), P_B(x), P_C(x)$ are polynomial in x. Then $ELie(\Sigma_\epsilon)$ $mod(\epsilon^r)$ is finite dimensional and solvable for all r.

14.3. Example.
ELie (weak cubic sensor) $mod(\epsilon^2)$ is 14 dimensional with basis

$$\frac{1}{2}\frac{d^2}{dx^2}-1\frac{1}{2}x^2-\epsilon x^4,\ x,\ \epsilon x^3, \frac{d}{dx},\ 1,\ \epsilon,\ \epsilon x^2\frac{d}{dx},\ \epsilon x$$
$$\epsilon x\frac{d}{dx},\ \epsilon\frac{d^2}{dx^2},\ \epsilon\frac{d}{dx},\ \epsilon\frac{d^3}{dx^3},\ \epsilon x\frac{d^2}{dx^2},\ \epsilon x^2$$

The next question is: do these finite dimensional solvable "quotients" of $ELie(\Sigma_\epsilon)$ calculate anything. Let us do the following. The solution of the DMZ equation will also depend on ϵ. Let us look for formal power series in ϵ solutions of the form

$$\rho(x,t,\epsilon) = \rho_0(x,t)+\epsilon\rho_1(x,t)+\epsilon^2\rho_2(x,t)+\ldots \tag{14.4}$$

Then the Wei-Norman equations for $ELie(\Sigma_\epsilon)$ $mod(\epsilon^r)$ precisely compute the first r coefficients of (14.4), i.e. $\rho_0(x,t),\ldots,\rho_{r-1}(x,t)$. This is quite simple and is in fact related of the second group of ideas re infinite dimensional Wei-Norman theory as discussed in section 13 above. (It also has aspects of the first group of ideas and is in fact a sort of amalgam of both).

The next question is whether the formal series (14.4) will converge. This is again a matter which still needs a great amount of investigation. The series does converge for the weak cubic sensor and the weak quadratic one. It is also pleasing to note that the resulting $mod(\epsilon^2)$ filter for the weak quadratic sensor sensor already performs much better than the extended Kalman filter [35,48]. Further it can be shown that the series (14.4) is always an asymptotic series in the technical sense of the word. On the other hand there are arguments indicating that the series will not converge for the weak quartic sensor and higher. Thus there appears much to do and it may well be fruitful to take into account that $\rho(x,t,\epsilon)$ only matters up to a normalization factor $r(t,\epsilon)$, which can be chosen as one pleases.

15. Identification of linear dynamical systems

Suppose now that we are faced with a somewhat different problem. Namely suppose one has reason to believe, or simply does not know anything better to do, that a given phenomenon, say a time series, is modeled by a linear dynamical system

$$dx = Axdt+Bdw,\ dy = Cxdt+dv \tag{15.1}$$

Now, however, the coefficients in A,B,C are unknown and also have to be estimated from the observations $y(t)$. That is the system (15.1) has to be identified. It is easy to turn this into a filtering problem by adding the (stochastic) equations

$$dA = 0,\ dB = 0,\ dC = 0 \tag{15.2}$$

(or just $dr_{ij}=0$ whether the r_{ij} run through the coefficients which are unknown, if A,B,C are partly known; for example because of structural considerations). The resulting filtering problem is nonlinear.

15.3. Observation

The estimation Lie algebra of the system (15.1)-(15.2) is a sub-Lie-algebra of the current Lie algebra $ls_n \otimes \mathbf{R}[A,B,C]$ where $\mathbf{R}[A,BC]$ stands for the ring of polynomials in the indeterminates a_{ij},b_{kl},c_{rs}. Here ls_n is the $(2n^2+3n+1)$-dimensional Lie algebra with basis (9.14); i.e. the Lie algebra of all differential operators of total degree $\leqslant 2$ in x_i and $\frac{\partial}{\partial x_j}$, so that $ls_n = \{\Sigma c_{\alpha,\beta} x^\alpha \frac{\partial^{|\beta|}}{\partial x^\beta} : c_{\alpha,\beta}=0$ if $|\alpha|+|\beta|>2\}$. The Lie algebra ls_n contains a subalgebra isomorphic to sp_n (cf section 9 above), so this does not yet prove that *ELie*((15.1)-(15.2)) is solvable. But as a matter of fact it is. Thus the ideas and considerations of the previous two sections can be brought into play. Some initial results exploiting the current algebra based ideas briefly discussed in the second half of section 13 above are contained in [42]. In this rather special case it turns out that the higher approximations (the zero-th approximation is simply the family of Kalman-Bucy filters parameterized by A,B,C also discussed in section 9 above) have to do with sensitivity equations: sensitivities of the output $y(t)$ with respect to changes in the parameters A,B,C.

As stated above, though, the problem is degenerate and likely to cause all kind of difficulties. The problem is that the conditional density $\rho(x,A,B,C,t)$ will be degenerate because the A,B,C are not uniquely determined by the observations. Indeed if S is an invertible $n \times n$ matrix then the system (15.1) given by the matrices SAS^{-1},SB,CS^{-1} instead of A,B,C gives exactly the same input-output behaviour. Thus we should really be considering this problem on a suitable quotient space $\{(A,B,C)\}/GL_n$. These quotient spaces as a rule are not diffeomorphic to open sets in some $\mathbf{R}^n$ [32,33]. This is one way in which stochastic systems like (1.1)-(1.2) on nontrivial manifolds naturally arise and it leads to the necessity of finding a DMZ-equation in this more general context. Work in this direction has been done by Ji Dunmu and T.E. Duncan.

Let me add one more possible approach, which is in the spirit of the ideas of section 14 and the first half of section 13, rather than based on current algebra ideas. For the filters giving $\hat{x},\hat{A},\hat{B},\hat{C}$ for problem (15.1)-(15.2) one expects $\hat{x}$ to move fast relative $\hat{A},\hat{B},\hat{C}$. Thus it would make sense to consider a system

$$dx = (A_0+\epsilon A_1)xdt+(B_0+\epsilon B_1)dw,\ dy = -\ (C_0+\epsilon C_1)dt+dv \tag{15.4}$$

$$dA_1 = 0,\ dB_1 = 0,\ dC_1 = 0$$

where A_0,B_0,C_0 are assumed known) and apply the ideas of section 11 above to find optimal directions of change (i.e. the A_1,B_1,C_1)).

16. Asymptotic expansions and approximate homomorphisms. The marking approach.

The ideas to be outlined below in this section are still speculative but there are quite a number of positive signs.

First however let me point out that the procedures based on Wei-Norman techniques as described in sections 13 and 14 above clearly indicate that existence, uniqueness and regularity results for solutions of the DMZ-equation have a lot to do with the existence of asymptotic expansions ([48]). for regularity results etc. cf e.g. [3,12,43,52] and references in these papers.

Let us consider a control system of the form

$$\dot{x} = f(x)+\Sigma u_i g_i(x) \tag{16.1}$$

where the f and g_i are vectorfields. To make thinking easier assume that 0 is a stable and asymptotically stable equilibrium for the unforced equation. A system like (16.1) is intended as a model of something and as such one can argue that say the values of $f(x), g_i(x)$ are relatively well known, the values of their (partial) derivatives (w.r.t. the x_i) will be less known, the second partial derivations are still less well determined etc..

Thus, intuitively, for systems which represent or model real (stable) things one would expect that in may cases the behaviour of (16.1) will depend primarily on the first few terms which appear in the Lie algebra generated by f and the g_i. The higher brackets should matter less and less.

That means that instead of looking at $Lie\{f, g_1, \ldots, g_m\}$, the Lie algebra generated by $f, g_1, \ldots, g_m$, as a Lie algebra without further structure we should look at it as a Lie algebra with a given set of generators and sort of keep track of how often these generators are used to generate further elements of the algebra. For each time a bracket is taken a differentiation is applied, and thus the higher brackets of the $f, g_1, \ldots, g_m$ depend only on the deeper parts of the Taylor expansions of $f, g_1, \ldots, g_m$. More precisely brackets of order n of a nearby changed system differ by terms of the form $\Delta(n_1)^{i_1} \ldots \Delta(n_r)^{i_r}$ with $i_1 n_1 + \ldots + i_r n_r \geqslant n$ wher $\Delta(n_k)$ symbolizes an upper bound for the uncertainty, i.e., the changes, at level n_k in the Taylor expansions. Let me also stress that, in spite of the word 'Taylor expansion' in the previous sentences I am attempting to formulate global ideas of approximation and definitely not local ones around one point. If f is a function of one variable depicted as a graph in the plane, then a piecewise linear approximation of f with rounded corners would be a low order illustration of what is intended here. Spline approximation takes us up (at least) one order higher.

Personally I would also say that having noises rather than precise deterministic controls u_i would enhance this type of (structural?) stability.

A precise way to keep track of how often the generators are used is to introduce one extra counting indeterminate z and to consider instead of $L = Lie\{f, g_1, \ldots, g_m\}$ the Lie algebra generated by the vectorfields $\{zf, zg_1, \ldots, zg_m\}$. This Lie algebra L_z is topologically nilpotent, i.e. if $L_z^{(n)} = [L_z, L_z^{(n-1)}]$, $L_z^{(0)} = L_z$, then $\cap L_z^{(m)} = \{0\}$. And a homomorphism $L_z \to V(M)$ into the vectorfields on M with kernel $L_z^{(n)}$ precisely means "respecting the structure of the Lie algebra L up to brackets of order n". All this is very much related to the ideas of nilpotent approximation introduced in the study of hypoellipticity [22,61], which are now also being investigated in control and system theoretic contexts [15,60].

Let me explain the context partly. Consider a homomorphism of a system of type (16.1) into an other one. That means a differentiable map $\phi: M \to M'$ where M and M' are the state space manifolds of (16.1) and (16.1)$'$ such that ϕ takes the vectorfields $f, g_1, \ldots, g_m$ into the vectorfields $f', g'_1, \ldots, g'_m$. (If there is also an output map $h: M \to \mathbb{R}^p$, then of course we must also have $h' \circ \phi = h$). Inversely if $Lie(\Sigma)$ is the Lie algebra generated by $f, g_1, \ldots, g_m$ and $\alpha: Lie(\Sigma) \to Lie(\Sigma')$ is a homomorphism of Lie algebras taking isotropy subalgebras into isotropy subalgebras, then, at least locally, there exists a ϕ such that '$d\phi = \alpha$'.

A first idea of an approximate homomorphism α of level m is that if σ resp. τ are elements in $Lie(\Sigma)$ which can be obtained by taking iterated brackets of the $f, g_1, \ldots, g_m$ at most a_σ resp. a_τ times and $a_\sigma + a_\tau \leqslant m$, then $\alpha[\sigma, \tau] = [\alpha(\sigma), \alpha(\tau)]$. This corresponds precisely to introducing markers, i.e. writing $zg_1, \ldots, zg_m$ and saying that a map $\alpha: Lie_z(\Sigma) \to Lie(\Sigma')$ is an approximate homomorphism of level m if it induces a homomorphism of Lie algebras $Lie_z(\Sigma)\, mod z^{m+1} \to Lie_z(\Sigma')\, mod z^{m+1}$

Thus in filtering theory, which can be seen as the theory of trying to find (approximate) homomorphism of the infinite dimensional system given by the DMZ equation (5.6) or (6.2) and the output map (4.4) to finite dimensional systems, it would seem natural to look at the Lie algebra of operators $ELie_z(\Sigma)$ generated by the operators

$$z_0 \pounds,\ z_1 h_1, \ldots, z_p h_p$$

where the $z_0, z_1, \ldots, z_p$ are additional variables (so as to give, if desired, certain observations more weight than others and to be able to set certain of them, especially z_0, equal to 1). The idea would be then to study the filters produced by Wei-Norman type techniques for the various finite dimensional quotients and to see whether this produces viable expansions.

Let me conclude this section with an argument indicating that the approximation scheme indicated above should work. Let the true system be Σ and assume it is stable in the sense that modifying the $f, g_1, \ldots, g_m$ in the manner indicated does not change the input-output behaviour much. Models of real systems are expected to be like this simply because they must still do a reasonable job if some of the measured coefficients are slightly wrong, which is inevitable. Now suppose moreover that I can modify the higher derivatives of the $f, g_1, \ldots, g_n$ in such a way that there exists an exact finite dimensional filter for the modified system Σ'. Thus the situation is as depicted below

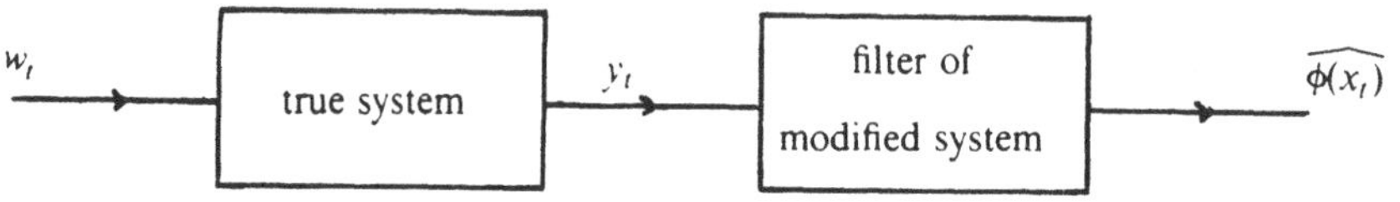

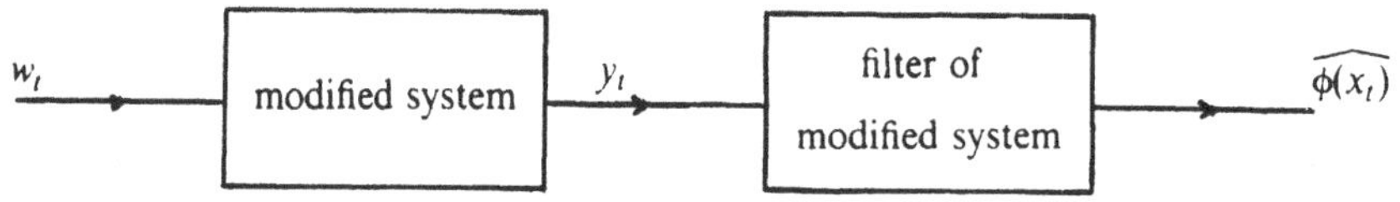

Now the filter of the modified system is also expected to be (input-output) stable. Indeed it will almost have to be that in order to do its job. Then the two composed systems shown above will be close in input-output sense, which means precisely that we have constructed an approximate filter for the true system.

Now, as far as I can see, for a given system Σ, there will as a rule not exist an approximation (in the given sense) which suddenly has a finite dimensional solvable estimation Lie algebra. Or even an infinite dimensional solvable one. In that case there certainly are lots of filters but it is less clear what quantities they filter for and it also remains to be investigated thoroughly whether they give usuable approximation to a $\rho(x,t)$, cf section 13, 14 above.

Thus it does not seem that the argument given above can be used to prove that the marking approach gives good approximate filters, but the argument certainly provides positive indications.

17. Removing outliers

A final idea in much the same spirit as before is the following. Suppose we are again dealing with a system

$$dx = f(x)dt + G(x)dw, \quad dy = h(x)dt + dv. \tag{17.1}$$

Suppose also, to make thinking easier, that the thing is more or less stable, so that x tends to remain in some bounded partion of $\mathbf{R}^n$ (f asymptotically stable), and maybe suppose also that h is proper, so that large y observations are exceedingly rare and should probably be discounted. Suppose that $e^{-\|x\|^2}$ is differential algebraically independent of f, G, h. This is for example the case if f, G, h are polynomial and also if they are of compact support. In other cases other functions with similar properties can presumably be found. Now instead of (17.1) consider the modified system

$$dx = f(x)dt + Gdw, \quad y = e^{-a\|x\|^2} h(x)dt + dv \tag{17.2}$$

where $a > 0$ is a small parameter. Note that the only thing which (17.2) does with respect to (17.1) is to discount large y observations.

Now consider the estimation Lie algebra of the system (17.2).

17.3. Theorem
If $e^{-a\|x\|^2}$ is differentially algebraically independent of f,G,h then the estimation Lie algebra of (17.2) is (infinite dimensional) solvable. To be more precise it is finite dimensional and solvable mod($a^i e^{-ja\|x\|^2}$, $i+j \geq n$) for all n.

Thus the yoga of the previous sections can again be applied and the behaviour of the resulting 2-parameter family of filters as a goes to zero and n goes to infinity could be studied.

1. L. ARNOLD, *Stochastische Differentialgleichungen,* Oldenbourg Verlag, 1972. (Translation: Stochastic differential equations, Wiley, 1974).
2. J.S. BARAS, Group invariance methods in nonlinear filtering of diffusion processes, In: [34], 565-572.
3. J.S. BARAS, G.L. BLANKENSCHIP, W.E. HOPKINS JR., Existence, uniqueness and asymptotic behaviour of solutions to a class of Zakai equations with unbounded coefficients, IEEE Trans. **AC-28** (1983), 203-214.
4. V.E. BENES, Exact finite dimensional filters for certain diffusions with nonlinear drift, *Stochastics* **5** (1981), 65-92.
5. G.L. BLANKENSCHIP, C.-H. LIU, S.I. MARCUS, Asymptotic expansions and Lie algebras for some nonlinear filter problems, IEEE Trans, **AC-28** (1983), 787-797.
6. R.W. BROCKETT, Nonlinear systems and nonlinear extimation theory, In: [34], 441-478.
7. R.W. BROCKETT, Classification and equivalence in estimation theory, *Proc, 1979 IEEE Conf. on decision and Control,* Ft. Lauderdale, December 1979.
8. R.W. BROCKETT, Remarks on finite dimensional nonlinear estimation, In: C. Lobry (ed.), Analyse des systèmes, Astérisque **76** (1980), Soc. Math. de France.
9. R.W. BROCKETT, J.M.C. CLARK, The geometry of the conditional density equation, In: O.L.R. Jacobs et al. (eds.), *Analysis and optimization of stochastic systems,* New York, 1980, 299-309.
10. R.S. BUCY, J.M.F. MOURA (eds.), Nonlinear stochastic problems, Reidel, 1983.
11. R.S. BUCY, J. PAGES, A priori error bounds for the cubic sensor problem, *IEEE Trans. Automatic Control* **AC-24** *(1979), 948-953.*
12. M. CHALEYAT-MAUREL, D. MICHE., Hypoellipticity theorems and conditional laws, *Z. Wahrsch. und verw. Geb.* **65** (1984), 573-597.
13. S.D. CHIKTE, J.T.-H. LO, Optimal filters for bilinear systems with nilpotent Lie algebras, *IEEE Trans, Automatic Control* **AC-24** (1979), 948-953.
14. J.M.C. CLARK, The design of robust approximations to the stochastic differential equations of nonlinear filtering, In: J.K. Skwirzynski (ed.), *Communication systems and random process theory,* Sijthoff & Noordhoff, 1978.
15. P.E. CROUCH, Solvable approximations to control systems, *SIAM J. Control and Opt.* **32** (1984), 40-54.
16. F.E. DAUM, Exact nonlinear recursive filters and separation of variables, *Proc. 25-th CDC,* Athens, Dec. 1986.
17. M.H.A. DAVIS, Exact finite dimensional nonlinear filters, *Proc. 24-th CDC, Las Vegas December 1985.*
18. M.H.A. DAVIS, Pathwise nonlinear filtering, In: [34], 505-528.
19. M.H.A. DAVIS, S.I. MARCUS, An introduction to nonlinear filtering, In: [34], 565-572.
20. T.E. DUNCAN, Probability densities for diffusion processes with applications to nonlinear filtering theory, *Ph. D. thesis,* Stanford, 1967.
21. M. FUJISAKI, G. KALLIANPUR, H. KUNITA, Stochastic differential equations for the nonlinear filtering problem, *Osaka J. Math.* **1** (1972), 19-40.
22. R.W. GOODMAN, Nilpotent Lie groups: structure and applications to analysis, LNM **562,**

Springer, 1976.
23. M. HAZEWINKEL, On deformations, approximations and nonlinear filtering, *Systems Control Lett.* **1** (1982), 29-62.
24. M. HAZEWINKEL, The linear systems Lie algebra, the Segal-Shale-Weil representation and all Kalman-Bucy filters, *J. Syst. Sci. & Math. Sci.* **5** (1985), 94-106.
25. M. HAZEWINKEL, S.I. MARCUS, On Lie algebras and finite dimensional filtering, *Stochastics* **7** (1982), 29-62.
26. M. HAZEWINKEL, S.I. MARCUS, H.J. SUSSMANN, Nonexistence of finite dimensional filters for conditional statistics of the cubic sensor problem, *Systems Control Lett.* **3** (1983), 331-340.
27. M. HAZEWINKEL, A tutorial introduction to differentiable manifolds and calculus on differentiable manifolds, *This volume.*
28. M. HAZEWINKEL, A tutorial on Lie algebras, *This volume.*
29. M. HAZEWINKEL, Lie algebraic methods in filtering and identification, In: M. Mebkhout, R. Sénéor (eds.), *Proc. VIII-th Int. Cong. Math. Physics,* Luming 1986, World Scientific, 1987, 120-137 (also to appear in *Proc. 1-st World Cong. Bernouilli Society,* Tashkent, 1986, VNU, 1988).
30. M. HAZEWINKEL, Identification, filtering, the symplectic group, and Wei-Norman theory. To appear in *Proc. 2-nd workshop on the Road-vehicle system and related mathematics,* Torino, 1987, Reidel-Teubner, 1988.
31. M. HAZEWINKEL, On filtering and polynomials in *n* independent Brownian motions, *Acta Scientifica,* Caracas, to appear.
32. M. HAZEWINKEL, Moduli and canonical forms for linear dynamical systems II: the topological case, *Math. Systems Th.* **10** (1977), 363-385.
33. M. HAZEWINKEL, (Fine) moduli (spaces) for linear systems: what are they and what are they good for, In: C.I. Byrnes, C.F. Martin (eds.), *Geometric methods for linear systems theory* (Harvard, June 1979), Reidel, 1980, 125-193.
34. M. HAZEWINKEL, J.C. WILLEMS, (eds.), Stochastic systems: the mathematics of filtering and identification and applications, Reidel, 1981.
35. M. HAZEWINKEL, An introduction to nilpotent approximation filtering. To appear in *Proc. 2-nd European Symp. on Math. in Industry* (ECMI 2), Oberwolfach, 1987, Reidel-Teubner, 1988.
36. R.C. HERMANN, Lie algebras and quantum mechanics, Benjamin, 1970.
37. R.C. HERMANN, Infinite dimensional Lie algebras and current algebras, In: *Lect. Notes in Physics* **6**, Springer, 1969.
38. R.C. HERMANN, Current algebras, the Sugawara model and differential geometry, *J. Math. Phys.* **11**: 6, (1970), 1825-1829.
39. O.B. HIJAB, Finite dimensional causal functionals of brownian motion, In [6], 425-436.
40. A.A. KIRILOV, Local Lie algebras, *Usp. Mat. Mauk.* **81**, 57-76.
41. A.J. KRENER, On the equivalence of control systems and the linearization of nonlinear systems, *SIAM J. Control* **11**, (1973), 670-676.
42. P.S. KRISHNAPRASAD, S.I. MARCUS, M. HAZEWINKEL, Current algebras and the identification problem, *Stochastics* **11** (1983), 65-101.
43. TH.G. KURTZ, D. OCONE, A martingale problem for conditional distributions and uniqueness for the nonlinear filtering equations, *Lect. Notes Control and Inf. Sci.* **69** (1985), 224-235.
44. H.J. KUSHNER, On the dynamical equations of conditional probability functions with application to optimal stochastic control theory, *J. Math. Anal. Appl.* **8** (1964), 332-344.
45. H.J. KUSHNER, Dynamical equations for optimal nonlinear filtering, *J. Diff. Equations* **3** (1967), 179-190.
46. R.S. LIPTSER, A.N. SHIRYAYEV, *Statistics of Random Processes I.* New York: Springer-Verlag, 1977.
47. C.-H. LIU, S.I. MARCUS, The Lie algebraic structure of a class of finite dimensional nonlinear filters, In: C.I. Byrnes, C.F. Martin (eds.), Algebraic and geometric methods in linear systems theory, *Amer. Math. Soc.,* (1980), 277-279.

48. S.I. MARCUS, Algebraic and geometric methods in nonlinear filtering, *SIAM J. Control Opt.* **22** (1984), 817-844.
49. S.I. MARCUS, A.S. WILLSKY, Algebraic structure and finite dimensional nonlinear estimation, *SIAM J. Math. Anal.* **9** (1978), 312-327.
50. S.I. MARCUS, Optimal nonlinear estimation for a class of discrete-time stochastic systems, *IEEE Trans. Automatic Control* **AC-24** (1979), 297-302.
51. S.I. MARCUS, S.K. MITTER, D. OCONE, Finite dimensional nonlinear estimation for a class of systems in continuous and discrete time, In: *Analysis and Optimization of Stochastic Systems,* O.L.R. Jacobs et al. eds, Academic Press, 1980, 387-406.
52. D. MICHEL, Régularité des lois conditionelles en théorie du filtrage non-linéaire et calcul des variations stochastiques, *J. Funct. Anal.* **14** (1981), 8-36.
53. S.K. MITTER, On the analogy between mathematical problems of non-linear filtering and quantum physics, *Richerche di Automatica* **10** (1980), 163-216.
54. S.K. MITTER, Nonlinear filtering and stochastic machanics, In: [34], 479-504.
55. R.E. MORTENSEN, Optimal control of continuous time stochastic systems, *Ph. D thesis,* Berkeley, 1966.
56. D. OCONE, *Topics in non-linear filtering,* Thesis, M.I.T., June 1980.
57. E. PARDOUX, D. TALAY, Discretization and simulation of stochstic differential equations, *Acta Appl. Math.* **3** (1982), 182-203.
58. K.R. PARTHASARATHY, K. SCHMIDT, A new method for constructing factorizable representations for current group and current algebras, *Comm. Math. Phys.* **50** (1976), 167-175.
59. L.E. PURSELL, M.E. SHANKS, The Lie algebra of a smooth manifold, *Proc. Amer. Math. Soc.* **5** (1954), 468-472.
60. CH. ROCKLAND, Intrinsic nilpotent approximation, *Acta Appl. Math.* **8** (1987), 213-270.
61. L..P. ROTHSCHILD, E.M. STEIN, Hypoelliptic differential operators and nilpotent groups, *Acta Math.* **137** (1976), 247-320.
62. S. STEINBERG, Application of the Lie algebraic formulas of Baker, Campbell, Hausdorff and Zassenhaus to the calculation of explicit solutions of partial differential equations, *J. Diff. Equations* **26** (1977), 404-434.
63. H.J. SUSSMANN, Existence and uniqueness of minimal realizations of nonlinear systems, *Math. Syst. Theory* **10** (1977), 349-356.
64. J. WEI, E. NORMAN, On the global representation of the solutions of linear differential equations as a product of exponentials, *Proc. Amer. Math. Soc.* **15** (1964), 327-334.
65. M. ZAKAI, On the optimal filtering of diffusion processes, *Z. Wahrsch. verw. Geb.* **11** (1969), 230-243.

STRUCTURAL PARAMETER IDENTIFICATION TECHNIQUES

F. Kozin
Polytechnic University, Farmingdale, New York, USA

ABSTRACT

In these lectures we review certain results on the problem of parameter estimation and system identification as applied to structural engineering. The importance of this problem in structural engineering has steadily increased in recent years, primarily motivated by the desire to have a more accurate description of the structure and its dynamical characteristics for purposes of predicting its response to environmental excitations such as earthquakes and wind generated pressure loadings, being able to assess aging or damage through changes in the salient structural parameters, and finally for purposes of applying controllers to structures that can reduce unwanted responses to the environmental excitations.

Naturally, since the field is quite broad in its scope, it will be quite impossible to cover all aspects, or all significant results. Instead, we present a summary of some significant results that have applications to the structural field.

Rather than merely present a brief discussion of results, we state in some detail what can be said concerning convergence and accuracy of the techniques available, along with examples to illustrate the ideas presented. We discuss time domain techniques, and concentrate on continuous time models.

1. INTRODUCTION

Our purpose in these lectures is to review certain results on the problem of parameter estimation and system identification as applied to structural engineering. The importance of this problem in structural engineering has steadily increased in recent years, primarily motivated by the desire to have a more accurate description of the structure and its dynamical characteristics for purposes of predicting its response to environmental excitations such as earthquakes and wind generated pressure loadings, being able to assess aging or damage through changes in the salient structural parameters, and finally for purposes of applying controllers to structures that can reduce unwanted responses to the environmental excitations. Certainly for these reasons and, perhaps, others that one may think of, an accurate model of the real structure must be available to the engineer.

The general topic of estimation, identification and modelling has grown greatly during this past decade to the extent that many basic problems have been solved, and techniques are available to study dynamical systems. To a great extent, the topic has developed as a result of the needs of the system sciences and more particularly, those requirements of control theory and applications. This activity has been developing for two decades (starting with the space programs), but only in the past decade has structural engineering looked to this development and recognized its significance. One can see this trend in recent survey papers such as [1], [2], and recent proceedings such as [3]. Previous surveys such as [4] were concerned with more traditional concepts of structural dynamics such as frequency response, eigenvalues, etc. But, now we recognize that those tools used by systems and control analysts have entered the structural identification literature.

Naturally, since the field is quite broad in its scope, it will be quite impossible to cover all aspects, or all significant results. We shall, however, attempt to present a summary of significant topics that have applications to the structural field.

We shall present rather practical approaches as well as discuss some theoretical questions that arise in connection with these approaches. We shall also present examples in brief form to illustrate the techniques discussed.

The general problem of modelling and identification can be divided into three parts:

I. Postulate a class of models for the data.
II. Estimate the unknown parameters of the model from the data.
III. Assess the quality or fit of the model to the given data.

To a great extent the models for structural systems are obtained from classical dynamical properties (Newton's Laws, etc.). Therefore, the class of models is usually well defined. However, there are cases where an accepted model is not available (biological systems, various nonlinearities, geophysical phenomena, soil structure interation) and a class of models must be postulated.

Once a class of models is postulated, the problem of identifying exactly that particular model from the class is basic. Generally, the class of models is a parametric class. That is, the class is known up to some set of parameters. These parameters will include coefficients of the variables of the model as well as parameters effecting the structure of the model (e.g. the order of the dynamical equations).

A specific model from the postulated class is selected according to some optimal criterion, based upon an analysis of the observed data. The procedures available for specific model selection (system identification) separate into time domain techniques and frequency domain techniques. In the time domain one is concerned with parameter estimates based upon least squares, maximum likelihood, recursive techniques and related techniques. In the frequency domain one is concerned with spectral estimates and identifying frequency response characteristics as well as decomposition into modal functions using frequency measurements to identify the structural parameters.

Those approaches that have appeared in the Structural Engineering literature are comprehensively reviewed for example in [5] which includes an excellent bibliography. A somewhat more recent series of papers on the topic of structural identification can be found in [6].

Rather than merely present a brief discussion of results, we shall state in some detail what can be said concerning convergence and accuracy of the techniques available, along with examples to illustrate the ideas presented. We also include an extensive bibliography. We approach the subject from the point of view to analytical tools and techniques rather than from the point of view of specific structural applications. We take this line since the ideas presented are broad in their applicability. A certain part of this material has already appeared in [68]. We discuss time domain techniques, only.

2. TIME DOMAIN IDENTIFICATION TECHNIQUES

2.1 General Discussion

In a general setting, the problem can be stated as follows, [7], we have selected a class of models

$$\mathfrak{M} = \{M(\theta) \mid \theta \varepsilon H\} , \qquad (2.1.1)$$

where each specific model $M(\theta)$ is determined by the parameter vector θ.

We have a set of data

$$Z_{\mathcal{J}} \equiv \{Z(t),\ t\varepsilon \mathcal{J}\} \qquad (2.1.2)$$

where $\mathcal{J}$ may be a discrete or continuous time domain. It is discrete if the structural dynamics are observed as sampled data; it is continuous if the structural dynamics are observed as analog curves. The Z-data comprises both the input as well as the output data of the structure.

The general problem is; given $Z_{\mathcal{T}}$, select $M(\theta)$ from $\mathfrak{M}$ based upon some optimal criterion of selection.

The method of optimization is generally in terms of an error based upon the "true", i.e. observed structural dynamics, and the model predicted structural dynamics. Thus, if $y(t)$ denotes the true structural output data (deflections, velocities, accelerations, etc., etc.) and $\hat{y}(t|\theta)$ denotes the predicted structural output from the model $M(\hat{\theta})$ determined by the estimated coefficient parameter $\hat{\theta}$ we can define the prediction error corresponding to the model $M(\hat{\theta})$ as

$$\varepsilon(t,\theta) = y(t) - \hat{y}(t|\theta) \tag{2.1.3}$$

The problem of optimal selection of a model $M(\theta)$ is that of computing the error $\varepsilon(t,\theta)$ over the set of $t\varepsilon\mathcal{T}$ and choosing that value θ which will make $[\varepsilon(t,\theta),\ t\varepsilon\mathcal{T}]$ small for some preselected criterion. If we introduce some scalar measure $C(t,\theta,\varepsilon(t,\theta))$, we can form a criterion such as the minimization of

$$V_{\mathcal{T}}(\theta) = \begin{cases} \dfrac{1}{N}\displaystyle\sum_{i=1}^{N} C(t_i,\theta,\varepsilon(t_i,\theta)), & \mathcal{T}\text{ discrete} \\[2ex] \dfrac{1}{T}\displaystyle\int_0^T C(t,\theta,\varepsilon(t,\theta))dt, & \mathcal{T}\text{ continuous} \end{cases} \tag{2.1.4}$$

That is, we choose θ so that $V_{\mathcal{T}}(\theta)$ is minimized. Perhaps the most significant special cases of this general formulation are to least squares estimation as well as to maximum likelihood estimation.

In the least squares case, we choose

$$C(t,\theta,\varepsilon) = |\varepsilon|^2, \tag{2.1.5}$$

and therefore choose the model $M(\theta)$ that minimizes

$$V_{\mathcal{T}}(\theta) = \begin{cases} \dfrac{1}{N}\displaystyle\sum_{i=1}^{N} (y(t_i) - y(t_i|\theta))^2, & \text{discrete} \\[2ex] \dfrac{1}{T}\displaystyle\int_0^T (y(t) - y(t|\theta))^2 dt, & \text{continuous} \end{cases} \tag{2.1.6}$$

In the case that $\varepsilon(t,\theta)$, given the data $Z_{\mathcal{T}}$ possesses the conditional density function $f(t,\theta,\varepsilon|Z_{\mathcal{T}})$, where

$$\mathrm{Prob}\{\varepsilon(t,\theta)\varepsilon A|Z_{\mathcal{T}}\} = \int_A f(t,\theta,\varepsilon|Z_{\mathcal{T}})d\varepsilon \tag{2.1.7}$$

we can write

$$y(t) = \hat{y}(t|\theta) + \varepsilon(t,\theta) \quad , \tag{2.1.8}$$

where from (2.1.7), the conditional probability density of y(t) given Z is

$$p(y(t)|Z_{\mathcal{T}}) = f(t,\theta,y(t)-\hat{y}(t|\theta)|Z_{\mathcal{T}}) \quad . \tag{2.1.9}$$

In the case that the errors $\varepsilon(t,\theta)$ are independent for distinct values of t_i for the discrete data problem, we can easily see that the $V_{\mathcal{T}}(\theta)$ criterion function becomes

$$V_{\mathcal{T}}(\theta) = -\sum_{i=1}^{N} \log f(t_i,\theta,\varepsilon(t_i,\theta)) \quad , \tag{2.1.10}$$

for $C(t,\theta,\varepsilon) = -\log f(t,\theta,\varepsilon)$, $\hat{\theta}$ is chosen so that $V_{\mathcal{T}}(\theta)$ is maximized.

Quite clearly there is the problem of the actual numerical minimization (or maximization) of the $V_{\mathcal{T}}(\theta)$ function. For the linear model, where

$$\hat{y}(t|\theta) = \theta^T\phi(t) \tag{2.1.11}$$

the least squares approach yields (θ^T denotes transpose)

$$V_{\mathcal{T}}(\theta) = \begin{cases} \frac{1}{N}\sum_{i=1}^{N} \| y(t_i) - \theta^T\phi(t_i)\|^2 \\ \frac{1}{T}\int_0^T \| y(t) - \theta^T\phi(t)\|^2 dt \end{cases} . \tag{2.1.12}$$

which is quadratic in θ, and can be explicitly minimized by

$$\hat{\theta} = \begin{cases} \left[\sum_{i=1}^{N} \phi(t_i)\phi^T(t_i)\right]^{-1} \sum_{i=1}^{N} \phi(t_i)y^T(t_i) \\ \left(\int_0^T \phi(t)\phi^T(t)dt\right)^{-1} \int_0^T \phi(t)y^T(t)dt \end{cases} \tag{2.1.13}$$

in the discrete time data and continuous time data cases, respectively.

For more general systems, it may be quite difficult to obtain the optimal θ explicitly. In that case a gradient search technique is required which is usually formulated as,

$$\hat{\theta}_{\mathcal{T}}(i+1) = \hat{\theta}_{\mathcal{T}}(i) - \mu_{\mathcal{T}}^{(i)} R_{\mathcal{T}}^{(i)} V' (\theta_{\mathcal{T}}^{(i)}) \tag{2.1.14}$$

where $V'_{\mathcal{T}}$ is the vector

$$V'_{\mathcal{T}}(\theta_{\mathcal{T}}^{(i)}) = \frac{1}{N}\sum_{i=1}^{N} [\psi(t_i,\theta)\frac{\partial \ell^T}{\partial \varepsilon}(t_i,\theta,\varepsilon(t_i,\theta)) - \frac{\partial \ell^T}{\partial \theta}(t_i,\theta,\varepsilon(t_i;\theta))] \tag{2.1.15}$$

and

$$\psi^T(t,\theta) = -\frac{d\varepsilon(t,\theta)}{d\theta} .$$

Here ψ is a matrix of order (dim θ, dim y).

The coefficient $\mu_{\mathcal{T}}^{(i)}$ is a convergence factor chosen so that the scalar $V_{\mathcal{T}}(\theta_{\mathcal{T}}^{(i+1)}) < V_{\mathcal{T}}(\theta_{\mathcal{T}}^{(i)})$. The matrix R is chosen so as to change the gradient search direction if so desired.

If data is continually being gathered, and an on-line identification is desired, one may also define recursive schemes to achieve this goal.

The most important questions to ask of these parameter estimates for increasing data sets (i.e. as $N\uparrow\infty$ or $T\uparrow\infty$) are

I. do the estimates converge?

II. in what statistical sense do they converge? (In probability, in mean square, with probability one?)

III. to what values do they converge?

IV. what is the asymptotic distribution of the estimates?

From the point of view of engineering application, it is certainly important that the estimates converge. Furthermore, it is clear that the estimates should converge with probability one, since that means we are assured the value obtained will be close to the true limiting value. We would like to know that the estimates $\hat{\theta}$ converge to the "true" value, so that we are indeed identifying the true parameters, and therefore the "true" model. This of course depends upon a number of factors.

If the $V_{\mathcal{T}}(\theta)$ function has a unique minimum value θ^*, then the estimator $\hat{\theta}_{\mathcal{T}}$ will converge to θ^*. But, in general, if $V_{\mathcal{T}}(\theta)$ possesses more than one minimum value, we can only be assured that the estimator will converge to one of the possible minima of the $V_{\mathcal{T}}(\theta)$ function. Perhaps, however, there is an even more basic philosophical question here. The engineering problem is that of identifying the analytical model $M(\theta)$ describing a real structure. If the assumed class of models

$\mathfrak{M}$ does not, in fact, contain the "true" model of the structure, what is the meaning of the value θ^* and the corresponding model $M(\theta^*)\varepsilon\mathfrak{M}$?

It is known that the model $M(\theta^*)$ that is selected by procedures discussed above is that model in $\mathfrak{M}$ that is "closest", in the sense of a norm or distance in a suitable function space (i.e. Hilbert space), to the "true" model of the structure in the realm of its excitations.

Thus, if the structure is oscillating in a non-linear manner, and the class of models $\mathfrak{M}$ are linear, then the parameter estimate $\hat{\theta}_{\mathcal{T}}$ will converge to the "best" linear approximation to the true dynamics in the sense of minimum distance between the true dynamics and the linear dynamics in Hilbert space. One can look at this also as a projection of the true model onto the subspace of linear models.

We can make a general statement concerning convergence of parameter estimates, which must be tested in specific cases. The general result is stated in the following theorem, in the notations above.

<u>Theorem [8]</u>

Let $Z_{\mathcal{T}}$ denote the observed data set, let $V_{\mathcal{T}}(\theta)$ be the functional of the data set and let $P(Z_{\mathcal{T}}|\theta^*)$ be the conditional probability distributions (measures) of the observed data given the true parameter θ^*. It is assumed that $\theta^*\varepsilon\Theta$, where Θ is a compact set in R_n, and

$$\mathcal{T}_1 \subset \mathcal{T}_2 \subset \ldots \subset \mathcal{T}_n \subset \ldots ; \bigcup_n \mathcal{T}_n = \begin{cases} \text{positive integers in discrete case} \\ \\ [0,\infty] \text{ in continuous case} \end{cases}$$

If $V_{\mathcal{T}}(\theta)$ satisfies

1. $V_{\mathcal{T}}(\theta)$ is measurable with respect to $P(Z_{\mathcal{T}}|\theta)$

2. $\inf_{\mathcal{T}_n \supset \mathcal{T}_N} [V_{\mathcal{T}_n}(\theta) - V_{\mathcal{T}_n}(\theta^*)]$ converges uniformly on Θ

   to $\lim_{n\uparrow\infty} \inf_{\mathcal{T}_n \supset \mathcal{T}_N} [V_{\mathcal{T}_n}(\theta) - V_{\mathcal{T}_n}(\theta^*)]$ with probability one

3. $\lim_{n\uparrow\infty} \inf_{\mathcal{T}_n \supset \mathcal{T}_N} [V_{\mathcal{T}_n}(\theta) - V_{\mathcal{T}_n}(\theta^*)]$ is continuous on Θ with

   probability one.

4. $\lim_{n\uparrow\infty} \inf_{\mathcal{T}_n \supset \mathcal{T}_N} [V_{\mathcal{T}_n}(\theta) - V_{\mathcal{T}_n}(\theta^*)] \geq 0$ with probability one.

   with equality iff $\theta = \theta^*$,

then

$$\lim_{n\uparrow\infty} \hat{\theta}_{\mathcal{T}_n} = \theta^* \text{ with probability one.}$$

Hence, we have our desired probability one convergence.

The question of the limiting statistical properties still remains. The importance of the limiting distribution of the error in the estimate $\hat{\theta}_{\mathcal{T}} - \theta^*$, is that it will allow a measure of the error variance in terms of the length of the observation interval (or number of observations in the discrete time case). In general this limiting distribution is gaussian with zero mean, and covariance matrix given in terms of the parameters of the structure. We shall see these results for specific cases below. We shall now look at a simple approach to estimation of parameters in continuous time, which will illustrate the ideas involved.

2.2 A Simple Approach for Estimation of System Parameters

In order to motivate the problem at hand, we shall describe an approach that was discussed a number of years ago [9] for purposes of identification of space structures. The technique has recently been found to be of significant practical use [10] for analysis of bridge motions, since it can identify model properties for structures with closely spaced natural frequencies and [11] for study of a three story reinforced concrete frame and panel building in USSR. As we shall see, the method though motivated by probabilistic ideas makes sense in the deterministic setting as well. The only requirement on the system model is that the unknown coefficienets in the dynamical equations appear linearly. The technique was "re-discovered", again recently in [12] and applied to non-linear systems.

We assume that the dynamical model is known via a set of appropriate differential equations. Furthermore, it is assumed that the state information (displacement, etc.) as well as the input is observable and noise free. Thus, at least the acceleration is known from which velocity and displacement information can be obtained by on-line integration.

The estimation occurs in the time domain from simple correlation averages of the observed input and output data.

In classical statistics we are confronted with the problem of representing a given random variable Z in terms of, say, a linear combination of observables y_i, $i = 1,\dots,n$ as

$$Z \approx \sum_{i=1}^{n} a_i y_i \tag{2.2.1}$$

There is usually no exact relation, so that a choice of the a_i's must be made on some optimal basis. The very important common criterion is least squares, where the coefficients a_i are chosen so that

$$E\left\{\left(Z-\sum_{i=1}^{n} a_i y_i\right)^2\right\}$$

is minimized.

Thus in the case that the y_i's are orthogonal random variables, we are led to the equations

$$a_i = \frac{E\{Zy_i\}}{E\{y_i^2\}}, \quad i=1,\ldots,n \tag{2.2.2}$$

for the least squares representation. One meets this type of estimation in applications of harmonic analysis as well. For, in the theory of generalized Fourier series one has a collection of functions Q_i, $i=1,\ldots,n$ defined on the interval $[a,b]$, which satisfy

$$\int_a^b Q_i(t)Q_j(t)dt = 0 \quad \text{for} \quad i \neq j \ .$$

In this situation we must represent a function f defined on [a,b] in some optimal fashion as the linear combination

$$f(t) \approx \sum_{i=1}^{n} a_i Q_i(t) \ . \tag{2.2.3}$$

Again the classical technique is least squares approximation where the a_i's are chosen so as to minimize

$$\int_a^b \left(f(t) - \sum_{i=1}^{n} a_i Q_i(t)\right)^2 dt \ .$$

This leads to the formulas,

$$\hat{a}_i = \frac{\int_a^b f(t)Q_i(t)dt}{\int_a^b Q_i^2(t)dt}, \quad i=1,\ldots,n \tag{2.2.4}$$

which are the deterministic analogue of (2.2.2).

If it was known beforehand that (2.2.3) was an equality, then the formulas (2.2.4) would yield the correct values for the coefficients a_i, $i=1,\ldots,n$. One would obtain the n equations

$$\int_a^b f(t)Q_k(t)dt = \sum_{i=1}^n a_i \int_a^b Q_i(t)Q_k(t)dt$$

$$k = 1,\ldots,n \tag{2.2.5}$$

whose solutions are given by (2.2.4) if the Q_i's are orthogonal on [a,b]. However, we hasten to add that the equations (2.2.5) hold whether or not the Q_i's are orthogonal on [a,b].

It is merely the application of the ideas above that allows us to identify system parameters. We can illustrate this point easily from the following situation.

Let $D \equiv \frac{d}{dt}$ and consider the differential equation.

$$\sum_{i=0}^n a_i D^i y(t) = f(t) . \tag{2.2.6}$$

where the $D^i y(t)$, $i = 1,\ldots,n$, and f(t) are known. Then we have an exact regression equality where the only difference is that the functions are all dynamically related via the linear operations of the ordinary calculus.

Thus we immediately obtain on an arbitrary interval, say [0,T],

$$\sum_{i=0}^n a_i \int_0^T [D^i y(t) D^j y(t)]dt = \int_0^T D^j y(t) f(t) dt$$

$$j = 0,\ldots,n . \tag{2.2.7}$$

In particular, if one defines a time average as the operator,

$$< > \equiv \lim_{T\to\infty} \frac{1}{T} \int_0^T (\ ) \, dt ,$$

then the equations (2.2.7) become

$$\sum_{i=1}^n a_i < D^i y D^j y> = <f \, D^j y> ,$$

$$j = 0,\ldots,n \tag{2.2.8}$$

from which the a_i's are exactly obtainable. This example simply and straightforwardly illustrates the idea for identification that we propose. It will hold for linear as well as non-linear systems.

We notice that no assumptions have been made relative to the input excitation f. It is clear, however, that various conditions on the f function input will have their effect upon the nature of the estimators.

Let us assume, for example, that the excitation f is a sample function from some stationary second order differentiable process defined on the positive time axis, $[0,\infty)$. Then for a stable system the differential equation (2.2.6) admits a stationary solution process on $[0,\infty)$ for which the indicated derivatives $Dy, D^2y,\ldots,D^ny$ exist at least in the mean square sense. Indeed, this would hold true for the general n^{th} order linear constant coefficient system denoted by

$$\dot{y}(t) = Ay(t) + f(t) \ , \qquad (2.2.9)$$

where y(t) is an n vector, f(t) an n-vector stationary process with mean square differential components, and A is an $n \times n$ constant matrix.

We can immediately write a general solution to the A matrix in (2.2.9). For we merely multiply (2.2.9) by the transpose $y^T(t)$ and take expectations to yield

$$E\{\dot{y}y^T\} = AE\{yy^T\} + E\{fy^T\}, \quad \text{T denotes Transpose}, \qquad (2.2.10)$$

from which A is obtained as

$$[E\{\dot{y}y^T\} - E\{fy^T\}]E\{yy^T\}^{-1} = A \ . \qquad (2.2.11)$$

We note that $E\{yy^T\}^{-1}$ exists since the covariance matrix $E\{yy^T\}$ will be positive definite. That is,

$$E\left\{\left(\sum_{i=1}^{n} \alpha_i y_i(t)\right)^2\right\} > 0 \ , \qquad (2.2.12)$$

and equality in (2.2.12) can hold only if there is a linear relationship among the components.

Due to the stationary statistics of the solution process, it will immediately follow that

$$\frac{d}{dt} E\{y_i^{\nu}(t)\} = 0$$

$$\frac{d}{dt} E\{y_i^{\nu}(t)y_j^{s}(t)\} = 0 \qquad (2.2.13)$$

for $i, j = 1,\ldots,n$ and ν, s any integers for which the indicated moments exist.

Furthermore, due to the mean square differential properties of y, guaranteed via the input f, it will follow that the derivative operator in (2.2.13) can be taken inside the expectations to yield

$$E\{y_i^{\nu-1}(t)\dot{y}_i(t)\} = 0$$

$$\nu E\{y_i^{\nu-1}(t)\dot{y}_i(t)y_j^s(t)\} + sE\{y_i^\nu(t)y_j^{s-1}(t)\dot{y}_j(t)\} = 0$$

$$i, j = 1,\dots,n; \ \nu, s = 1,2,\dots \tag{2.2.14}$$

In particular,

$$E\{y_i(t)\dot{y}_i(t)\} = 0$$

$$E\{y_i(t)\dot{y}_j(t)\} + E\{\dot{y}_i(t)y_j(t)\} = 0 . \tag{2.2.15}$$

We note that this precludes the use of white noise inputs, since in that case the derivatives do not exist.

As an example of how the equations above may be used for identification purposes, let us consider the system

$$\ddot{Z}_1(t) + c_1[\dot{Z}_1(t) - \dot{Z}_2(t)] + k_1[Z_1(t) - Z_2(t)] = f(t)$$

$$\ddot{Z}_2(t) + c_2\dot{Z}_2(t) + k_2Z_2(t) - c_1[\dot{Z}_1(t) - \dot{Z}_2(t)] -$$

$$k_1[Z_1(t) - Z_2(t)] = 0 . \tag{2.2.16}$$

This system can be written in the form

$$\dot{y}_1(t) = y_2(t)$$

$$\dot{y}_2(t) = -c_1[y_2(t) - y_4(t)] - k_1[y_1(t) - y_3(t)] + f(t) \tag{2.2.17}$$

$$\dot{y}_3(t) = y_4(t)$$

$$\dot{y}_4(t) = -c_2y_4(t) - k_2y_3(t) + c_1[y_2(t) - y_4(t)] + k_1[y_1(t) - y_3(t)],$$

substituting the state vector $\begin{bmatrix} Z_1 \\ \dot{Z}_1 \\ Z_2 \\ \dot{Z}_2 \end{bmatrix}$ by $\begin{bmatrix} y_1 \\ y_2 \\ y_3 \\ y_4 \end{bmatrix}$

We can multiply the second equation of (2.2.17) respectively by y_1, y_2 and take averages. The fourth equation (2.2.17) is to be multiplied by y_3, y_4 and averaged. On the basis of equations (2.2.15) it will follow that

$$
\begin{aligned}
E\{y_2^2\} + E\{fy_1\} &= -c_1 E\{y_1 y_4\} + k_1(E\{y_1^2\} - E\{y_1 y_3\}) \\
E\{fy_2\} &= c_1(E\{y_2^2\} - E\{y_2 y_4\}) - k_1 E\{y_2 y_3\} \\
-E\{y_4^2\} &= -k_2 E\{y_3^2\} + c_1 E\{y_2 y_3\} + k_1(E\{y_1 y_3\} - E\{y_3^2\}) \\
0 &= -c_2 E\{y_4^2\} + c_1(E\{y_2 y_4\} - E\{y_4^2\}) + k_1 E\{y_1 y_4\}
\end{aligned}
\tag{2.2.18}
$$

This set of four linear algebraic equations is easily solved for the unknowns in order to determine the parameters, c_1, k_1 for example, as

$$
c_1 = \frac{\begin{pmatrix} E\{y_2^2\} + E\{y_1 f\} & E\{y_1^2\} - E\{y_1 y_3\} \\ E\{fy_2\} & -E\{y_2 y_3\} \end{pmatrix}}{B}
$$

$$
k_1 = \frac{\begin{pmatrix} -E\{y_1 y_4\} & E\{y_2^2\} + E\{y_1 f\} \\ E\{y_2^2\} - E\{y_2 y_4\} & E\{fy_2\} \end{pmatrix}}{B}
\tag{2.2.19}
$$

where

$$
B = \begin{pmatrix} -E\{y_1 y_4\} & E\{y_1^2\} - E\{y_1 y_3\} \\ E\{y_2^2\} - E\{y_2 y_4\} & -E\{y_2 y_3\} \end{pmatrix}
$$

Similar equations yield c_2, k_2.

Naturally, the equations of identification above hold in the case that the coefficients of $\ddot{Z}_1$, $\ddot{Z}_2$ in (2.2.16) are unity. If not, then in the case of unknown mass coefficients, two other equations will be necessary to identify the system.

Again as an example, consider the general chainlike system

$$
\begin{aligned}
&1)\quad m_1\ddot{Z}_1 = f_1 - c_1(\dot{Z}_1-\dot{Z}_2) - k_1(Z_1-Z_2) \\
&2)\quad m_2\ddot{Z}_2 = f_2 - c_2(\dot{Z}_2-\dot{Z}_3) - k_2(Z_2-Z_3) + c_1(\dot{Z}_1-\dot{Z}_2) + k_1(Z_1-Z_2) \\
&\ \vdots \qquad \vdots \\
&n)\quad m_n\ddot{Z}_n = f_n - c_n\dot{Z}_n - k_nZ_n + c_{n-1}(\dot{Z}_{n-1}-\dot{Z}_n) + k_{n-1}(Z_{n-1}-Z_n) \quad .
\end{aligned}
\tag{2.2.20}
$$

We multiply the first equation by Z_1, $\dot{Z}_1$, $\ddot{Z}_1$ respectively and average to yield the parameter identification equations,

$$m_1E\{Z_1\ddot{Z}_1\} = E\{Z_1f_1\} - c_1E\{Z_1(\dot{Z}_1-\dot{Z}_2)\} - k_1 E\{Z_1(Z_1-Z_2)\}$$

$$m_1E\{\dot{Z}_1\ddot{Z}_1\} = E\{\dot{Z}_1f_1\} - c_1E\{\dot{Z}_1(\dot{Z}_1-\dot{Z}_2)\} - k_1 E\{\dot{Z}_1(Z_1-Z_2)\}$$

$$m_1E\{\ddot{Z}_1^2\} = E\{\ddot{Z}_1f_1\} - c_1E\{\ddot{Z}_1(\dot{Z}_1-\dot{Z}_2)\} - k_1 E\{\ddot{Z}_1(Z_1-Z_2)\}$$

etc. (2.2.21)

the relations (2.2.15) can be used to simplify the equations above. However, we hasten to point out that the equations above will hold on a deterministic basis if the statistical expectation operation "E{ }" is replaced by the time average "< >". In that case stationarity is not a requirement. Thus if f is a sinusoidal or sweep sinusoidal excitation, the expectations will be replaced by time averages. The discussion above holds for non-linear systems as well, as we now see.

Let us consider

$$\ddot{Z} + \nu^2 Z + h(Z,\dot{Z}) = f \quad . \tag{2.2.22}$$

where h is a polynomial in Z, $\dot{Z}$ and the parameters ν^2 and those of h are unknown. We can easily identify by the same procedures above, as shown in the following example.

Consider the van der Pol oscillator

$$\ddot{Z} + \mu(Z^2 - 1)\dot{Z} + Z = f \quad , \tag{2.2.23}$$

where μ is unknown.

We simply multiply (2.2.23) by $\dot{Z}$ and take expectations using (2.2.15) in the stationary case, to yield

$$\hat{\mu} = \frac{E\{f\dot{Z}\}}{E\{(Z^2-1)\dot{Z}^2\}} \tag{2.2.24}$$

Similarly for the non-linear system [1], [12],

$$m\ddot{Z} + a_1Z + a_2Z^3 + a_3\dot{Z} + a_4\dot{Z}^3 = f \quad . \tag{2.2.25}$$

with unknown parameters a_1, a_2, a_3, a_4 we will obtain the equations

$$
\begin{aligned}
&mE\{\ddot{Z}Z\} + a_1E\{Z^2\} + a_2E\{Z^4\} + a_3E\{Z\dot{Z}\} + a_4E\{Z\dot{Z}^3\} = E\{Zf\} \\
&mE\{\ddot{Z}Z^3\} + a_1E\{Z^4\} + a_2E\{Z^6\} + a_3E\{Z^3\dot{Z}\} + a_4E\{Z^3\dot{Z}^3\} = E\{Z^3f\} \\
&mE\{\ddot{Z}\dot{Z}\} + a_1E\{Z\dot{Z}\} + a_2E\{Z^3\dot{Z}\} + a_3E\{\dot{Z}^2\} + a_4E\{\dot{Z}^4\} = E\{\dot{Z}f\} \qquad (2.2.26) \\
&mE\{\ddot{Z}\dot{Z}^3\} + a_1E\{Z\dot{Z}^3\} + a_2E\{Z^3\dot{Z}^3\} + a_3E\{\dot{Z}^4\} + a_4E\{\dot{Z}^6\} = E\{\dot{Z}^3f\}
\end{aligned}
$$

by which we can estimate the coefficients.

As we mentioned in the beginning of this section, these estimation equations can be looked upon as coming from a least squares criterion, so that (2.2.26) can be obtained from (2.2.25) by minimizing the quantity,

$$E\{(m\ddot{Z} + a_1Z + a_2Z^3 + a_3\dot{Z} + a_4\dot{Z}^3 - f)^2\}$$

with respect to a_1, a_2, a_3, a_4.

Furthermore, if we are in the stationary realm, terms such as $E\{\ddot{Z}\dot{Z}\}$, $E\{\ddot{Z}\dot{Z}^3\}$, $E\{Z^3\dot{Z}\}$, $E\{Z\dot{Z}\}$ are zero, (orthogonal random variables) making (2.2.26) somewhat simpler. [See also, [12], [13]].

We shall now consider the results of simulated experiments, to determine the practicality of this procedure.

Experiment I

The system simulated in this experiment is a 5-mass chain. The simulation rate was 5000 c.p.s. The force input f_1 was an actual taped excitation from a non-white Gaussian process with spectrum centered at 270 c.p.s. and bandwidth approximately 350 c.p.s.

Two sets of estimates are calculated, one for 1 second duration, the other for 2 second duration each starting at $t = 0$. In this case, the system was still transient hence time averages were used to estimate the coefficients as shown in Table I.

	True Value	Estimation Interval 1 sec	Estimation Interval 2 sec	True Value	Estimation Interval 1 sec	Estimation Interval 2 sec	True Value	Estimation Interval 1 sec	Estimation Interval 2 sec
i	m_i	$\hat{m}_i(1)$	$\hat{m}_i(2)$	k_i	$\hat{k}_i(1)$	$\hat{k}_i(2)$	c_i	$\hat{c}_i(1)$	$\hat{c}_i(2)$
1	.052	.052	.052	$.93\times10^6$	$.93\times10^6$	$.93\times10^6$	0.900	1.041	1.082
2	.052	.052	.052	$.83\times10^5$	$.83\times10^5$	$.83\times10^5$	1.900	1.900	1.903
3	.052	.052	.052	$.93\times10^6$	$.93\times10^6$	$.93\times10^6$	0.900	1.179	1.037
4	.052	.052	.052	$.83\times10^5$	$.83\times10^5$	$.83\times10^5$	1.900	1.865	1.907
5	.052	.052	.052	$.43\times10^2$	$.43\times10^2$	$.43\times10^2$	.100	.135	.088

TABLE I

It is clear from the table that the 2 second estimates of c_i are better than the 1 second estimates. The estimates of m_i and k_i are uniformly excellent. Presumably, longer periods of estimation will yield better estimates for c_i. Although, as we shall see in Section 2.3, there is a theoretical reason for greater errors in damping coefficient estimates.

Experiment II

In this example we considered the identification of a two-dimensional multi-degree of freedom system as defined by the differential equations below.

$$m_1\ddot{x}_1 + 2(c_1 + c_3)\dot{x}_1 - 2c_3\dot{x}_4 + 2(k_1 + k_3)x_1 - 2k_3x_4 = f_1$$

$$m_1\ddot{x}_2 + 2(c_2 + c_4)\dot{x}_2 - 2c_4\dot{x}_5 + 2(k_2 + k_4)x_2 - 2k_4x_5 = f_2$$

$$I_1\ddot{x}_3 + 2(c_1b^2 + c_2a^2 + c_3c^2 + c_4e^2)\dot{x}_3 - 2(c_3cd + c_4e^2)\dot{x}_6$$
$$+ 2(k_1b^2 + k_2a^2 + k_3c^2 + k_4e^2)x_3 - 2(k_3cd + k_4e^2)x_6 = f_3$$

$$m_2\ddot{x}_4 - 2c_3\dot{x}_1 + 2c_3\dot{x}_4 - 2k_3x_1 + 2k_3x_4 = f_4$$

$$m_2\ddot{x}_5 - 2c_4\dot{x}_2 + 2c_4\dot{x}_5 - 2k_4x_2 + 2k_4x_5 = f_5$$

$$I_2\ddot{x}_6 - 2(c_3cd + c_4e^2)\dot{x}_3 + 2(c_3d^2 + c_4e^2)x_6$$
$$- 2(k_3cd + k_4e^2)x_3 + 2(k_3d^2 + k_4e^2)x_6 = f_6$$

The constants a, b, c and d represent the various vertical and horizontal distances between the two-mass centers and the spring-damper components. This system was simulated and driven by random excitations as well as sinusoidal excitations. The system was simulated and sampled at intervals of .0005 secs. Observations commenced at $t = 0$ and were made at intervals of 0.0 - 1.0 secs and 0.0 - 2.0 secs. In the random case, f_4, f_5 and f_6 were independent random excitations generated by passing white noise through a filter to yield a process with spectrum having center frequency 70 c.p.s. and bandwidth 20 c.p.s. Furthermore, f_1, f_2, f_3 were identically zero. For the sinusoidal cases f_1, f_2, and f_3 again were zero, f_4, f_5 and f_6 were given as

$$f_4 = 25 \sin 100\pi t, \quad f_5 = 25 \sin(100\pi t + \frac{\pi}{2})$$

$$f_6 = 25 \sin(100\pi t + \frac{3}{4}\pi).$$

The results are presented in the two following tables first for random Table II(a) then for sinusoidal, Table II(B).

Again, to the first few places, the estimated parameters are identical with the actual system parameters. It is only at the fifth and sixth places where deviation occurs.

	True Value	ESTIMATION INTERVALS 0.0-1.0sec	1.0-2.0sec	True Value	0.0-1.0sec	1.0-2.0sec	True Value	0.0-1.0sec	1.0-2.0sec	True Value	0.0-1.0sec	1.0-2.0sec
i	k_i	$\hat{k}_i(1)$	$\hat{k}_i(2)$	c_i	$\hat{c}_i(1)$	$\hat{c}_i(2)$	m_i	$\hat{m}_i(1)$	$\hat{m}_i(2)$	I_i	$\hat{I}_i(1)$	$\hat{I}_i(2)$
1	$.99\times10^5$	$.99\times10^5$	$.10\times10^6$	10.0	10.0	10.0	.26	.26	.26	70.0	70.0	70.0
2	$.40\times10^6$	$.40\times10^6$	$.40\times10^6$	20.0	20.0	20.0	.26	.26	.26	70.0	70.0	70.0
3	$.99\times10^5$	$.10\times10^6$	$.10\times10^6$	10.0	10.0	10.0						
4	$.40\times10^6$	$.40\times10^6$	$.40\times10^6$	20.0	20.0	20.0						

TABLE II(a) Random Excitation

i	k_i	$\hat{k}_i(1)$	$\hat{k}_i(2)$	c_i	$\hat{c}_i(1)$	$\hat{c}_i(2)$	m_i	$\hat{m}_i(1)$	$\hat{m}_i(2)$	I_i	$\hat{I}_i(1)$	$\hat{I}_i(2)$
1	$.99\times10^5$	$.10\times10^6$	$.10\times10^6$	10.0	10.0	10.0	.26	.26	.26	70.0	70.0	70.0
2	$.40\times10^6$	$.40\times10^6$	$.40\times10^6$	20.0	20.0	20.0	.26	.26	.26	70.0	70.0	70.0
3	$.99\times10^5$	$.99\times10^5$	$.99\times10^5$	10.0	10.0	10.0						
4	$.40\times10^6$	$.40\times10^6$	$.40\times10^5$	20.0	20.0	20.0						

TABLE II(b) Sinusoidal Excitation

Experiment III

The system (2.2.25) was studied in [1]. The forcing factor used to excite the system was the North-South component of the 1934 El-Centro earthquake. Furthermore, it was assumed that the observed dynamical variable data was corrupted by uniformly distributed purely random noise samples, with zero mean and variance $\alpha^2/3$. Thus, we write

$$\bar{Z} = Z(1 + r_1), \ \bar{\dot{Z}} = \dot{Z}(1 + r_2), \ \bar{\ddot{Z}} = \ddot{Z}(1 + r_3) \ , \qquad (2.2.27)$$

where (r_1, r_2, r_3) are the purely random observation noise, $(Z, \dot{Z}, \ddot{Z})$ are the true dynamical variables and $(\bar{Z}, \bar{\dot{Z}}, \bar{\ddot{Z}})$ are the noisy observation data.

Three cases were considered,

Case A. $(\bar{Z}, \bar{\dot{Z}}, \bar{\ddot{Z}})$ are all observed, and given by (2.2.27);

Case B. $(\bar{Z}, \bar{\ddot{Z}})$ are observed and $\bar{\dot{Z}}$ is obtained by integration, hence,

$$\bar{\dot{Z}} = \dot{Z}(0) + \int \bar{\ddot{Z}}\, dt$$

Case C. $\bar{\ddot{Z}}$ is observed and $\bar{Z}$, $\bar{\dot{Z}}$ is obtained by integration as

$$\bar{\dot{Z}} = \dot{Z}(0) + \int \bar{\ddot{Z}}\, d\tau, \ \bar{Z} = Z(0) + \iint \bar{\ddot{Z}}\, d\tau_1 d\tau_2$$

The results for different levels of observation noise are given in Table III. The observation time is T = 10 secs.

Estimated Parameters (1)	True Values (2)	A			B		C	
		$\alpha = \pm 2.5\%$ (3)	$\alpha = \pm 5.0\%$ (4)	$\alpha = \pm 10\%$ (5)	$\alpha = \pm 2.5\%$ (6)	$\alpha = \pm 5.0\%$ (7)	$\alpha = \pm 1\%$ (8)	$\alpha = \pm 2.5\%$ (9)
$\hat{a}_1$	25.0000	24.9897	25.0024	25.4312	25.0577	24.8789	24.6233	24.8168
$\hat{a}_2$	2.5000	2.4157	2.4120	2.4875	2.5616	2.2947	3.1181	3.2679
$\hat{a}_3$	1.0000	0.9957	1.0041	1.0621	1.0012	0.9833	1.0331	1.0094
$\hat{a}_4$	0.1000	0.0998	0.0978	0.0945	0.0991	0.0978	0.0975	0.0981

TABLE III

The estimated values are quite acceptable, although in case C, the error in $\hat{a}_2$ is large. Case C is, of course, more prone to error due to the fact that only $\ddot{Z}$ is observed and each integration to yield $\bar{Z}$, $\dot{Z}$ introduces computational noise. However, case C is important since it is the most common case for earthquake engineering application due to the fact that buildings have accelerometers permanently attached (as required by law in California).

In a comprehensive study of strong motion earthquake records from highway bridges, Raggett and Rojahn [10] applied these techniques to estimate the structural parameters of the bridges, [11]. In particular they looked at continuous two span and three span bridges described by a system of ordinary differential (lumped parameter) equations.

The common form of this "lumped mass" model is

$$M\ddot{\vec{u}} + C\dot{\vec{u}} + K\vec{u} = -\vec{M}_x \ddot{f}_{xg} - \vec{M}_y \ddot{f}_{yg} - \vec{M}_z \ddot{f}_{zg} , \tag{2.2.28}$$

where u is the vector of bridge displacements relative to the ground at the preselected points; M,C,K are the mass, damping and stiffness matrices; $\ddot{f}_{xg}$, $\ddot{f}_{yg}$, $\ddot{f}_{zg}$ accelerations and $\vec{M}_x$, $\vec{M}_y$, $\vec{M}_z$ are participation factors for the orthogonal ground motions.

The ith participation component, M_{xi}, is equal to the mass M_i if the ith transducer is directed parallel to $\ddot{f}_{xg}$ (x-direction), it is zero if it is orthogonal to the x-direction. The participation factors, M_{yi}, M_{zi} are similarly defined. If the ith mass is rotational all participation factors M_{xi}, M_{yi}, M_{zi} are zero.

$$M = \begin{bmatrix} 39.54 & & & & & & \\ & 39.54 & & & & \bigcirc & \\ & & 39.54 & & & & \\ & & & 118.62 & & & \\ & \bigcirc & & & 19.77 & & \\ & & & & & 19.77 & \\ & & & & & & 19.77 \end{bmatrix}$$

is the mass matrix for this seven degree of freedom model. with participation vectors,

$$\vec{M}_x = \begin{bmatrix} 0 \\ 0 \\ 0 \\ 118.62 \\ 0 \\ 0 \\ 0 \end{bmatrix}, \quad \vec{M}_y = \begin{bmatrix} 39.54 \\ 39.54 \\ 39.54 \\ 0 \\ 0 \\ 0 \\ 0 \end{bmatrix}, \quad \vec{M}_z = \begin{bmatrix} 0 \\ 0 \\ 0 \\ 0 \\ 19.77 \\ 19.77 \\ 19.77 \end{bmatrix}$$

From the geometry and anticipated structural behavior, the authors assumed that the lateral or transverse response will be independent of the longitudinal and vertical motion. Thus, the equations will split into subsystems $(u_1, u_2, u_3)^T$ and $(u_4, u_5, u_6, u_7)^T$. Further, the C,K matrices are assumed to be symmetric. The simulated bridge model was subjected to the Oroville earthquake whose orthogonal component accelerations are shown in Figure 1.

The linear system equations (2.2.28) were transformed into modal equations

$$\ddot{\eta}_i + 2\xi_i\omega_i\dot{\eta}_i + \omega_i^2\eta_i = -\gamma_{xi}\ddot{f}_{xg} - \gamma_{yu}\ddot{f}_{yg} - \gamma_{zi}\ddot{f}_{zg} \qquad (2.2.29)$$

The parameters used for simulation are shown in Table IV.

i	ω_i	$\xi(\%)$	γ_x	γ_y	γ_z
1	17.780	2.00	0	1.1888	0
2	36.040	2.00	0	.3291	0
3	64.310	2.00	0	-.4002	0
4	13.730	2.00	.9380	0	-.0446
5	19.660	2.00	.0413	0	.0639
6	23.360	2.00	.3240	0	.0546
7	31.740	2.00	-.6340	0	1.6910

TABLE IV

From these parameters, the corresponding state vector form of the equations (2.2.28), were obtained, and numerically integrated to yield $\vec{u}$, $\dot{\vec{u}}$, $\ddot{\vec{u}}$. The estimated damping and stiffness matrices C,K obtained are as follows

$$C = \begin{bmatrix} 64.1 & -24.5 & 6.82 & 0 & 0 & 0 & 0 \\ -24.5 & 61.4 & -4.74 & 0 & 0 & 0 & 0 \\ 6.82 & -4.74 & 53.7 & 0 & 0 & 0 & 0 \\ 0 & 0 & 0 & 54.1 & .288 & 6.16 & -7.35 \\ 0 & 0 & 0 & .288 & 14.3 & -6.18 & -5.41 \\ 0 & 0 & 0 & 6.16 & -6.18 & .193 & -.800 \\ 0 & 0 & 0 & -7.35 & -5.41 & -.800 & 1.82 \end{bmatrix}$$

$$K = \begin{bmatrix} 75{,}200 & -54{,}600 & 32{,}200 & 0 & 0 & 0 & 0 \\ -54{,}600 & 64{,}300 & -44{,}000 & 0 & 0 & 0 & 0 \\ 32{,}200 & -44{,}000 & 86{,}900 & 0 & 0 & 0 & 0 \\ 0 & 0 & 0 & 24{,}800 & -1330 & -2770 & 2500 \\ 0 & 0 & 0 & -1330 & 10{,}800 & 4510 & 291 \\ 0 & 0 & 0 & -2770 & 4510 & 17{,}300 & 4140 \\ 0 & 0 & 0 & 2500 & 291 & 4140 & 11{,}100 \end{bmatrix}$$

From the given matrix M and the estimated matrix K the natural frequencies, ω, of $K-\omega^2M$ were obtained.

We list the estimated parameters in Table V.

i	ω	ξ(%)	γ_x	γ_y	γ_z
1	17.714	2.84	0	1.248	0
2	36.028	1.96	0	.322	0
3	64.141	1.65	0	-.333	0
4	13.741	1.94	.949	0	-.032
5	19.584	1.01	.039	0	.054
6	23.461	1.15	.327	0	.096
7	32.697	-.28	-.099	0	1.353

TABLE V

Upon comparing the exact parameters of Table IV and the estimated parameters of Table V, we find the natural frequencies to be very well estimated, the participation parameters acceptable, and the damping ratios becoming more in error at higher modes. The first four modal damping estimates are certainly reasonable.

In the report the estimated mode shapes compare extremely well with the simulated model mode shapes.

The authors conclude from a number of simulations that modal damping errors will be greater for those modes that are not significantly excited. Their rule of thumb in time domain identification procedures is that if that response is particularly sensitive to a parameter, that parameter will be found with accuracy.

We have seen in this section that estimators for the unknown coefficients can be obtained through a least squares approach that can be applied for random as well as deterministic inputs. [See also,[14],[15]].

The natural questions that occur are, how good are the estimated parameters?, what are their errors?, what are their statistical properties?, will they converge to specific values as the time interval of observation becomes larger?, what is the significance of these "limiting" values, if they exist?

We shall take up a number of these questions in the next section.

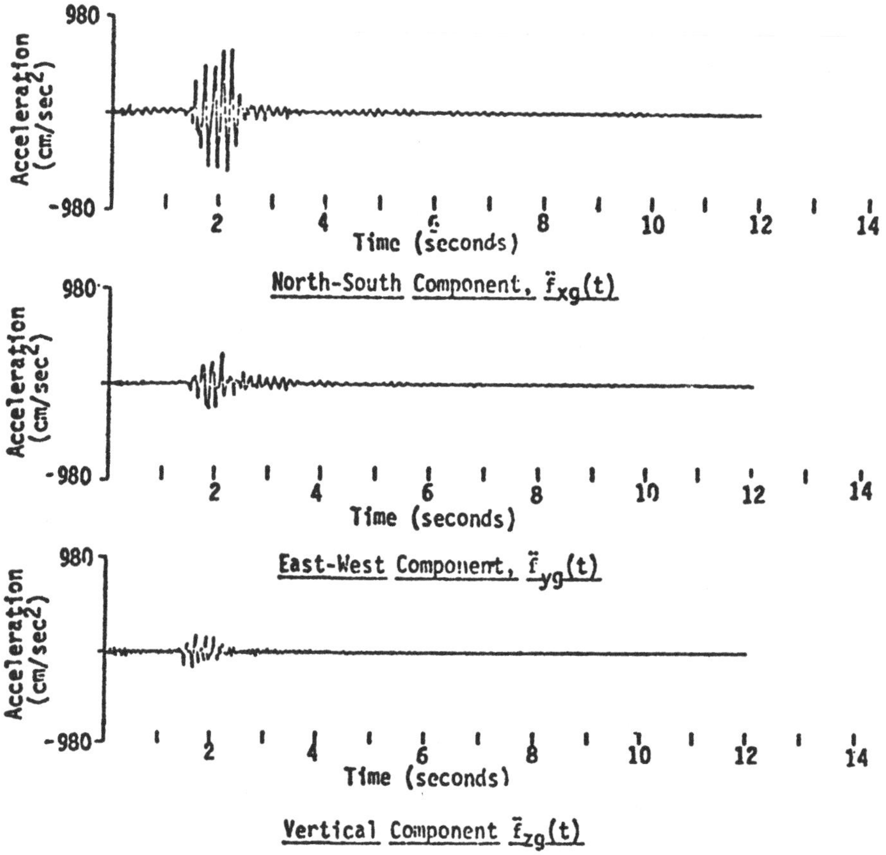

Figure 1.

So far in this section for the physical noise excitation case, we have not established convergence properties of the estimated parameters, nor have we considered the case of noise in the observed data. We shall see, however, in the next subsection that if the physical noise can be assumed to be generated by an Ito differential equation, that is, the physical noise process itself is a diffusion process (or, at least, a component of a diffusion process), then we can use the ideas to be developed in the next subsection to establish the desired statistical properties of the estimates in the physical noise case.

2.3 Identification of Parameters for Systems Subjected to White Noise Excitation

The problem that we encounter in this section concerns the estimation of unknown parameters of Ito differential systems. In particular we are interested in systems which can be written in the Ito form

$$dx_t = \sum_{j=1}^{k} a_j^o \phi_j(x_t)dt + dB_t \quad , \tag{2.3.1}$$

where x_t is an n-vector, $\phi_j(x)$, $j = 1,\dots,k$ are given n-vector functions, $\{a_j^o\}$ are the "true" scalar coefficients and $\{B_t, t \geq 0\}$ is an n-vector of independent standard Brownian motion processes.

That is, each component of B_t is a gaussian process with stationary independent increments for which $E\{B_t\} = 0$ and $E\{B_t B_t'\} = \text{diag}(t)$, an $n \times n$ diagonal matrix.

The question we ask is the following: given that we observe the state vector x_t, under what conditions (if any) can we be assured that an estimator of the coefficients, as a functional of the observations, which we denote by $\hat{a}(T)$, will converge to the true coefficients a^o as the observation time T increases to infinity. That is, under what conditions can we be assured that in some limiting sense we can write

$$\lim_{T\uparrow\infty} \hat{a}(t) = a^o \tag{2.3.2}$$

Notice that we are talking about the limit properties of random variables. As such, therefore, there are a number of modes of convergence that we might consider such as, convergence in probability, convergence in the mean, and, finally, almost sure convergence [16].

The most practical form of convergence from the engineering point of view is almost sure convergence, since it tells us that for almost any time history of observed values of x_t the estimated coefficients $\hat{a}(T)$ will converge to the true coefficients a^o as the observation time approaches infinity.

The estimator that we shall be concerned with is the maximum likelihood estimator which we denote as $\hat{a}_{ML}$ and we wish to determine the conditions for which

$$\lim_{T\uparrow\infty} \hat{a}_{ML}(T) = a^o, \text{ almost surely.} \tag{2.3.3}$$

Then, it will allow us to determine the parameters of the structure being subjected to white noise inputs. We will have a firm statistical foundation upon which to base our estimates. This is somewhat different from the situation in Section 2.2. Furthermore, as we shall see, we can even obtain some idea of the range within which the estimate lies after a given observation time T, which will allow us to have some idea of the error bounds for our estimated parameters. We recall that the likelihood function for a given set of discrete observations $x_1,\ldots,x_N$, with parameter vector $\vec{a}$, is given by the joint density of the random variables $x_1,\ldots,x_N$.

$$f(x_1,\ldots,x_N:\vec{a}) \equiv L(\vec{x}:\vec{a}) \tag{2.3.4}$$

The maximum likelihood estimate of $\vec{a}$ is that value of $\vec{a}$ that maximizes (2.3.4) relative to the given observations. Thus $\hat{a}_{ML}$ satisfies

$$\max_{\vec{a}} L(\vec{x}:\vec{a}) = L(\vec{x}:\hat{a}_{ML}) \tag{2.3.5}$$

This usually may be obtained simply by taking the partial derivatives of L with respect to the unknown parameters and solving

$$\frac{\partial L(\vec{x}:\vec{a})}{\partial \vec{a}} = 0 \tag{2.3.6}$$

We may recall that in the gaussian case, the maximum likelihood estimators are identical to least squares estimators.

Let us consider a simple illustration of the maximum likelihood procedure for a discrete parameter system.

Let us assume that a discrete model of linear dynamics can be written in the so-called auto-regressive form as

$$x_j = \sum_{k=1}^{K} a_k x_{j-k} + w_j , \tag{2.3.7}$$

where $\{w_j\}$ is a sequence of independent identically distributed gaussian random variables with zero mean and variance σ^2.

We assume that we have N observed data values $x_1,\ldots x_N$, $N \geq K$. Clearly from (2.3.7) and the assumptions on w_j, we know that

$$x_j - \sum_{k=1}^{K} a_k x_{j-k} = w_j$$

is gaussian, with density function

$$f(w_j) = f\left(x_j - \sum_{k=1}^{K} a_k x_{j-k}\right) = \frac{1}{\sqrt{2\pi}\,\sigma}\, e^{-\frac{(x_j - \sum a_k x_{j-k})^2}{2\sigma^2}} \tag{2.3.8}$$

Since w_j's are independent, we have

$$L(x_1,\ldots,x_N:\vec{a}) = \frac{1}{|J|}\prod_{j=1}^{N} f(w_j) = \frac{1}{|J|}\prod_{j=1}^{N} f\left(x_j - \sum_{k=1}^{K} a_k x_{j-k}\right)$$

$$= \frac{1}{|J|}\prod_{j=1}^{N}\frac{1}{\sqrt{2\pi}\,\sigma}\, e^{-\frac{(x_j-\sum a_k x_{j-k})^2}{2\sigma^2}} \qquad (2.3.9)$$

$$= \frac{1}{(2\pi\sigma)^{N/2}}\, e^{-\frac{1}{2\sigma^2}\sum_{j=1}^{N}\left(x_j - \sum_{k=1}^{K} a_k x_{j-k}\right)^2},$$

using the fact that J, the Jacobian of the transformation, is unity since $\partial x_i/\partial w_i = 1$, and $\partial x_i/\partial w_j = 0$, $i < j$. The maximum likelihood estimates of the unknown coefficients are obtained by solving $\partial L/\partial a_i = 0$ or equivalently for the log likelihood, $\ell \equiv \log L$, we solve $\partial\ell/\partial a_i = 0$, $i = 1,2,\ldots,K$.

From (2.3.9), the equations are obtained as,

$$\left.\begin{aligned} &\text{(a)}\quad \frac{\partial \ell}{\partial a_i} = 0 = \sum_{j=1}^{N}\left(x_j - \sum_{k=1}^{K} a_k x_{j-k}\right) x_{j-i}, \quad i = 1,\ldots,K. \\ &\qquad\qquad \text{or} \\ &\text{(b)}\quad \sum_{j=1}^{N} x_j x_{j-i} = \sum_{k=1}^{K} a_k \sum_{j=1}^{N} x_{j-k} x_{j-i}, \quad i = 1,\ldots,K. \end{aligned}\right\} \quad (2.3.10)$$

The values a_k satisfying (2.3.10) are the maximum likelihood estimates of the unknown coefficients.

It is easily recognized that (2.3.10(b)) are exactly the same equations for the least squares estimates of the unknown coefficients.

How many of the ideas above carry over into the system defined by (2.3.1)? Note that we are concerned with a continuous time parameter, possibly non-linear, formulation of the system dynamics. We shall obtain the likelihood function for the system (2.3.1) in a highly heuristic fashion which contains the essential ideas but was first made rigorous by Girsanov, [17] in 1960 and is fully discussed in [18].

We shall consider a simple scalar version of the Ito equation (2.3.1). On the interval [0,T] we form the partition $t_i = i(T/N) \equiv i\Delta_N$. Therefore, $\Delta_i t = t_{i+1} - t_i = \Delta_N$, $x_i = x_{t_i}$, $\Delta_i x = x_{i+1} - x_i$, and finally

$\Delta_i B = B_{i+1} - B_i$. Therefore, for large N we can interpret (2.3.1) simply as,

$$\Delta_i x = \sum_{k=1}^{K} a_j \phi_j (x_i) \Delta_N + \Delta_i B \ . \tag{2.3.11}$$

But since $\Delta_i B$, $\Delta_j B$ are on disjoint intervals for $i \neq j$, it follows that they are independent random variables, and their joint density can be written as

$$\prod_{i=0}^{N-1} \frac{1}{\sqrt{2\pi\Delta_n}} e^{-\frac{1}{2\Delta_N}(\Delta_i B)^2}$$

$$= \frac{1}{(2\pi\Delta_N)^{N/2}} e^{-\frac{1}{2\Delta_N}\sum_{i=0}^{N-1}(\Delta_i B)^2} = p_B \ . \tag{2.3.12}$$

Substitution of the variable x from (2.3.11) into (2.3.12) will give us the joint density

$$\frac{1}{(2\pi\Delta_N)^{N/2}} e^{-\frac{1}{2\Delta_N}\sum_{i=0}^{N-1}\left(\Delta_i x - \sum_{j=1}^{K} a_j \phi(x_i)\Delta_N\right)^2}$$

$$= \frac{1}{(2\pi\Delta_N)^{N/2}} e^{-\frac{1}{2\Delta_N}\left[\sum_{i=0}^{N-1}(\Delta_i x)^2 - 2\sum_{i=1}^{N-1}\Delta x_i \sum_{i=1}^{K} a_j \phi(x_i)\Delta_N + \sum_{i=0}^{N-1}\left(\sum_{j=1}^{K} a_j \phi(x_i)\right)^2 \Delta_N^2\right]}$$

$$= \frac{1}{(2\pi\Delta_N)^{N/2}} e^{-\left[\frac{1}{2\Delta_N}\sum_{i=0}^{N-1}(\Delta_i x)^2 - \sum_{i=1}^{N-1}\Delta x_i \left(\sum_{j=1}^{K} a_j \phi(x_i)\right) + \frac{1}{2}\sum_{i=0}^{N-1}\left(\sum_{j=1}^{K} a_j \phi(x_i)\right)^2 \Delta_N\right]}$$

$$= p_x \ , \tag{2.3.13}$$

again since the Jacobian of the transformation (2.3.11) is unity.

The so-called likelihood ratio of the (discrete) x-process with respect to the Brownian motion process is the ratio, p_x/p_B, where x is substituted for B in (2.3.12).

Thus, we obtain

$$\frac{p_x}{p_B} = e^{\left[-\sum_{i=1}^{N-1} \Delta x_i \left(\sum_{j=1}^{K} a_j \phi(x_i)\right) + \frac{1}{2} \sum_{i=C}^{N-1} \left(\sum_{j=1}^{K} a_j \phi(x_i)\right)^2 \Delta_N\right]} \quad (2.3.14)$$

The two sums can be interpreted as stochastic integrals to yield in the limiting case (for continuous time parameter), the proper likelihood function,

$$L(a_1,\ldots,a_K:T) = e^{-\frac{1}{2}\int_0^T \left(\sum_{j=1}^{K} a_j \phi(x_t)\right)^2 dt + \int_0^T \left(\sum_{j=1}^{K} a_j \phi(x_t)\right) dx_t} \quad (2.3.15)$$

Again we repeat, though correct in the sequence of steps, this derivation is only formal. All limiting operations must be made rigorous. Modern Martingale representation theorems [18] make this derivation quite clean and straightforward.

If we set $\mu(x_t) \equiv \sum a_j \phi_j(x_t)$ in (2.3.1) for the general vector case, we have

$$L(\vec{a},T) = e^{\left[-\frac{1}{2}\int_0^T \mu'(x_t)\mu(x_t)dt + \int_0^T \mu'(x_t)dx_t\right]} \quad (2.3.16)$$

$$\equiv e^{\ell(\vec{a},T)},$$ where ℓ is the log likelihood

and "'" denotes transpose.

There is one more result that is required. It is known that

$$\lim_{T\uparrow\infty} \frac{1}{T} \int_0^T g'(x_t)dB_t = 0 \quad (2.3.17)$$

almost surely for any function g, satisfying $E\{g'(x_t)g(x_t)\} < \infty$, for the x_t-process defined by (2.3.1).

The result follows from simple Borel-Cantelli arguments for the integral martingale (2.3.17). [See Khazminskii [19]].

With these preliminaries at hand, we can now state our main result, [20].

We assume that there is a statistically stationary solution process for (2.3.1) with density $f_o(x)$ for which

$$\int \phi_j'(x)\phi_j(x)f_o(x)dx < \infty,\ j = 1,\ldots,k \quad (2.3.18)$$

The main theorem follows.

Theorem. Let $a_{ML}(T) = (a_1(T),\ldots,a_k(T))'$ be the estimate of $(a_1^o,\ldots,a_k^o)'$ that is obtained by maximizing (2.3.16) for the observed sample solution to (2.3.1).

It follows that

$$\lim_{T\uparrow\infty} a_{ML}(T) = (a_1^o,\ldots,a_k^o)' \text{ almost surely.}$$

Proof.

$$\ell(a,T) = -\frac{1}{2}\int_o^T \mu'(x_s)\mu(x_s)ds + \int_o^T \mu'(x_s)dx_s$$

$$= -\frac{1}{2}\int_o^T \left[\sum_{i=1}^k a_i\phi_i(x_s)\right]'\left[\sum_{j=1}^k a_j\phi_j(x_s)\right] ds + \sum_{j=1}^k a_j \int_o^T \phi'_j(x_s)dx_s$$

$$= -\frac{1}{2}\sum_{i,j=1}^k a_i a_j \int_o^T \phi_i'(x_s)\phi_j(x_s)ds + \sum_{j=1}^k a_j \int \phi_j'(x_s)dx_s$$

$$= -\frac{1}{2}\, a'Ca + a'\rho \,, \tag{2.3.19}$$

where

$$\left.\begin{aligned} c_{ij} &= \int_o^T \phi_i'(x_s)\phi_j(x_s)ds, \quad C = (c_{ij}) \\ \rho_j &= \int_o^T \phi_j'(x_s)dx_s \,. \end{aligned}\right\} \tag{2.3.20}$$

The matrix C, above, is non-singular for sufficiently large T with probability one, since the $\{\phi_j(x)\}$ are linearly independent. Hence, from (2.3.19) we have

$$a_{ML}(T) = C^{-1}\rho \quad . \tag{2.3.21}$$

Furthermore,

$$\rho_i = \int_o^T \phi_i'(x_s)dx_s = \int_o^T \phi_i'(x_s)\left[\sum_{j=1}^k a_j^o\phi_j(x_s)\right] ds + \int_o^T \phi_i'(x_s)dB_s$$

$$= \sum_{j=1}^k a_j^o \int_o^T \phi_i'(x_s)\phi_j(x_s)ds + \int_o^T \phi_i'(x_s)dB_s \;;$$

or

$$\rho = Ca^o + Z \quad , \tag{2.3.22}$$

where Z_1 is defined as

$$Z_i = \int_o^T \phi_i'(x_s) dB_s \ .$$

Hence, from (2.3.22) we have

$$a^o = C^{-1}\rho - C^{-1}Z \ ,$$

and combining this result with (2.3.21) yields

$$\hat{a}_{ML}(T) - a^o = C^{-1}Z \ . \tag{2.3.23}$$

Now, since we assume that there is a stationary solution to the stochastic equation (2.3.1) with stationary density $f_o(x)$, it is known that

$$\lim_{T\uparrow\infty} \frac{1}{T} \int_o^T g(x_t) dt = \int_{-\infty}^{\infty} g(x) f_o(x) dx$$

almost surely [21], therefore,

$$\lim_{T\uparrow\infty}(C/T) = \Gamma, \text{ with probability one,} \tag{2.3.24}$$

where

$$\Gamma_{ij} = \int_o^T \phi_i'(x)\phi_j(x) f_o(x) dx \quad . \tag{2.3.25}$$

It is shown below that Γ is non-singular; thus Γ^{-1} exists and

$$\lim_{T\uparrow\infty} TC^{-1} = \Gamma^{-1} \quad \text{with probability one} \quad . \tag{2.3.26}$$

Finally, from (2.3.23) we have, by application of (2.3.17)

$$\lim_{T\uparrow\infty}(\hat{a}_{ML}(T) - a^o) = \lim_{T\uparrow\infty}(TC^{-1}(Z/T))$$

$$= \Gamma^{-1} \lim_{T\uparrow\infty}(Z/T) = 0. \tag{2.3.27}$$

This completes the proof.

To show that Γ is non-singular, we simply note that the $k \times k$ matrix $G(x)$, with elements

$$G_{ij}(x) = \phi_i'(x)\phi_j(x) \ ,$$

is positive definite.

This follows from the fact that $\phi_1(s),\ldots,\phi_k(x)$ are linearly independent vectors. Therefore, there is no non-trivial relationship of the form

$$\sum_{i=1}^{k} c_i\phi_i(x) = 0 \ ,$$

for any non-identically zero vector of constants $(c_1,\ldots,c_k)$.

Hence, the magnitude

$$\|\sum c_i\phi_i(x)\| = \left(\sum_{i=1}^{k} c_i\phi_i(x), \ \sum_{j=1}^{k} c_j\phi_j(x)\right)^{1/2}$$

$$= c'(\phi_i'(x)\phi_j(x))c \geq 0 \ ,$$

is only equal to zero for $c \equiv 0$.

Thus, the matrix $G(x) = (\phi_i'(x)\phi_j(x))$, usually called the Gram matrix is non-negative definite.

Thus, for any non-zero k-vector, b, we have for the probability density $f_o(x)$,

$$0 < \int_{-\infty}^{\infty} dx[b'G(x)b]f_o(x) = b'\left[\int_{-\infty}^{\infty} dxG(x)f_o(x)\right]b$$

$$= b'\Gamma b \ ,$$

and the non-singularity follows.

Perhaps the most significant assumption allowing our theorem to go through, is the existence of an invariant (stationary distribution) measure for the system (2.3.1).

The problem of existence of invariant measures for Ito systems was studied rigorously by Khazminskii [21]. Using Lyapunov ideas, Wonham [22] modified the conditions in [21] somewhat.

Extensions were studied by Zakai [23] and more recently Blankenship [24] has related their existence to asymptotic moment properties. It should also be mentioned that H. Dym [25] characterized all linear constant coefficient systems possessing invariant measures, and recently Arnold and his co-workers have obtained conditions for existence of stationary solutions [26].

It is known that for certain regularity conditions on $\phi(x)$ the solutions to (2.3.1) is a Markov diffusion process. In terms of notation for Markov processes we write the transition probability function as

$$P(t,x,S) = \mathrm{Prob}\{x(t) \varepsilon S \mid x(0) = x\} .$$

If $\mu_o(\cdot)$ is the distribution of the initial value $x(0) = x_o$, then we write

$$\mu_t(S) = \int_{R^n} P(t,x,S)\mu_o(dx) \equiv (U_t\mu_o)(S) ,$$

which is the probability distribution for $x(t)$.

If μ_o satisfies the property,

$$U_t\mu_o = \mu_o \quad \text{for all } t \geq 0,$$

then μ_o is called an invariant measure.

In the usual terms of transition densities, if there is a density $f_o(x)$, for the Markov diffusion process with transition probability density, $p(x,t|\xi,s)$, $s < t$, that satisfies,

$$f_o(x) = \int_{R^n} p(x,t|\xi,0)f_o(\xi)d\xi \tag{2.3.28}$$

then $f_o(x)$ is a stationary density for the process.

Examples

(a) Ornstein-Uhlenbeck Process

$$p(x,t|\xi,0) = \frac{1}{\sqrt{2\pi\frac{\sigma^2}{2\beta}(1-e^{-2\beta t})}}\, e^{-\frac{1}{2(\frac{\sigma^2}{2\beta})}\frac{(x-\xi e^{-\beta t})^2}{1-e^{-2\beta t}}}$$

It is easy to verify (2.3.28) for

$$f_o(x) = \frac{1}{\sqrt{2\pi\frac{\sigma^2}{2\beta}}} e^{-\frac{x^2}{2(\frac{\sigma^2}{2\beta})}}$$

(b) Non-linear Diffusion Processes

Let the system be defined by

$$dx_1 = x_2 dt$$

$$dx_2 = (-x_2 - h(x_1))dt + dB_t \quad ,$$

then

$$f_o(x_1,x_2) = C\, e^{-[\frac{1}{2} x_2^2 + H(x_1)]} \quad ,$$

where

$$H(x_1) = \int_o^{x_1} h(x)dx, \text{ C is a normalizing constant.}$$

For the arbitrary Ito system, $dx(t) = m(x(t))dt + G(x(t))dB_t$, where x,m are n-vectors, G is $n \times q$, and B_t is a q-vector of independent Brownian motions.

The associated Backward operator is given by

$$\mathscr{L} = \sum_{i=1}^{n} m_i(x) \frac{\partial}{\partial x_i} + \frac{1}{2} \sum_{i,j=1}^{n} g_{ij}(x) \frac{\partial^2}{\partial x_i \partial x_j} \, , \tag{2.3.29}$$

where

$$g_{ij} = (GG')_{ij} \quad .$$

The Fokker-Planck equation $\frac{\partial u}{\partial t} = \mathscr{L}^* u$, where $\mathscr{L}^*$ is the adjoint operator of $\mathscr{L}$ (usually referred to as the forward operator) can often be used to obtain the stationary density as the normalized positive solution to $\mathscr{L}^* u = 0$.

The result of Wonham is as follows: Let V(x) be a function (analogous to Lyapunov functions) satisfying

(1) V is defined for all x in $\bar{D}_R$, $D_R = \{x: \|x\| > R\}$ for any $R > 0$.

(2) V is continuous in $\bar{D}_R$, and twice continuously differentiable in D_R.

(3) $V(x) \geq 0$, $x \varepsilon \bar{D}_R$ and $V(x) \uparrow \infty$ as $\|x\| \uparrow \infty$.

Theorem [22]

If there exists a function V satisfying (1)-(3) for which

$$\mathscr{L}V(x) \leq -1, \; x \varepsilon D_R,$$

then the x-process possesses an invariant measure.

It is known that the almost sure, ergodic property [21]

$$\text{Prob}\left\{\lim_{T\uparrow\infty} \frac{1}{T}\int_0^T g(x_t)dt = \int_{R^n} g(x)f_o(x)dx\right\} = 1 ,$$

holds.

We now look at two numerical examples illustrating maximum likelihood estimation for the unknown coefficients for systems satisfying the conditions of our theorem.

We present two examples of stochastic Luré equations, which are of the following form,

$$dx(t) = Fx(t)dt - b\phi(\sigma)dt + dB(t)$$

$$\sigma = c'x(t) \tag{2.3.30}$$

where F is a n x n constant matrix, b and c are constant n-vectors, ϕ is a scalar function and B(t) is a n-vector standard Brownian motion with independent components. Applying his result [22] to the Luré equation, Wonham proved that if

(i) all eigenvalues of F have negative real parts;

(ii) $\sigma\phi(\sigma) > 0$ for large $|\sigma|$; $\Phi(\sigma)$ is continuously differentiable and $|d\Phi(\sigma)/d\sigma| < \infty$ for all σ;

(iii) there are α,β non-negative such that $\alpha + \beta > 0$ and $\text{Re}(\alpha + i\omega\beta)c'(i\omega I-F)^{-1}b > 0$ for all real ω; then the process x(t) defined by (2.3.30) has a unique invariant measure.

Example 1

$$\begin{bmatrix} dx_1(t) \\ dx_2(t) \end{bmatrix} = a_o \begin{bmatrix} 0.1x_2(t) \\ -5x_1(t) - x_2(t) \end{bmatrix} dt + \begin{bmatrix} dB_1(t) \\ dB_2(t) \end{bmatrix} . \tag{2.3.31}$$

with

$$a^o = 5 .$$

(2.3.31) is rewritten as a form of (2.3.30) with

$$F = \begin{bmatrix} 0 & 0.5 \\ -24 & -5 \end{bmatrix}$$

$$b' = (0,1) \text{ and } c' = (1,0)$$

$$\Phi(\sigma) = \sigma \quad .$$

Condition (i) is satisfied because the eigenvalues of F are

$$\frac{1}{2}(-5 \pm \sqrt{23}i) \quad .$$

Condition (ii) on $\Phi(\sigma)$ is easily checked. To see Condition (iii), let $\beta > 0$ and $\alpha = 0$.

$$i\omega\beta(1 \quad 0)\begin{pmatrix} i\omega & -0.5 \\ 24 & i\omega + 5 \end{pmatrix}^{-1}\begin{pmatrix} 0 \\ 1 \end{pmatrix} = 1\omega\beta \cdot \frac{+0.5}{5i\omega+(12-\omega^2)} \quad \mathrm{Re} \frac{0.5i\omega\beta}{5i\omega+(12-\omega^2)}$$

$$= \frac{2.5\omega^2\beta}{(12-\omega^2)^2+25\omega^2} > 0 \text{ for all real } \omega \quad .$$

Sample functions of (2.3.31) are approximated by the discrete form

$$\begin{bmatrix} x_1(t+\Delta t) \\ x_2(t+\Delta t) \end{bmatrix} = \begin{bmatrix} x_1(t) \\ x_2(t) \end{bmatrix} + \left\{ a^o \begin{bmatrix} 0.1x_2(t) \\ -5x_1(t) - x_2(t) \end{bmatrix} \cdot \Delta t \right\}$$

$$+ \begin{bmatrix} B_1(t+\Delta t) - B_1(t) \\ B_2(t+\Delta t) - B_2(t) \end{bmatrix} \qquad (2.3.32)$$

for $\Delta t = 0.01$ and $E(B_i(t+\Delta t) - B_i(t))^2 = \Delta t$, $i = 1,2$.

The independent random increments, $B_i(t+\Delta t) - B_i(t)$ are generated by the IBM standard Gauss subroutine as independent Gaussian random variables with zero mean and variance equal to 0.01. The estimates are

	$\hat{a}$
500 sample points	4.819
1000 sample points	5.051
1500 sample points	4.943

Example 2

$$\begin{bmatrix} dx_1(t) \\ dx_2(t) \end{bmatrix} = \left\{ a^o \begin{bmatrix} x_2(t) \\ x_1(t) + \sin x_1(t) \end{bmatrix} + b^o \begin{bmatrix} 0 \\ 2x_1(t) + x_2(t) \end{bmatrix} \right\} dt + \begin{bmatrix} dB_1(t) \\ dB_2(t) \end{bmatrix} \tag{2.3.33}$$

with $a^o = 1$ and $b^o = -2$. (2.3.33) is written as a form of (2.3.30) with

$$F = \begin{bmatrix} 0 & 1 \\ -2 & -2 \end{bmatrix}$$

$$b' = (0,1) \quad \text{and} \quad c' = (1,0)$$

$$\Phi(\sigma) = \sigma \sin \sigma .$$

Sample functions of (2.3.33) are approximated by

$$\begin{bmatrix} x_1(t+\Delta t) \\ x_2(t+\Delta t) \end{bmatrix} = \begin{bmatrix} x_1(t) \\ x_2(t) \end{bmatrix} + \left\{ a^o \cdot \begin{bmatrix} x_2(t) \\ x_1(t)+\sin x_1(t) \end{bmatrix} + b^o \cdot \begin{bmatrix} 0 \\ 2x_1(t)+x_2(t) \end{bmatrix} \right\} \cdot \Delta t$$

$$+ \begin{bmatrix} B_1(t+\Delta t)-B_1(t) \\ B_2(t+\Delta t)-B_2(t) \end{bmatrix}$$

where $t = 0.01$ and $E(B_i(t+\Delta t) - B_i(t))^2 = \Delta t$ for $i = 1,2$. The estimates are

	$\hat{a}$	$\hat{b}$
500 sample points	1.060	-1.689
1000 sample points	1.633	-2.293
1500 sample points	1.141	-2.006

Except for the $\hat{a}$ value in Example 2, at 1000 points, we note that the estimates are quite good.

In general we have found the estimates to be consistently close to the true values for other examples of second and third order stochastic differential equations, using various sample functions.

The next question that naturally comes to mind concerns the error bounds on the estimated variables. We shall approach this question by first showing that the maximum likelihood estimators of the coefficients in (2.3.1) are asymptotically normal [see [20] and references], with zero mean and known variance. This will allow us to obtain a measure of the error to be expected in our estimators.

We state, but do not prove, the basic theorem establishing the asymptotic normality. This is an extension of a result in [27].

Theorem [20]

Let v be a vector in R^k, u(x) a k-dimension vector function of the x-process with stationary density $f_o(x)$ for which

$$\int u(x)u'(x)f_o(x)dx = \Gamma_u \quad . \tag{2.3.34}$$

Then if B(t) is a k-vector of independent standard Brownian motions, we have

$$T^{-\frac{1}{2}} \int_o^T \sum_{i=1}^k v_i u_i(x_s) dB_i(s) \overset{D}{\to} N\left(0,\ \sum_{i=1}^k v_i^2 \Gamma_{ii}\right) \tag{2.3.35}$$

as T approaches infinity, where D denotes convergence in distribution. Here, $N(m,\sigma^2)$ denotes normality with mean m and variance σ^2.

For our specific application, we need the following.

Theorem [20]

$$T^{1/2}(\hat{a}(T) - a^o) \overset{D}{\to} N(o,\ \Gamma^{-1}) \tag{2.3.36}$$

as $T\uparrow\infty$.

Proof. From the equality (2.3.23) we can write,

$$T^{1/2}(\hat{a}(T) - a^o) = T\ C^{-1} \frac{Z}{T^{1/2}} \tag{2.3.37}$$

For (2.3.36), we can consider the limit of (2.3.37)

$$\begin{aligned} \lim_{T\uparrow\infty} T^{1/2}(\hat{a}(T) - a^o) &= \lim_{T\uparrow\infty} T\ C^{-1} \frac{Z}{T^{1/2}} \\ &= \Gamma^{-1} \lim_{T\uparrow\infty} \frac{Z}{T^{1/2}} \end{aligned} \tag{2.3.38}$$

where C,Z are defined by (2.3.20), (2.3.22).

Thus, by the Theorem above, we have

$$T^{1/2} v'Z \xrightarrow{D} N\left(0, \sum_{j=1}^{n} E_o\left\{\left[\sum_{i=1}^{k} v_i \phi_{ij}(x)\right]^2\right\}\right) \qquad (2.3.39)$$

as T approaches infinity.

It is easy to check that

$$\sum_{j=1}^{n} E_o\left\{\left[\sum_{i=1}^{k} v_i \phi_{ij}(x)\right]^2\right\} = v'\Gamma v \ .$$

Hence,

$$T^{-1/2} Z \xrightarrow{D} N(0,\Gamma) \quad \text{as} \quad T\uparrow\infty \qquad (2.3.40)$$

Finally, from (2.3.38), (2.3.39), (2.3.40), we have our Theorem.

As a result of the Theorem above, we can assess the error statistics of the maximum likelihood estimates in the examples.

For Example 1, there is only one parameter, a_o. Therefore Γ defined by (2.3.25) is a scalar given as

$$\begin{aligned}\Gamma &= E\{(0.1x_2)^2 + (5x_1 + x_2)^2\} \\ &\approx 2.70, \text{ and } \Gamma^{-1} \approx 0.37 \ .\end{aligned} \qquad (2.3.41)$$

In order to obtain (2.3.41) explicitly, we must calculate the stationary moments $E\{x_1^2\}$, $E\{x_2^2\}$, $E\{x_1x_2\}$. This is easily accomplished by Dynkin's theorem, using the generator $\mathscr{L}$ of the process defined by (2.3.29). The generator for (2.3.31) is

$$\mathscr{L} = (0.5)x_2 \frac{\partial}{\partial x_1} - (25x_1 + 5x_2) \frac{\partial}{\partial x_2} + \frac{1}{2}\left(\frac{\partial^2}{\partial x_1^2} + \frac{\partial^2}{\partial x_2^2}\right). \qquad (2.3.42)$$

Dynkin's theorem gives us

$$\begin{aligned}E\{\mathscr{L}\, x_1^2\} &= E\{x_1x_2\} + 1 = 0 \\ E\{\mathscr{L}\, x_1x_2\} &= (0.5)E\{x_2^2\} - 5E\{x_1x_2\} - 25E\{x_1^2\} = 0 \\ E\{\mathscr{L}\, x_2^2\} &= -\ 10E\{x_2^2\} - 50E\{x_1x_2\} + 1 = 0\end{aligned} \qquad (2.3.43)$$

with solution

$$E\{x_2^2\} = 5.1,\ E\{x_1x_2\} = -\ 1,\ E\{x_1^2\} = 0.302 \qquad (2.3.44)$$

which leads to (2.3.41). Therefore, the asymptotic standard deviaiton of $T^{1/2}(\hat{a}(t) - a_o)$ is approximately 0.609.

In the examples, $\Delta t = 0.01$, thus the values of T in the Theorem are T = 5, 10, 15. From Example 1, we have the values of $T^{1/2}(\hat{a}(t) - a_o)$ as

T	$T^{1/2}(a(T) - a_o)$
5	- 0.4047
10	0.1612
15	- 0.22076

which are well within one standard deviation.

Proceeding in the same way for Example 2, Γ is given as the matrix

$$\Gamma = \begin{pmatrix} E\{x_2^2 + (x_1 + \sin x_1)^2\} & E\{(2x_1 + x_2)(x_1 + \sin x_1)\} \\ E\{(2x_1 + x_2)(x_1 + \sin x_1)\} & E\{(2x_1 + x_2)^2\} \end{pmatrix} \tag{2.3.45}$$

and the generator of (2.3.33) is

$$\mathscr{L} = x_2 \frac{\partial}{\partial x_1} + (\sin s_1 - 3x_1 - 2x_2) \frac{\partial}{\partial x_2} + \frac{1}{2}\left(\frac{\partial^2}{\partial x_1^2} + \frac{\partial^2}{\partial x_2^2}\right). \tag{2.3.46}$$

From (2.3.46) as before, we obtain

$$E\{\mathscr{L}x_1^2\} = E\{x_1x_2\} + 0.5 = 0$$

$$E\{\mathscr{L}x_1x_2\} = E\{x_2^2\} + E\{x_1 \sin x_1\} - 3E\{x_1^2\} - 2E\{x_1x_2\} = 0$$

$$E\{\mathscr{L}x_2^2\} = 2E(x_2 \sin x_1) - 6E\{x_1x_2\} - 4E\{x_2^2\} + 1 = 0. \tag{2.3.47}$$

Except for the explicit value $E\{x_1x_2\} = -0.5$, we cannot solve these equations exactly. In view of this fact, we generated sample solutions of (2.3.33) and estimated the desired expectations by taking time averages for T = 100, 200, 300, 400. For T = 400, one obtains

$$\overline{x_1^2} = 0.798, \quad \overline{x_2^2} = 0.845, \quad \overline{x_1 x_2} = -0.495$$

$$\overline{\sin^2 x_1} = 0.401, \quad \overline{x_1 \sin x_1} = 0.540, \quad \overline{x_2 \sin x_1} = -0.330.$$

Notice that the estimated value of $\overline{x_1 x_2}$ compares favorably with the exact value - 0.5 obtained from the first equation of (2.3.47). Placing the estimated values in the equations of (2.3.43) yields relative errors of approximately 1%. Placing these values in (2.3.45) gives

$$\Gamma = \begin{pmatrix} 3.124 & 1.851 \\ 1.851 & 2.057 \end{pmatrix} \qquad \Gamma^{-1} = \begin{pmatrix} 0.686 & -0.617 \\ -0.617 & 1.041 \end{pmatrix}, \tag{2.3.48}$$

Hence, we have the standard deviations σ_a, σ_b of $T^{1/2}(\hat{a}(T) - a_o)$, $T^{1/2}(b(T) - b_o)$ respectively are approximately $\sqrt{0.686} = 0.828$ and $\sqrt{1.041} = 1.02$. From Example 2, we have,

T	$T^{1/2}(\hat{a}(T) - a_o)$	$T^{1/2}(\hat{b}(T) - b_o)$
5	0.134	-0.695
10	2.00	-0.927
15	0.546	-0.023

We now present a few examples that allow us to determine, analytically, from (2.3.36), the asymptotic error properties of the parameter estimates.

Example 3 Non-Linear Spring

Consider the system

$$\left.\begin{aligned} dx_1 &= x_2 dt \\ dx_2 &= -[a^o x_2 + f(x_1)]dt + dB \end{aligned}\right\} . \tag{2.3.49}$$

For simple odd degree polynomials $f(x_1)$ it can be shown that (2.3.49) possesses a stationary density. We let $dF(x_1)/dx_1 = f(x_1)$.

We can rewrite (2.3.49) as

$$\begin{pmatrix} dx_1 \\ dx_2 \end{pmatrix} = \left[-a^o \begin{pmatrix} 0 \\ x_2 \end{pmatrix} - \begin{pmatrix} 0 \\ f(x_1) \end{pmatrix} + \begin{pmatrix} x_2 \\ 0 \end{pmatrix} \right] dt + \begin{pmatrix} 0 \\ dB \end{pmatrix} \tag{2.3.50}$$

The only unknown is the coefficient of $\begin{pmatrix} 0 \\ x_2 \end{pmatrix}$. Therefore it follows that we need only consider a single term,

$$\phi_1(x_1,x_2) = \begin{pmatrix} 0 \\ x_2 \end{pmatrix} ,$$

which yields through (2.3.20), (2.3.21)

$$\hat{a}_{ML}(t) = \left(\int_0^T x_2^2(s)ds \right)^{-1} \int_0^T x_2(s)dx_2(s) . \qquad (2.3.51)$$

By (2.3.36) we simply have $T^{1/2}(\hat{a}_{ML}(T) - a^o) \to N(0, E\{x_2^2\}^{-1})$.

The generator of (2.3.49) is $\mathscr{L} = x_2 \frac{\partial}{\partial x_1} - (a^o x_2 + f(x_1)) \frac{\partial}{\partial x_2} + \frac{1}{2} \frac{\partial^2}{\partial_{x_2}^2}$.

From the equations $\mathscr{L} x_2^2 = 0, \mathscr{L} F(x_1) = 0$, we obtain

$$-2a^o E\{x_2^2\} - 2E\{x_2 f(x_1)\} + 1 = 0, \; E\{x_2 f(x_1)\} = 0. \qquad (2.3.52)$$

Thus, we find $E\{x_2^2\}^{-1} = 2a^o$. Hence for the system (2.3.49) for small damping we obtain a small variance in the estimate of a^o. However, the coefficient of variation [see example 4] is large.

Example 4 2nd Order Oscillator

In this example we consider the general simple damped oscillator driven by gaussian white noise, which in scalar form, we write as

$$\ddot{x}_t + 2\zeta\omega x_t + \omega^2 x_t = W_t \qquad (2.3.53)$$

where as usual, ζ is the damping ratio, ω is the natural frequency and W is the gaussian white noise. It is easily shown that for positive ζ, a stationary solution to (2.3.53) exists. We can write (2.3.53) in the vector Ito form (2.3.1) as

$$\begin{pmatrix} dx_1 \\ dx_2 \end{pmatrix} = \left[-\omega^2 \begin{pmatrix} 0 \\ x_1 \end{pmatrix} - 2\zeta\omega \begin{pmatrix} 0 \\ x_2 \end{pmatrix} + \begin{pmatrix} x_2 \\ 0 \end{pmatrix} \right] dt + \begin{pmatrix} 0 \\ dB \end{pmatrix} . \qquad (2.3.54)$$

We estimate the parameters, $(\omega^2, \zeta\omega)$. Notice that the coefficient of $\binom{x_2}{0}$ is known (unity) and is not to be estimated.

We find from (2.3.20), (2.3.21), for finite observation time T, the estimators to be

$$\begin{pmatrix} \hat{\omega}^2 \\ 2\hat{\zeta\omega} \end{pmatrix}_{ML} = \begin{pmatrix} \int_0^T x_1^2(s)ds & \int_0^T x_1(s)x_2(s)ds \\ \int_0^T x_1(s)x_2(s)ds & \int_0^T x_2^2(s)ds \end{pmatrix}^{-1} \begin{pmatrix} \int_0^T x_1(s)dx_2(s) \\ \int_0^T x_2(s)dx_2(s) \end{pmatrix} \tag{2.3.55}$$

The asymptotic error variances as given by (2.3.36) are

$$\Gamma^{-1} = \begin{pmatrix} E\{x_1^2\}^{-1} & 0 \\ 0 & E\{x_2^2\}^{-1} \end{pmatrix} \tag{2.3.56}$$

The constant moments in (2.3.56) may be evaluated by Dynkin's Theorem through the backward operator of (2.3.54), which from (2.3.29) is found to be

$$\mathscr{L} = x_2 \frac{\partial}{\partial x_1} - (\omega^2 x_1 + 2\zeta\omega x_2) \frac{\partial}{\partial x_2} + \frac{1}{2} \frac{\partial^2}{x_2^2} \tag{2.3.57}$$

From the three equations $\mathscr{L}x_1^2 = 0$, $\mathscr{L}x_1x_2 = 0$, $\mathscr{L}x_2^2 = 0$, we obtain the stationary moments

$$E\{x_1^2\}^{-1} = 4\zeta\omega^3 = \sigma^2_{\omega^2}, \quad E\{x_2^2\}^{-1} = 4\zeta\omega = \sigma^2_{2\zeta\omega} \;. \tag{2.3.58}$$

and thus the error variances. The coefficients of variation are defined as the ratio of the standard deviation of errors to the mean value. Thus, we have respectively,

$$\delta_{\omega^2} = \frac{\sqrt{4\zeta\omega^3}}{\omega^2} = 2\sqrt{\frac{\zeta}{\omega}} \;; \quad \delta_{\zeta\omega} = \frac{\sqrt{4\zeta\omega}}{2\zeta\omega} = \sqrt{\frac{1}{\zeta\omega}} \tag{2.3.59}$$

But, since the damping ratio ζ is generally quite small, we have that relative error in the estimate of $\zeta\omega$ is much greater than the relative error in the estimate of ω^2. Indeed from (2.3.59) we see that the relative error for $\zeta\omega^2$ will be $1/\zeta$ times the relative error for ω^2. Hence, even with accurate observed data, we see theoretically that there will always be a greater error in the estimate of the damping coefficient as compared with the error in the spring constant.

Example 5 nth Order Linear System

$$\text{For } dx_t = Ax_t dt + dB_t \quad , \tag{2.3.60}$$

where $A = (a_{ij})$ is a stability matrix, and the components of the B-vector are independent Brownian motions. We can write (2.3.60) as

$$\begin{pmatrix} x_1 \\ \vdots \\ x_n \end{pmatrix} = \left[a_{11}\begin{pmatrix} x_1 \\ 0 \\ \vdots \\ 0 \end{pmatrix} + \cdots + a_{1n}\begin{pmatrix} x_n \\ 0 \\ \vdots \\ 0 \end{pmatrix} + \cdots + a_{ni}\begin{pmatrix} 0 \\ \vdots \\ 0 \\ x_n \end{pmatrix} + \cdots + a_{nn}\begin{pmatrix} 0 \\ \vdots \\ 0 \\ x_n \end{pmatrix} \right] dt + \begin{pmatrix} dB_1 \\ \vdots \\ dB_n \end{pmatrix} \tag{2.3.61}$$

Since A is a stability matrix, we know that (2.3.60) possesses a stationary solution. The estimates of the a_{ij} are given by

$$\hat{a}_{ML}(T) = \mathrm{Diag}(C^{-1}, C^{-1}, \ldots, C^{-1}) \begin{pmatrix} \rho_1 \\ \rho_2 \\ \vdots \\ \rho_n \end{pmatrix} \tag{2.3.62}$$

$$\text{where all } C = \left(\int_0^T x_k(s) x_j(s) ds \right) \quad \text{and } \rho_i = \begin{pmatrix} \int_0^T x_1(s) dx_i(s) \\ \vdots \\ \int_0^T x_n(s) dx_i(s) \end{pmatrix} \tag{2.3.63}$$

We find furthermore, that $\Gamma = \mathrm{diag}(R, R, \ldots, R)$, where

$$R = (E\{x_i x_j\}) = \int_0^\infty e^{As} e^{A's} ds \tag{2.3.64}$$

since A is a stability matrix. Therefore, to obtain the asymptotic error estimates from Theorem 2, we can use the integral in (2.3.64) or we can in simpler cases use the generator as was done in Example 4. In either case we will obtain the desired estimate properties.

As we stated at the end of Section 2.2, the convergence properties of the parameter estimators, when the excitation is physical noise, requires further assumptions over those stated in Section 2.2. In fact, if the physical noise is generated as the solution of an Ito differential equation, then we can establish convergence using the ideas we have just developed.

In particular, let us assume that we can write our system as

$$\dot{x}_t = Ax_t + n_t \tag{2.3.65}$$

where A is an unknown stability matrix and $\{n_t,\ t\varepsilon[0,\infty]\}$ is a stationary gaussian vector process which is generated as the Ito equation

$$dn_t = Dn_t dt + dB_t \tag{2.3.66}$$

We shall assume that D is a stability matrix, so that the n-process is stationary gaussian. Furthermore, we must assume that the statistics of the n-process, in particular the D matrix, is known.

Furthermore, since the D matrix is a stability matrix (in order to obtain the stationary n-process), it follows that D must be non-singular. We see this easily from the fact that the characteristic polynomial

$$\phi_D(s) = \det(sI-D)$$

of the D matrix cannot have the solution $s = 0$. Hence, it follows that

$$\phi_D(0) = \det D \neq 0$$

Therefore, the matrix D must be non-singular.

Now, from (2.3.65), (2.3.66) it follows, by taking differentials, that

$$\begin{aligned} d\dot{x}_t &= A\dot{x}_t dt + dn_t \\ &= A\dot{x}_t dt + Dn_t dt + dB_t \\ &= A\dot{x}_t dt + D(\dot{x}_t - Ax_t)dt + dB_t \\ &= (A+D)\dot{x}_t dt - DAx_t dt + dB_t \end{aligned} \tag{2.3.67}$$

Upon taking the transpose of (2.3.67) multiplying by x_t and integrating, we obtain

$$\frac{1}{T}\int_0^T x_t d\dot{x}_t' = \frac{1}{T}\int_0^T x_t \dot{x}_t' dt(A+D)' - \frac{1}{T}\int_0^T x_t x_t' dtA'D'$$

$$+ \frac{1}{T}\int_0^T x_t dB_t' \tag{2.3.68}$$

Since the x-process is stationary, it follows that there is finite k, such that $E\{x_t'x_t\} < k^2$, uniformly in t.

Therefore, from the stationary ergodic measure of the x-process, we find, almost surely

$$\left.\begin{array}{l} \text{(a)}\ \lim\limits_{T\uparrow\infty} \frac{1}{T}\int_0^T x_t\dot{x}_t' dt = E\{x_t\dot{x}_t'\} = (0) \\ \\ \text{(b)}\ \lim\limits_{T\uparrow\infty} \frac{1}{T}\int_0^T x_t x_t' dt = E\{x_t x_t'\} \\ \\ \text{(c)}\ \lim\limits_{T\uparrow\infty} \frac{1}{T}\int_0^T x_t dB_t' = 0,\ \text{(from (2.3.17))} \end{array}\right\} \tag{2.3.69}$$

Thus, from (2.3.68), using the equalities (2.3.69), we obtain

$$\lim_{T\uparrow\infty} \widehat{DA}(T) = -\lim_{T\uparrow\infty}\left[\left(\frac{1}{T}\int_0^T x_t d\dot{x}_t'\right)'\left(\frac{1}{T}\int_0^T x_t x_t' dt\right)^{-1}\right]$$

$$= -(-E\{x_t x_t'\}A'D')'(E\{x_t x_t'\})^{-1}$$

$$= DA \text{ almost surely.} \tag{2.3.70}$$

Finally, since D is known and D^{-1} exists, $\lim_{T\uparrow\infty} D^{-1}\, DA(T) = A$ almost surely, and we find that

$$D^{-1}\left(\frac{1}{T}\int_{o}^{T} x_t dx_t'\right)'\left(\frac{1}{T}\int_{o}^{T} x_t x_t' dt\right)^{-1} \qquad (2.3.71)$$

is the consistent estimator for the unknown matrix A.

This greatly extends the ideas developed in [28], for linear systems.

We can in fact make further extensions to non-stationary, even unstable models, noisy observation models, and physical noise excitation models. In general, we are concerned with special cases of the following system

$$\begin{cases} \dot{X}_t = Q(t,A) + N_t \\ Y_t = X_t + M_t \end{cases}, \qquad (2.3.72)$$

where N_t and M_t are noise processes, A is the set of unknown parameters, and Q(t,A) is a stochastic process. We wish to estimate A from the continuous observation Y_t over the time interval [0,T]. We are interested in the strong consistency (i.e., the estimator converges to the true parameter almost surely), and the convergence rates of the strong consistent estimators. The model given by (2.3.72) is too general to discuss, thus we confine ourselves to some special cases.

Model 1. n^{th} order linear system:

$$dX_t = AX_t dt + dW_t\ , \qquad (2.3.73)$$

where W_t is an n-dimensional standard Wiener process. It is assumed that X_t can be continuously observed directly, i.e., without observation noise. A is an $n \times n$ matrix to be estimated. For the nth order case, when all the eigenvalues of A are negative, we have the strong consistent estimator of A. For nearly arbitrary A, without the assumption of negative eigenvalues, Bellach [28] in 1980, gave a weak consistent estimator of A. The question is that "Is it possible to give a strong consistent estimator of A without the assumption of negative eigenvalues? The answer is yes, as we shall see.

Model 2. n^{th} order nonlinear, physical noise excitation system:

$$\begin{cases} \dot{X}_t = Q(X_t)H + N_t \\ dN_t = BN_t dt + dW_t \end{cases}, \qquad (2.3.74)$$

where H is an m-dimensional column vector of unknown parameters. Q is an $n \times m$ known function of X. X_t is assumed observable, N_t is not a white noise. It is an n-dimensional diffusion process generated by an nth order linear Ito differential equation. For the first order case with $Q(X_t) = X_t$, Bellach also discussed this model and gave a weak consistent estimator of H. We will study the general model (2.3.74) and give a strong consistent estimator.

Model 3. n^{th} order linear, noisy observation system:

$$\begin{cases} dX_t = AX_t dt + dW_t \\ Y_t = X_t + N_t \\ dN_t = BN_t dt + dV_t \end{cases} \tag{2.3.75}$$

where W_t and V_t are independent n-dimensional standard Wiener processes. Here Y_t is the observation. B is assumed to be known. When N_t is a white noise, Bagchi [44] in 1980 discussed this problem. He showed there exists a strong consistent estimator of A which makes a q function minimized. But this q function is too complicated for one to give an explicit solution for the estimator.

Model 4. n^{th} order nonlinear, nonstationary system:

$$dX_t = Q(t)Hdt + dW_t \tag{2.3.76}$$

where H is an m-dimensional column vector of unknown parameters. Q(t) is a stochastic process. For arbitrary n and $Q(t) = Q(X_t)$, we remove the stationary condition and obtain a strong consistent estimator for the first order case.

We first require a few preliminaries.

Let $M_2^{n,m}[t_o,t]$ denote the set of those $n \times m$ nonanticipating functions G_t with respect to the m-dimensional standard Wiener process W_s, $t_o \leq s \leq t$, such that $\int_{t_o}^{t} |G_s|^2 ds$ exists a.s.. $M_2[t_o,t]$ denotes $M_2^{n,m}[t_o,t]$ when $n = m = 1$, where $|\cdot|$ is the matrix norm defined as $|A|^2 = \sum_{i=1}^{n} \sum_{j=1}^{m} |a_{ij}|^2$.

Lemma (Khazminskii) If G_t is in $M_2[0,T]$ with $EG_t^2 \leq (K > 0$ is a constant), then $\lim_{T\uparrow\infty} \frac{1}{T} \int_0^T G_t dW_t = 0$.

The elegant proof of this lemma can be found in [19] on page 222.

If G_t is a stationary process and its second moment exists, then the second moment is a constant. Therefore this lemma holds for every stationary process. But for nonstationary processes $EG_t^2 \leq K$ is not generally true. In our generalization we allow EG_t^2 to be an increasing (or even more general) function of t, then study how much we must increase the order of the T function in 1/T.

Consider an Ito integral $X_t = \int_0^t G_s dW_s$ where W_s is a one dimensional standard Wiener process and G_s belongs to $M_2[0,t]$ for $t \geq 0$. Let $u(t) = \int_0^t G_s^2 ds$ and let $u^{-1}(t)$ be the left-continuous inverse function: $u^{-1}(t) = \min(s{:}u(s) = t)$ defined for $t < u(\infty)$.

Lemma $\overline{W}_t = X_{u^{-1}(t)}$ is a Wiener process for $t < u(\infty)$. This is the same as saying that $X_t = \overline{W}_{u(t)}$ with a new Wiener process $\overline{W}_t$.

The proof can be found in [45], page 29. This lemma tells us that any Ito integral is essentially (that is, up to within a proper transformation of the time axis) a piece of the Wiener process.

We shall now state two important results for our further development.

In the proof of Khazminskii's Lemma, the fact that $\int_0^T G_t dW_t$ is a martingale is very important. We try to generalize Khazminskii's lemma in two directions. One is to generalize the factor 1/T in $\frac{1}{T}\int_0^T G_t dW_t$ to a more general T function. The other is to generalize the Ito integral $\int_0^T G_t dW_t$ to any martingale.

Theorem [46] - Let X_t be an n-dimensional martingale such that $E|X_t|^p$ exists for some integer $p > 0$. And let $F_1(t)$, $F_2(t)$ be two non-decreasing deterministic functions. If there exists a positive, increasing function f(r) which goes to infinity such that $\sum_{r=1}^{\infty} \frac{F_1^p(f(r+1))}{F_2^p(f(r))} E|X_{f(r+1)}|^p < \infty$, then $\lim_{t\to\infty} \frac{F_1(t)}{F_2(t)} X_t = 0$, a.s.. Where "a.s." is the abbreviation of "almost surely".

Proof - It is sufficient if we have,

$$P\left(\bigcap_{k=1}^{\infty} \bigcup_{j=k}^{\infty} \left(\sup_{t>f(j)} \frac{F_1(t)}{F_2(t)} |X_t| \geq c\right)\right) = 0 \tag{2.3.77}$$

for any $c > 0$, where $f(r)$ is any positive, increasing function which goes to infinity. Since $f(j)$ is increasing, we have the left side of (2.3.77) equal to

$$P\left(\bigcap_{k=1}^{\infty} \left(\sup_{t\geq f(k)} \frac{F_1(t)}{F_2(t)} |X_t| \geq c\right)\right) = P\left(\bigcap_{k=1}^{\infty} \bigcup_{r=k}^{\infty} \left(\sup_{f(r+1)\geq t\geq f(r)} \frac{F_1(t)}{F_2(t)} |X_t| \geq c\right)\right).$$

By the Borel-Cantelli lemma, we need to show that

$$\sum_{r=1}^{\infty} P\left(\sup_{f(r+1)\geq t\geq f(r)} \frac{F_1(t)}{F_2(t)} |X_t| \geq c\right) < \infty . \tag{2.3.78}$$

Since $F_1(t)$, $F_2(t)$ are non-decreasing functions and X_t is a martingale, by virtue of the martingale inequality, we have the left side of (2.3.78) $\leq \sum_{r=1}^{\infty} \dfrac{E|X_{f(r+1)}|^p}{\left(\dfrac{cF_2(f(r))}{F_1(f(r+1))}\right)^p}$.

The proof is completed.

The following corollary is more convenient for our use.

<u>Corollary</u> Let G_t denote an $n \times m$ matrix function in $M_2^{n,m}[0,T]$ such that $\int_0^t E|G_s|^2 ds$ exists for any $t \leq T$. And let $F(T)$ denote a positive, non-decreasing function. If there exists a positive, increasing function $f(r)$ which goes to infinity such that

$$\sum_{r=1}^{\infty} \frac{1}{F^2(f(r))} \int_0^{f(r+1)} E|G_t|^2 dt < \infty , \text{ then } \lim_{T\uparrow\infty} \frac{1}{F(T)} \int_0^T G_t dW_t = 0, \text{ a.s..}$$

This corollary is true since $\int_0^T G_t dW_t$ is a martingale.

<u>Theorem [46]</u> - For G in $M_2[0,T]$, if $\lim_{t\uparrow\infty} \int_0^t G_s^2 ds = \infty$, a.s., then, for any $f(t)$ such that $\lim_{t\uparrow\infty} \dfrac{(2t\log(\log t))^{1/2}}{f(t)} = 0$, we have

$$\lim_{t\uparrow\infty} \frac{\int_0^t G_s dW_s}{(f\int_0^t G_s^2 ds)} = 0, \text{ a.s.}$$

Proof - From the iterated logarithm of Wiener process, we know

$$\limsup_{t\uparrow\infty} \frac{|W_t|}{(2t\log(\log t))^{1/2}} = 1, \text{ a.s., thus we have } \lim_{t\uparrow\infty} \frac{W_t}{f(t)} = 0, \text{ a.s.}$$

Let $u(t) = \int_0^t G_s^2 ds$. Then from the lemma, we know there exists a new Wiener process $\bar{W}_t$ such that $\int_0^t G_s dW_s = \bar{W}_{u(t)}$. Hence, we have

$$\lim_{t\uparrow\infty} \frac{\int_0^t G_s dW_s}{(f\int_0^t G_s^2 ds)} = \lim_{t\uparrow\infty} \frac{\bar{W}_{u(t)}}{f(u(t))} = \lim_{u\uparrow\infty} \frac{\bar{W}_u}{f(u)} = 0, \text{ a.s..}$$

Now we establish strong consistent parameter estimators for the systems discussed before. The main tools we adopt for our proofs are the two theorems and their corollary established above. We shall state the results without proofs.

Model 1

Let us consider the nth order linear system given by (2.3.73) defined on $[0,T]$ and with $X_o = 0$.

For the system given by (2.3.73), suppose there exists a non-singular (full rank) matrix S such that $SAS^{-1} = D = \text{diag}(d(1), d(2),\ldots, d(n))$, where all $d(i)$'s are real. If 1). $d(1) > 0$, for every $i = 1,2,\ldots,n$, and $d(i) \neq d(j)$ for $i \neq j$; or 2). $d(i) < 0$, for every $i = 1,2,\ldots n$; then $\hat{A} = (\int_0^T X_t dX_t')'(\int_0^T X_t X_t' dt)^{-1}$ is a strong consistent estimator of A, i.e., $\lim_{T\uparrow\infty} \hat{A} = A$, a.s..

Remark - Under our assumptions 1). if $d(i) > 0$, $i = 1,2,\ldots,n$, then as T goes to infinity, $|\hat{A} - A|$ almost surely converges to zero faster than

$T^{(1/2)+c}e^{-dT}$, for any $c > 0$, where $d = \min(d(1), d(2),\ldots,d(n))$; 2). If $d(i)$ 0, $i = 1,2,\ldots,n$, then as T goes to infinity, $|\hat{A} - A|$ almost surely converges to zero faster than $T^{-1(1/2)-c)}$, for any $c > 0$.

Model 2

For the nth order physical noise excitation system given by (2.3.74), if

C.1: $\frac{\partial Q}{\partial X}$ (the partial derivative with respect to X) exists,

C.2: the columns of $Q(X_t) - BQ(X_t)$ are linearly independent on $[0,T]$,

C.3: $\lim_{T\uparrow\infty} \frac{1}{T} \int_0^T (\dot{Q}(X_t) - BQ(X_t))'(\dot{Q}(X_t) - BQ(X_t))dt = P$, a.s.,

where P is a positive definite matrix,

C.4: $E|\dot{Q}(X_t) - BQ(X_t)|^2$ is bounded by a constant, then

$$\hat{H} = \left(\int_0^T (\dot{Q}(X_t) - BQ(X_t))'(\dot{Q}(X_t) - BQ(X_t))dt\right)^{-1}$$

$$\cdot \int_0^T (\dot{Q}(X_t) - BQ(X_t))'d(\dot{X}_t - BX_t)$$

is a strong consistent estimator of H.

As a special case, if X_t is a stationary and ergodic process, then C.3 and C.4 hold as long as $E|\dot{Q}(X_t) - BQ(X_t)|^2$ exists, for any $t < T$.

Model 3

Consider the nth order linear, noisy observation system given by (2.3.75). B is assumed to be known.

For the system given by (2.3.75) with $X_0 = N_0 = 0$, suppose there exist nonsingular matrices S and Z such that $SAS^{-1} = D = \text{diag}(d(1), d(2),\ldots,d(n))$, $ZBZ^{-1} = H = \text{diag}(h(1), h(2),\ldots,h(n))$, where all $d(i)$'s and $h(i)$'s are real. If $d(i), h(i) < 0$, for $i = 1,2,\ldots,n$, then

$$\hat{A} = [(\frac{1}{T} \int_0^T Y_t dY_t')' - BF]\cdot(\frac{1}{T} \int_0^T Y_t Y_t' dt - F)^{-1}$$

is a strong consistent estimator of A, where $F = \int_0^\infty e^{Bt}e^{B't}dt$.

Model 4

Now, consider the first order case of the nonlinear, nonstationary system given by (2.3.76), i.e., we investigate the parameter estimation problem of the following model

$$dx_t = aq_t dt + dw_t, \quad 0 \le t \le T, \tag{2.3.79}$$

where q_t is a stochastic process, w_t is a first order standard Wiener process, and a is the unknown parameter to be estimated based on the observations to x_t and q_t.

For the system given by (2.3.76), if q_t is in $M_2[0,T]$, and $\lim_{T\uparrow\infty} \int_0^T q_t^2 dt = \infty$, a.s., then $\hat{a} = \int_0^T q_t dx_t \cdot (\int_0^T q_t^2 dt)^{-1}$ is a strong consistent estimator of a.

Remark - for system (2.3.76), under our assumptions as T goes to infinity, the convergence rate of $|\hat{a} - a| \to 0$ is the same as that of $((1/u)\log\log u)^{1/2} \to 0$, where $u = \int_0^T q_t^2 dt$.

As a special case, if $q_t = q(x_t)$ and if x_t is a stationary ergodic process, then the result follows.

For Model 2, we assumed rather strong conditions on X_t in order to get the strong consistent estimators of H. For the first order case it is possible to remove those constraints.

For the system

$$\begin{cases} x_t = aq_t + n_t \\ \\ dn_t = bn_t dt + dw_t, \end{cases} \quad 0 \le t \le T. \tag{2.3.80}$$

where q_t is a stochastic process which can be observed, and b is a known number.

If

C.1: q_t is nonanticipating with respect to w_t,

C.2: $\dot{q}_t$ (derivative with respect to t) exists, almost surely,

C.3: there is no T_o such that $\dot{q}_t = bq_t$ for all t in $[0,T_o]$,

C.4: $\int_0^T (\dot{q}_t - bq_t)^2 dt$ exists, for any finite T,

C.5: $\lim_{T\uparrow\infty} \int_0^T (\dot{q}_t - bq_t)^2 dt = \infty$, a.s.,

then

$$\hat{a} = \int_0^T (\dot{q}_t - bq_t) d(\dot{x}_t - bx_t) \left(\int_0^T (\dot{q}_t - bq_t)^2 dt \right)^{-1}$$

is a strong estimator of a.

As a special case, if $q_t = q(x_t)$ and if x_t and n_t are stationary and ergodic processes, then C.5 always holds.

If system (2.3.80) is linear, it is possible to find strong consistent estimators for both a and b.

For the system

$$\begin{cases} x_t = ax_t + n_t \\ dn_t = bn_t dt + dw_t, \end{cases} \qquad 0 \le t \le T, \text{ with } x_o = n_o = 0.$$

if $a \neq b$ and $ab > 0$, then

$$\begin{bmatrix} 0 & 1 \\ -\hat{b}\hat{a} & \hat{a}+\hat{b} \end{bmatrix} = \left(\int_0^T Y_t dY_t' \right)' \left(\int_{0_-}^T Y_t Y'_t dt \right)^{-1}$$

is a strong consistent estimator of $\begin{bmatrix} 0 & 1 \\ -ba & a+b \end{bmatrix}$, where $Y_t = (x_t, \dot{x}_t)'$.

In Model 3, we discussed the nth order system given by (2.3.75). But, there we required the stability condition for the processes X_t and N_t, and we assumed B is known. Actually for the first order case, it is possible to remove these restrictions.

For the system

$$\begin{cases} dx_t = ax_t dt + dw_t \\ y_t = x_t + n_t \qquad 0 \leq t \leq T, \text{ with } x_o = n_o = 0. \\ dn_t = bn_t dt + dv_t, \end{cases}$$

where w_t and v_t are two independent standard Wiener processes, if $\max(a,b) > 0$, then $\int_0^T y_t dy_t (\int_0^T y_t^2 dt)^{-1}$ is a strong consistent estimator of $\max(a,b)$.

Let us consider the following system [(see (2.3.65)]

$$\begin{cases} \dot{X}_t = AX_t + N_t \\ \qquad\qquad \text{with } X_o = N_o = 0 , \\ dN_t = BN_t dt + dW_t , \end{cases} \tag{2.3.81}$$

where W_t is an n-dimensional standard Wiener process, B is an $n \times n$ known matrix, and A is an $n \times n$ matrix to be estimated based on the observations to X_t over [0,T]. Actually this is a special case of (2.3.74), but for this linear model, it is possible to give a strong consistent estimator for both stable and unstable cases.

For the system given by (2.3.81), if $AB = BA$, then under the conditions for Model 1,

$$\hat{A} = (\int_0^T (\dot{X}_t - BX_t) d(\dot{X}_t - BX_t)')'(\int_0^T (\dot{X}_t - BX_t)(\dot{X}_t - BX_t)' dt)^{-1}$$

is a strong estimator of A.

We shall now turn to an interesting connection between parameter estimation and the often used method of statistical linearization.

2.4 Application to Statistical Linearization

The idea of statistical linearization was developed in 1953 [47], motivated by the desire to study non-linear control systems with random inputs. The method was further developed due to problems in non-linear random vibrations [48], stochastic control [49], [50], and more recently in non-linear structural response [51], [52], [53], [54]. The fundamental idea is quite straight forward. For the system, for which we assume a stationary, ergodic probability measure exists,

$$\dot{\vec{y}}(t) + \vec{f}(\vec{y}(t)) = \vec{n}(t), \tag{2.4.1}$$

An approximate linearized form is to be obtained simply by adding and subtracting $A\vec{y}$ in (2.4.1) to yield

$$\dot{\vec{y}}(t) + A\vec{y}(t) + [\vec{f}(\vec{y}(t)-A\vec{y}(t)] = \vec{n}(t) . \tag{2.4.2}$$

The difference term in (2.4.2) chosen to be as small as possible in the sense of least squares. Thus, the constant matrix A is determined by

$$\min_{A} E\{||\vec{f}(\vec{y}(t))-A\vec{y}(t))||^2\} \tag{2.4.3}$$

which leads to

$$A = E\{\vec{f}(\vec{y})\vec{y}'\}E\{\vec{y}\,\vec{y}'\}^{-1} \tag{2.4.4}$$

where "'" denotes transpose.

The expectations in (2.4.4) are taken with respect to the stationary measure generated by the system (2.4.1). This measure is, in general, unknown. Thus, the statistics used to define A are taken instead from the linear system

$$\dot{\vec{x}}(t) + A\vec{x}(t) = \vec{n}(t) . \tag{2.4.5}$$

In the case that the n-vector has gaussian components, or when the f-vector contains only polynomials in the components of $\vec{y}$, equations can be written that will yield a solution for the A-matrix.

The rationale here, of course, is that if (2.4.3) can be made arbitrarily small, then the y-process statistics will be "close" to the statistics of the x-process defined by (2.4.5). Thus, there are two approximations here, one coming from the approximate statistics. Obviously, if the true probability measure generated by (2.4.1) were known, we would not have to linearize in the first place. Although there are a number of approximations made, the idea has been used in many applications over the past 30 years since it was developed. The method as it has been studied, is strictly analytical and has no data requirements as long as one knows or assumes the statistics of the excitation process as well as the explicit form of the non-linear $\vec{f}$-function in (2.4.1).

We shall now see that for the "true" linearized coefficient matrix A, given by (2.4.4), the second moments of the solution process (2.4.5) will be identical to the second moments of the original non-linear system (2.4.1). We will assume that the excitation process is the gaussian white noise. Thus, we shall interpret (2.4.1) as the Ito differential equation

$$d\vec{y}(t) + \vec{f}(\vec{y}(t))dt = d\vec{B}(t) \qquad (2.4.6)$$

where the components of the $\vec{B}$-vector process are independent Brownian motions with $E\{B_i(t)\} = 0$, $E\{B_i^2(t)\} = \sigma^2 t$, $i = 1,\ldots,n$.

The associated statistically linearized form will be obtained from (2.4.5) as

$$d\vec{x}(t) + A\vec{x}(t)dt = d\vec{B}(t) \; . \qquad (2.4.7)$$

In order to study the second moments of (2.4.7) and their relation to the second moments of (2.4.1) it is convenient to write the generator of the joint $\vec{x}$-$\vec{y}$ process. This can immediately be seen to be

$$\mathscr{L} = -\sum_{i=1}^{n} f_i(\vec{y})\frac{\partial}{\partial y_i} - \sum_{i=1}^{n}\Big(\sum_{k=1}^{n} a_{ik}x_k\Big)\frac{\partial}{\partial x_i} + \frac{\sigma^2}{2}\sum_{i=1}^{n}\left[\frac{\partial^2}{\partial x_i^2} + 2\,\frac{\partial^2}{\partial x_i \partial y_i} + \frac{\partial^2}{\partial y_i^2}\right] \qquad (2.4.8)$$

Upon applying Dynkin's formula for the generator (2.4.8) to $x_i x_j$ and to $y_i y_j$, we obtain, again assuming stationary statistics, the following equations satisfied by the resulting expectations.

For $y_i y_j$, we obtain for $i,j = 1,2,\ldots,n$

$$\begin{cases} -2E\{y_i f_i(\vec{y})\} + \sigma^2 = 0 & i = j \\ E\{y_j f_i(\vec{y})\} + E\{y_i f_j(\vec{y})\} = 0, & i \neq j \end{cases} \qquad (2.4.9)$$

For $x_i x_j$, we obtain for $i, j = 1,2,\ldots,n$

$$\begin{cases} -2\sum_{k=1}^{n} a_{ik}E\{x_k x_i\} + \sigma^2 = 0, & i = j \\ \sum_{k=1}^{n} a_{ik}E\{x_k x_j\} + \sum_{k=1}^{n} a_{jk}E\{x_k x_j\} = 0, & i \neq j \end{cases} \qquad (2.4.10)$$

Using the notation $\Gamma_x = E\{\vec{x}(t)\vec{x}'(t)\}$, for the second moment matrix of the x-process, equations (2.4.10) is equivalent to the, so-called, Lyapunov equation

$$A\Gamma_x + \Gamma_x A^T = \frac{\sigma^2}{2} I \; . \qquad (2.4.11)$$

It is known that equation (2.4.11) possesses a unique solution, Γ_x, when A is a stability matrix [55]. But, this is exactly our case.

For the x-process, defined by (2.4.5) to be stationary, the coefficient matrix A must be a stability matrix. That is, all initial conditions must die out. If A is given as (2.4.4) using the statistics of the true solution process, y, then it must follow that the unique solution to (2.4.11) is $\Gamma_x \equiv \Gamma_y$ where $\Gamma_y = E\{\vec{y}(t)\vec{y}'(t)\}$ is the second moment matrix for the y-process. Hence, the second moments of the approximate x-process will be exactly identical to the second moments of the y-process. To establish this identity, we write (2.4.11) as

$$E\{\vec{f}(\vec{y})\vec{y}'\}E\{\vec{y}\,\vec{y}'\}^{-1}\Gamma_x + \Gamma_x E\{\vec{y}\,\vec{y}'\}^{-1}E\{\vec{y}\vec{f}'(\vec{y})\} = \frac{\sigma^2}{2}I. \quad (2.4.12)$$

Upon replacing Γ_x with $E\{\vec{y}\,\vec{y}'\}$ $(\equiv\Gamma_y)$, in (2.4.12), the resulting equality is

$$E\{\vec{f}(\vec{y})\vec{y}'\} + E\{\vec{y}'\vec{f}(\vec{y})\} = \frac{\sigma^2}{2}I. \quad (2.4.13)$$

But, (2.4.13) is an identity equivalent to the equations (2.4.9). Thus, we have established the identity $\Gamma_x \equiv \Gamma_y$ for A given by the true statistics (2.4.4).

The fundamental questions, however, still remain due to the fact that the probability density for the $\vec{y}$ process is not known exactly, therefore we cannot calculate any expectations analytically. Instead, let us look at the problem from a more contemporary view based upon observed response data, and parameter estimation.

We shall now establish the very interesting connection between parameter estimation and the method of statistical linearization. It has been well accepted and understood for centuries that a mathematical model of a real dynamical system yields, at best, an approximation to the "true" response of the system due to external as well as internal forcing. Under the assumption of small oscillations the linear model has shown itself to be a quite accurate representation for non-linear systems.

In the general problem of system identification, a prespecified class of models is assumed from which the specific model that best describes the response of the dynamical system is obtained through some model fitting and estimation procedure. In some cases, it can be established that the best fitting model from the prespecified class is, in a function space sense, the projection of the true system model onto the prespecified class of models. In any case, once a class of models is selected, then the optimal choice is made based upon observations on the true system. Let us look at the statistical linearization problem from this point of view.

We assume that the true system is governed by the equation (2.4.1), but we are assuming that the system is linear governed by the equation

$$\dot{\vec{x}}(t) = -A\vec{x} + \vec{w}(t) \tag{2.4.14}$$

where A is a constant coefficient matrix and $\vec{w}(t)$ is the vector of independent gaussian white noise components. Upon writing (2.4.14) in Ito differential form, we have simply,

$$d\vec{x}(t) = -A\vec{x}(t)dt + d\vec{B}(t) \ . \tag{2.4.15}$$

Based upon the equations (2.3.19), (2.3.20), we find that the maximum likelihood estimate of A from observations $\{\vec{x}(t);\ t \in [0,T]\}$, is given as

$$-\hat{A}(T) = \left[\int_0^T \vec{x}(s)d\vec{x}'(s)\right]' \left[\int_0^T \vec{x}(s)\vec{x}'(s)ds\right]^{-1} \tag{2.4.16}$$

However, since we are observing the true response $\{\vec{y}(t),\ t \in [0,T]\}$, then the integrals in (2.4.16) will be evaluated as,

$$-\hat{\hat{A}}(T) = \left[\int_0^T \vec{y}(s)d\vec{y}'(s)\right]' \left[\int_0^T \vec{y}(s)\vec{y}'(s)ds\right]^{-1} \tag{2.4.17}$$

We must now determine the limit value,

$$\lim_{T\uparrow\infty} \hat{\hat{A}}(T) \ .$$

Since we are assuming that a stationary (ergodic) measure exists it follows that each term in (2.4.17) will possess an almost sure limit. In particular, from (2.4.6), we have

$$\lim_{T\uparrow\infty} \frac{1}{T}\int_0^T \vec{y}(s)d\vec{y}'(s) = \lim_{T\uparrow\infty}\frac{1}{T}\int_0^T -\vec{y}(s)\vec{f}'(y(s))ds + \lim_{T\uparrow\infty}\frac{1}{T}\int_0^T \vec{y}(s)d\vec{B}'(s), \tag{2.4.18}$$

which finally yields

$$-\lim_{T\uparrow\infty} \frac{1}{T}\int_0^T \vec{y}(s)d\vec{y}'(s) = E\{\vec{y}\,\vec{f}'(\vec{y})\} \tag{2.4.19}$$

almost surely. We note that the second integral on the right of (2.4.18) possesses a zero limit by Khazminskii's lemma. Similarly,

$$\lim_{T\uparrow\infty} \frac{1}{T} \int_0^T \vec{y}(s)\vec{y}'(s)ds = E\{\vec{y}\,\vec{y}'\} \tag{2.4.20}$$

almost surely. Upon combining (2.4.19), (2.4.20), we find from (2.4.17)

$$\lim_{T\uparrow\infty} \hat{A}(T) = E\{\vec{f}(\vec{y})\vec{y}'\}E\{\vec{y}'\vec{y}\}^{-1} \tag{2.4.21}$$

almost surely.

But, most surprising is that the expectation terms in (2.4.21) are exactly the constants for statistical linearization given by (2.4.4).

There is, moreover, an extremely important point to stress here. We need only observe the response vector $\{\vec{y}(t),\ t\in[0,T]\}$ and evaluate (2.4.16) with $\vec{x}(s)$ replaced by $\vec{y}(s)$, that is we simply evaluate (2.4.17). Thus, in the white noise case, the specific form of the non-linearity does not have to be known. Indeed, we can obtain our statistically equivalent linear system without any assumptions on the nature of the true non-linearity. All the information that is required is contained in the response for the white noise excited non-linear system.

Clearly, if we are investigating the response of a real system, the true form of the non-linearities are generally unknown. By estimating the coefficients of a linear model of the response vector $\vec{y}(t)$ to white noise excitation we are, in fact, actually performing statistical linearization. This apparently has always been true. For linear models of hysteretic restoring forces, what is required is that the restoring force be observable and measured to allow calculation of the required integrals (2.4.17). We note also, that for purely analytical models of the non-linear response straight-forward computer simulation for a single record will yield the desired linear coefficient estimates from equation (2.4.17).

An important question that remains is how much of the preceding analysis goes through for the physical noise excitation case. Thus, in equations (2.4.1), (2.4.5) we interpret $\vec{n}(t)$ as a statistically stationary physical excitation source. We estimate the matrix A in (2.4.5) by least squares defined as

$$\min_{A} \frac{1}{T} \int_0^T \| \dot{\vec{x}}(t) + A\vec{x}(t) - \vec{n}(t) \|^2 dt\ . \tag{2.4.22}$$

Minimization of (2.4.22) leads to $\hat{A}(T)$ given as,

$$A(T) = \left[\frac{1}{T}\int_0^T \vec{n}(t)\vec{x}'(t)dt\right]\left[\frac{1}{T}\int_0^T \vec{x}(t)\vec{x}'(t)dt\right]^{-1}$$
$$- \left[\frac{1}{T}\int \dot{\vec{x}}(t)\vec{x}'(t)dt\right]\left[\frac{1}{T}\int_0^T \vec{x}(t)\vec{x}'(t)dt\right]^{-1} \qquad (2.4.23)$$

Again as in the white noise case we are actually observing the response vector $\vec{y}$. Hence, we replace $\vec{x}$, by $\vec{y}$ in (2.4.23) and take the limit as T approaches infinity. Based upon our assumption of a stationary (ergodic) solution process y for (2.4.1), and using the fact that

$$E\{\dot{\vec{y}}(t)\vec{y}'(t)\} \equiv 0 , \qquad (2.4.24)$$

we will obtain the estimator for A as, the almost sure limit,

$$\lim_{T\uparrow\infty} \hat{A}(T) = E\{\vec{n}(t)\vec{y}'(t)\}\ E\{\vec{y}(t)\vec{y}'(t)\}^{-1} \qquad (2.4.25)$$

We note that there is a significant difference between (2.4.23), (2.4.25) and the equivalent expression for the white noise case, (2.4.16), (2.4.17), (2.4.21). This difference is that for the physical excitation case we <u>must</u> observe the excitation process $\vec{n}(t)$, as well as the response in order to estimate the coefficient matrix for the linear model. From (2.4.1), (2.4.24), it follows that

$$E\{\vec{n}(t)\vec{y}'(t)\} = E\{\vec{f}(\vec{y}(t))\vec{y}'(t)\}. \qquad (2.4.26)$$

It immediately follows that (2.4.25) is equivalent to

$$\lim_{T\uparrow\infty} \hat{A}(T) = E\{\vec{f}(\vec{y}(t))\vec{y}'(t)\}\ E\{\vec{y}(t)\vec{y}'(t)\}^{-1}$$

with probability one.

Again we see, as in the white noise case, that the coefficient matrix estimate for the linear model in the physical noise case is identical to the coefficient matrix (2.4.4) determined by statistical linearization. Furthermore, the specific non-linear function $\vec{f}(\vec{y})$ is not required. All required information is contained in the joint $(\vec{y},\vec{n})$ process.

3. CONCLUSION

In these lectures we have attempted to present a few of the explicit results, and techniques that are available for the estimation of parameters for linear as well as non-linear systems subjected to random excitations. We have been concerned with the estimators, their asymptotic convergence properties to the "true" value, as well as the asymptotic statistical properties of the estimator that will allow some measure of the accuracy for finite observation times. To a great extent the theoretical results available rely heavily upon the white noise excitation case and the associated Ito differential equations. However, physical excitation cases can be treated theoretically, if they are generated by Ito systems.

Many examples and special cases of interest are presented to illustrate the insight obtained from identification methods. We have concentrated on parameter estimation, leaving out the very important question of modelling such as AIC criteria [58]. Furthermore, we have discussed only continuous time models which leaves out the vastly important case of ARMA models and the many identification and modelling techniques available, [see e.g. [31]-[43], [56]-[64]]. Finally, we have not discussed the recent adaptive algorithms that allow identification even when the statistical properties of the excitation process are completely unknown [65]-[67].

We hope that the interested reader will look further into the suggested references and see the many tools and techniques available for practical estimation procedures, as well as the many problems awaiting solution.

REFERENCES

1. Hart, G.C., Yao, J.T.P., System identification in structural dynamics, ASCE Journ. Eng. Mech. EMG, Dec. 1977, 1089-1104.
2. Gersch, W., Parameter identification: Stochastic process techniques, Shock and Vibration Digest, 1975.
3. Proceedings of 6th IFAC Symposium on Identification and System Parameter Estimation, Arlington, Va., 1982.
4. Pilkey, W.D., Cohen, R. (Editors), System identification of vibrating structures-Mathematical models from test data, ASME publications, 1972.
5. Ting, E.C., et al., System identification, damage assessment and reliability evaluation of structures, Report CE-STR-78-1, School of Civil Eng., Purdue Univ., 1978.
6. Natke, H.G., Editor, Identification of vibrating structures, CISM Lectures No. 272 Springer-Verlag, New York, 1982.
7. Ljung, L., Identification Methods, Proc. 6th IFAC Symposium on Identification and System Parameter Estimation, Arlington, Va., June 1982, 11-18.
8. Nakajima, F., Kozin, F., A characterization of consistent estimators, IEEE Trans. Auto. Cont. Vol. AC-24, No. 5, Oct. 1979, 758-765.

9. Kozin, F., Kozin, C.H., A moment technique for system parameter identification, Shock and Vibration Bulletin, No. 38, Part II, Aug. 1968, 119-131 (also NASA report No. CR98738, April 1968).
10. Raggett, J.D., Rojahn, C., Use and interpretation of strong-motion records for highway bridges, Federal Highway Administration Report No. FHWA-RD-78-158, October, 1978.
11. Raggett, J.D., Rojahn, C., Analysis of the three story polygon test structure vibration tests, Dushanbe Tajikistan-Preliminary report, Seismic Eng. Branch, U.S. Geological Survey.
12. Distefano, N., Todeschini, R., Modeling, identification and prediction of a class of non-linear visoelastic systems, Int. Journ. of Solids and Structures, Vol. 1, No. 9, 1974, 805-818.
13. Distefano, N. Rath, A., System identification in nonlinear structural seismic dynamics, Computer Meth. Appl. Mech. and Engr. Vol. 5, No. 3, 1975.
14. Udwadia, F.E., Sharma, D.K., Some uniqueness results related to building structural identification, SIAM Journ. Appl. Math. Vol. 34, No. 1, Jan. 1978, 104-118.
15. Udwadia, F.E., Some uniqueness results related to soil and building structural identification, SIAM Journ. Appl. Math. Vol. 45, No. 4, Aug. 1985, 674-685.
16. Fisz, M., Probability theory and mathematical statistics, Wiley & Sons, New York, 1963.
17. Girsanov, I.V., On transforming a certain class of stochastic processes by absolutely continuous substitution of measures, Theory of Prob. and Appl., Vol. 5, 1960, 285-301.
18. Wong, E., Stochastic processes in information and dynamical systems, McGraw-Hill, New York, 1971.
19. Khazminskii, R.Z., Stochastic stability of differential equations, (Eng. translation) (Chap. 6, Sec. 7) Sijthoff-Noordhoff, Alphen aan den Rijn, Holland, 1980.
20. Lee, T.S., Kozin, F., Almost sure asymptotic likelihood theory for diffusion processes, Jour. Appl. Prob., Vol. 14, 1977, 527-537.
21. Khazminskii, R.Z., Ergodic properties of recurrent diffusion processes and stabilization of the solution to the Cauchy problem for Parabolic equations, Theory of Prob. and Appl., Vol. 5, 1960, 179-196.
22. Wonham, W.M., Lyapunov criteria for weak stochastic stability, Jour. Diff. Equa., Vol. 2, 1966, 195-207.
23. Zakai, M., A Lyapunov criterion for the existence of stationary probability distribution for systems perturbed by noise, SIAM Jour. Contr. Vol. 7, 1969, 390-397.
24. Blankenship, G.L., Limiting distributions and the moment problem for nonlinear stochastic differential equations, Report, Systems Research Center, Case Western Reserve University, July 1975.
25. Dym, H., Stationary measures for the flow of a linear differential equation driven by white noise, Trans. Amer. Math. Soc., Vol. 123, 1966, 130-164.
26. Arnold, L., Wihstutz, V., Stationary solutions of linear systems with additive and multiplicative noise, Stochastics Vol. 7, 1982, 133.

27. Brown, B.M., Eagleson, G.K., Martingale convergence to infinitely divisible laws with finite variances, Trans. Amer. Math. Soc., Vol. 162, 1971, 449-453.
28. Bellach, B., Parameter estimators in linear stochastic differential equations and their asymptotic properties, Math. Opers. Forsch. Statis. Vol. 14, No. 1, 1983, 141-191.
29. Wen, Y.K., Equivalent linearization for hysteretic systems under random vibration, ASME Jour. Appl. Mech. Vol. 47, No. 1, March 1980.
30. Wedig, W., Fast algorithms in the parameter identification of dynamic systems, Proc. IUTAM Symp. Random vibrations and reliability, K. Hennig, Ed., Akademie-Verlag, Berlin, 1983, 217-227.
31. Gersch, W., Nielsen, N., Akaike, H., Maximum likelihood estimation of structural parameters from random vibrational data, Jour. Sound and Vibration, Vol. 31, 1973, 295-308.
32. Gersch, W., Luo, S., Discrete time series synthesis of randomly excited structural system response, Jour. Acous. Soc. of America. Vol. 51, 1972, 402-408.
33. Gersch, W., On the achievable accuracy of structural system parameter estimates, Jour. Sound and Vibration, Vol. 34, 1974, 63-79.
34. Gersch, W., Taoka, G.T., Liu, R., Estimation of structural system parameters by a two stage least squares method, ASCE Natl. Structural Engineering Convention, New Orleans, April 1975, preprint #2440.
35. Gersch, W., Foutch, D.A., Least squares estimates of structural system parameters using covariance data, IEEE Trans. Auto. Cont. AC-19, 1974, 898-903.
36. Gersch, W., Liu, S., Time series methods for synthesis of random vibration systems, ASME Trans. Appl. Mech. Vol. 43, March 1976, 159-165.
37. Bartlett, M.S., The theoretical specification and sampling properties of autocorrelated time series, Jour. Royal Stat. Soc. Series B., Vol. 8, 1946, 27-41.
38. Shinozuka, M., Samaras, E., ARMA model representation of random processes, Proc., 4th ASCE Specialty Conference. Probabilistic Mechanics and Structural Reliability, Berkeley, Jan. 1984, 405-409.
39. Akaike, H., Maximum likelihood identification of Gaussian auto-regressive moving average models, Biometrika, 1973.
40. Box, G.E.P., Jenkins, G.M., Time series, forecasting and control, Revised edition, Holden-Day, San Francisco, 1976.
41. Lee, D.T.L., et al, Recursive least squares ladder estimation algorithms, IEEE Trans. Acous., Speech & Signal Proc. Vol. ASSP-29, No. 3, June 1981.
42. Ljung, L., Soderstrom, T., Theory and practice of recursive identification, The MIT Press, Cambridge, Mass., 1983.
43. Lefkowitz, R., Evaluation of various methods of parameter estimation for ARMA processes, Eng. Degree Thesis, Systems Engineering, Polytechnic Institute of New York, January 1986.
44. Bagchi, A., Consistent estimates of parameters in continuous time systems, in: O. Jacobs et al., eds., Analysis and Optimization of Stochastic Systems, Academic Press, New York.

45. McKean, H.P., Stochastic Integrals, Academic Press, New York and London, 1969.
46. Chen, X.K., Strong consistent parameter estimation. Ph.D. dissertation, Dept. of Ele. Engineering, Polytechnic University, June 1987.
47. Booton, R.C., The analysis of non-linear control systems with random inputs, Proc. MRI symposium on Nonlinear Circuits, Polytechnic Inst. of Brooklyn, 1958, 341-344.
48. Caughey, T.K., Equivalent linearization techniques, Jour. Acous. Soc. Amer. Vol. 35, 1963, 1706-1711.
49. Kazakov, I.E., Approximate probability analysis of the operational precision of essentially nonlinear feedback control systems, Auto. and Remote Control, Vol. 17, 1956, 423-450.
50. Sunahara, Y., et al., Statistical studies in nonlinear control systems, Nippon Printing Co., Osaka, Japan, 1962.
51. Spanos, P.D., Stochastic linearization in structural dynamics, Appl. Mech. Rev., Vol. 34, No. 1, 1981, 1-8.
52. Wen, Y.K., Equivalent linearization for hysteretic systems under random loading, ASME Jour. Appl. Mech. Vol. 47, 1980, 150-154.
53. Casciati, F., Faravelli, L., Methods of nonlinear stochastic dynamics for assessment of structural fragility, Nucl. Eng. and Design Vol. 90, 1985, 341-356.
54. Hampl, N.C., Schuëller, G.I., Probability densities of the response of non-linear structures under stochastic dynamic excitation, Proc. U.S.-Austria seminar on Stoch. Struc. Dynam. Florida Atlantic Univ., Boca Raton, Fl., May, 1987.
55. Kailath, T., Linear Systems (Section 9.1), Prentice-Hall, New Jersey, 1980.
56. Mehra, R.K., Lainiotis, D.G., System identification advances and case studies, Math. in Sci., and Eng., Vol. 126, Academic Press, New York, 1976.
57. Goodwin, G.C., Sin, K.S., Adaptive filtering prediction and control, Prentice-Hall, New Jersey, 1984.
58. Akaike, H., A new look at statistical model identification, IEEE Trans. Auto Contr. Vol. 19, 1974, 716-723.
59. Yule, G.U., On a method of investigating periodicities in disturbed series with special reference to Wolfer's sunspot numbers, Phil. Trans., A-226, 267, 1927.
60. Walker, G., On periodicity in series of related terms, Proc. Royal Soc., A-131-518, 1931.
61. Levinson, N., The Wiener RMS Criterion in Filer Design and Prediction, Appendix B of Wiener, N., Extrapolation, Interpolation, and Smoothing of Stationary Time Series with Engineering Applications, John Wiley & Sons, New York, N.Y., 1949, 129-148.
62. Wiggins, R.A., and Robinson, E.A., Recursive solution to the multichannel filtering problem, Journal Geophysical Research, 70(8), April, 1965.
63. Lee, Daniel T.L., Friedlander, B., and Morf, M., Recursive ladder algorithms for ARMA modeling, IEEE Transactions on Automatic Control, AC-27 (4), August 1982.

64. Marquardt, D.W., An algorithm for least squares estimation of non-linear parameters, Journal Society of Industrial Applied Mathematics, 1963, 11, 431.
65. Widrow, B., et al., Stationary and non-stationary learning characteristics of the LMS adaptive filter, Proc. IEEE, Vol. 64, 1976, 1151-1162.
66. Nagumo, J., Noda, A., A learning method for system identification, IEEE Trans. Auto. Control, Vol. AC-12, 282-287, 1967.
67. Shi, D.H., Kozin, F., On almost sure convergence of adaptive algorithms, IEEE Trans. Auto Contr. Vol. AC-31, 1986, 471-474.
68. Kozin, F., Natke, H.G., System identification techniques, Structural Safety, Vol. 3, 1986, 209-316.

PARAMETRIC INSTABILITY AND PROCESS IDENTIFICATION

W. Wedig
University of Karlsruhe, Karlsruhe, FRG

ABSTRACT

The topic of parametric instability and process identification is treated in two papers as follows.

1. Stability of Parametric Systems

2. Parameter Identification of Road Spectra and Nonlinear Oscillators

STABILITY OF PARAMETRIC SYSTEMS

Walter V. Wedig
Institute for Technical Mechanics
University of Karlsruhe
D - 7500 Karlsruhe, Kaiserstr. 12

ABSTRACT

Linear time-variant dynamic systems can be reduced to time-invariant ones by means of state transformations based on the parametric excitation. The transformation defines an eigenvalue problem which is solved by functional analytic methods. These procedures are demonstrated for mechanical oscillator systems studying linear or quadratic state coordinates and their p-th means, respectively.

1. INTRODUCTION

In the special case that the parametric excitation of dynamic systems are modelled by deterministic harmonic functions, we are able to apply the well-established Floquet theory in order to check the stability behaviour in dependence on the intensity of the perturbations. However, this theory is restricted to the harmonic case and has to be extended to more general and more realistic excitations.

The contribution derives a generalization by means of state transformations defined on the stationary excitation processes for the classical investigation of linear and quadratic state coordinates. They are more generally defined on the stationary phase processes of the system if we investigate the p-th norm of its state vector. The transformations lead to deterministic eigenvalue problems which are solved by means of functional analytic methods.

Such methods are applicable for harmonic or almost periodic perturbations as well as for the stochastic case where the parametric excitations are assumed to be filtered or white noise. In dependence on the excitation models the applied orthogonal functions are Fourier expansions or Hermite polynomials leading to infinite determinants the eigenvalues of which decide the stability behaviour of the dynamic system and its state vector norm of interest.

2. STABILITY OF THE HARMONIC OSCILLATOR

To start we consider the example of an harmonically excited oscillator described by the following single differential equation:

$$\ddot{x}(t) + 2D\omega_1\dot{x} + \omega_1^2[1 + \sigma\, z(t)]\, x(t) = 0, \qquad z(t) = \cos(\omega_e t). \tag{1}$$

Herein, ω_1 is the natural frequency, D is a dimensionless damping measure and σ denotes the intensity of the parametric perturbation z(t). Dots denote derivations with respect to the time variable t.

2.1 Application of the Floquet theory

In the special case of the harmonic perturbation with the excitation frequency ω_e noted in (1) we apply the Floquet theory [1] for the stability investigation of the equilibrium position $x(t) \equiv 0$. According to [2], we insert the well-known Floquet setup

$$x(t) = \exp(\lambda t)\sum_{n=-\infty}^{\infty}\exp(-in\omega_e t)\, c_n, \qquad (\lambda = i\omega) \tag{2}$$

into the differential equation (1). Herein, the exponent λ is the stability deciding eigenvalue of the system. In the critical state, its real part is vanishing and its imaginary part is determined by the following infinite equation system derived by a simple coefficient comparison.

$$f_n c_n + \tfrac{1}{2}\sigma\,(c_{n+1} + c_{n-1}) = 0, \qquad \text{for } n = 0, \pm 1, \pm 2, \dots, \tag{3}$$

$$f_n = 1 - (\kappa - n\eta)^2 + i2D(\kappa - n\eta), \qquad \kappa = \omega/\omega_1, \quad \eta = \omega/\omega_e. \tag{4}$$

Non-trivial coefficients c_n require that the determinant (3) vanishes. This results in an infinite determinant of the form:

$$\cdot\,\Delta(\sigma,\kappa) = \begin{vmatrix} ..f_{-2} & \tfrac{1}{2}\sigma & 0 & 0 & 0 \\ \tfrac{1}{2}\sigma & f_{-1} & \tfrac{1}{2}\sigma & 0 & 0 \\ 0 & \tfrac{1}{2}\sigma & f_0 & \tfrac{1}{2}\sigma & 0 \\ 0 & 0 & \tfrac{1}{2}\sigma & f_{+1} & \tfrac{1}{2}\sigma \\ 0 & 0 & 0 & \tfrac{1}{2}\sigma & f_{+2}\,... \end{vmatrix}. \tag{5}$$

According to the sufficient condition of H. von Koch [3] the determinant (5) converges with $1/n^4$ in its normalized form. Equated to zero it determines the critical excitation intensity σ and associated unknown response frequency ω in dependence on given damping values and excitation frequencies.

In the special case of the oscillator (1) there exist integral periodic solutions and half-integral ones [4]. Correspondingly, we know strong solutions of the unknown response frequency for the stability boundary.

$$\kappa = k\eta, \quad \text{or} \quad \kappa = (k + 1/2)\eta, \qquad \text{for } k = 0, (\mp 1, \mp 2, \dots). \tag{6}$$

By (6) the original two-parametric determinant $\Delta(\sigma,\kappa)$ is reduced to the one-parametric eigenvalue problem $\Delta(\sigma) = 0$ for the determination of the critical intensity parameter σ. Finally, this leads to the following matrix recursion:

$$\mathbf{D}_{n+2} = \mathbf{D}_{n+1}\,\mathbf{B}_{n+2} - \mathbf{D}_n\,\sigma^2/4, \qquad \underset{n\geq 2}{\mathbf{B}_n} = \begin{bmatrix} 1-(\kappa-n\eta)^2 & -2D(\kappa-n\eta) \\ 2D(\kappa-n\eta) & 1-(\kappa-n\eta)^2 \end{bmatrix}. \quad (7)$$
$$n = 0,1,\ldots,\ \mathbf{D}_o = \mathbf{I},\ \mathbf{D}_1 = \mathbf{B}_1,$$

Herein, $\mathbf{D}_n$ is a 2x2 stability matrix which results in $|\mathbf{D}_n| = \Delta(\sigma)$ and therewith in the determinant value of interest. The matrix recursion (7) is started with the unit matrix $\mathbf{I}$ and the two $\mathbf{B}_1$ matrices

$$\mathbf{B}_1^o = \begin{bmatrix} 1-\eta^2-\sigma^2/2 & 2D\eta \\ -2D\eta & 1-\eta^2 \end{bmatrix}, \qquad \mathbf{B}_1^{1/2} = \begin{bmatrix} 1-\eta^2/4 + \sigma/2 & D\eta \\ -D\eta & 1-\eta^2/4 - \sigma/2 \end{bmatrix}. \quad (8)$$

The first one $\mathbf{B}_1^o$ belongs to the integral periodic solutions. The second $\mathbf{B}_1^{1/2}$ is applied for the determination of the half-integral periodic stability boundaries, both for k = 0. All other frequency values, noted in (6), lead to identical stability values if the original determinant (5) is symmetrically evaluated under the auxiliary condition of purely real-valued results.

2.2 State transformation of the oscillator

It is now interesting to derive the same results by means of a coordinate transformation. For this purpose, we rewrite the differential equation (1) into the form of a first order system.

$$\left.\begin{aligned} &\dot{\mathbf{x}}(t) = [\,\mathbf{A} + \sigma\,\omega_1^2\,z(\varphi)\,\mathbf{R}\,]\,\mathbf{x}(t), \qquad z(\varphi) = \cos\varphi, \qquad \varphi = \omega_e t, \\ &\mathbf{x}(t) = \begin{bmatrix} x(t) \\ \dot{x}(t) \end{bmatrix}, \quad \mathbf{A} = \begin{bmatrix} 0 & 1 \\ -\omega_1^2 & -2D\omega_1 \end{bmatrix}, \quad \mathbf{R} = \begin{bmatrix} 0 & 0 \\ -1 & 0 \end{bmatrix}. \end{aligned}\right\} (9)$$

Subsequently, we introduce the new state vector $\mathbf{y}(t)$ via the transformation matrix $\mathbf{M}(\varphi)$ defined on the dimensionless time $\varphi = \omega_e t$. It is provided that $\mathbf{M}(\varphi)$ is nonsingular and bounded so that both state vectors $\mathbf{x}(t)$ and $\mathbf{y}(t)$ have the same stability behaviour.

$$\mathbf{y}(t) = \mathbf{M}(\varphi)\,\mathbf{x}(t), \qquad \dot{\mathbf{y}}(t) = \mathbf{\Lambda}\,\mathbf{y}(t), \quad (10)$$

$$\omega_e\,\mathbf{M}_\varphi(\varphi) + \mathbf{M}(\varphi)\,(\mathbf{A} + \sigma\,\omega_1^2\cos\varphi\,\mathbf{R}) = \mathbf{\Lambda}\,\mathbf{M}(\varphi). \quad (11)$$

It is obvious that the resulting dynamic system (10) is time-invariant described by the constant matrix $\mathbf{\Lambda}$ if the transformation matrix $\mathbf{M}(\varphi)$ satisfies the eigenvalue equation (11).

Without any loss of generality the unknown system matrix $\mathbf{\Lambda}$ is assumed to be diagonal, i.e. $\mathbf{\Lambda} = \mathrm{diag}(\lambda_l)$ for $l = 1,2$. Therewith, the matrix differential equation (11) can be evaluated row-wise resulting in the following vector differential equation for each row vector $\mathbf{m}^{(l)}$ of the matrix $\mathbf{M}(\varphi)$ associated to the eigenvalues λ_l $(l=1,2)$.

$$\omega_e \mathbf{m}_\varphi(\varphi) + (\mathbf{A}^T + \sigma\omega_1^2 \cos\varphi\, \mathbf{R}^T - \lambda \mathbf{I})\mathbf{m}(\varphi) = 0, \quad \mathbf{m}(\varphi) = \sum_{-\infty}^{\infty} \exp(in\varphi)\mathbf{c}_n. \quad (12)$$

Finally, the Fourier series, noted in (12), yields the following system equations for the coefficient vectors $\mathbf{c}_n$.

$$[\mathbf{A}^T + (in\omega_e - \lambda)\mathbf{I}]\, \mathbf{c}_n + \frac{1}{2}\sigma\omega_1^2\, \mathbf{R}^T\, (\mathbf{c}_{n-1} + \mathbf{c}_{n+1}) = 0, \quad n = 0, \mp 1, . (13)$$

Evaluated in a scalar form, they lead to the same result (3) already derived by means of the Floquet theory. The further evaluation of (13) is described by (7) and (8). Typical results are given below (fig.1).

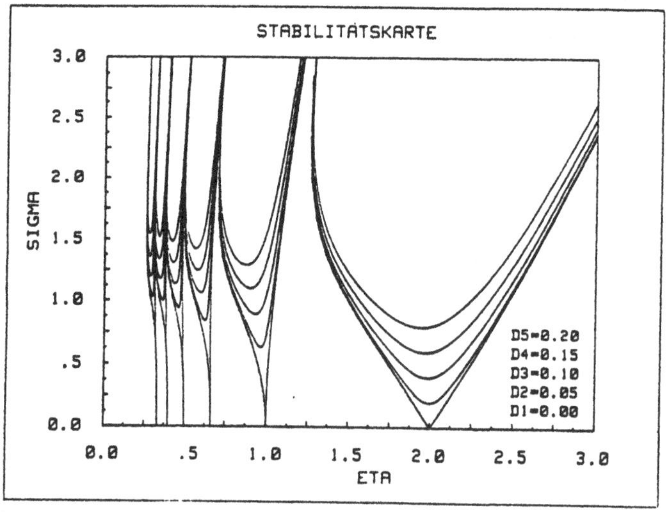

Fig. 1: Stability map of the harmonic oscillator
- critical intensity versus excitation frequency

Concluding it is worth to emphasize that the eigenvalue problem (12) is solved by a Fourier series which is simplier in comparison with the Floquet setup (2). Furthermore, we recognize that (12) is generated by the backward operator of the original dynamic system (9). Such properties are important for the following applications in the stochastic case.

3. MOMENTS STABILITY UNDER FILTERED NOISE

The modelling of more realistic parametric excitations by shaping filters under white noise has been started by Kozin and Bogdanoff [5] in 1962. This model implies a nonlinearity in form of a product of the filter excitation and the state process of the parametric system resulting in a sequence of moments equations. Hence, convergence problems are involved.

3.1 Formulation of the problem

To introduce the problem, mentioned above, let us consider the parametric oscillator with a filtered noise coefficient.

$$\ddot{X}_t + 2D\omega_1\dot{X}_t + (\omega_1^2 + \alpha Z_t)X_t = 0, \qquad (t \geqslant 0), \tag{14}$$

$$dZ_t = -\omega_g Z_t dt + \sigma dW_t, \qquad E[(dW_t)^2] = dt. \tag{15}$$

As before, ω_1 is the natural frequency of the system, D is a dimensionless damping measure and α denotes the intensity of the parametric excitation Z_t generated from normed white noise $\dot{W}_t$ via the scalar first order filter (15) with the band limiting frequency ω_g. The parameter σ gives the intensity of the Wiener process W_t. The index t in W_t, Z_t and X_t denotes the time dependency of these processes. Dots are abbreviations for time derivatives.

To (14) belong quadratic state processes which are described by the third order system, noted in (16).

$$\mathbf{X}_t^T = (X_t^2, X_t\dot{X}_t, \dot{X}_t^2), \qquad \dot{\mathbf{X}}_t = (\mathbf{A} + \alpha Z_t\mathbf{R})\mathbf{X}_t \tag{16}$$

$$A = \begin{bmatrix} 0 & 2 & 0 \\ -\omega_1^2 & -2D\omega_1 & 1 \\ 0 & -2\omega_1^2 & -4D\omega_1 \end{bmatrix}, \qquad R = \begin{bmatrix} 0 & 0 & 0 \\ -1 & 0 & 0 \\ 0 & -2 & 0 \end{bmatrix}. \tag{16}$$

Applying Ito's formula to both, the filter equation (15) and the quadratic state equation (16), we obtain a sequence of differential equations for the second order moments of interest.

$$E(Z_t^n X_t) = (A - n\omega_g I)E(Z_t^n X_t) + \alpha R E(Z_t^{n+1} X_t) + \tfrac{1}{2}\sigma^2 n(n-1)E(Z_t^{n-2} X_t). \tag{17}$$

The exponential set-up $E(Z_t^n X_t) = \exp(\lambda t)c_n$ $(n = 0,1,2,..)$ reduces (17) to an infinite set of homogeneous algebraic equations coupled by c_{n+1} and c_{n-2}. This is a consequence of the nonlinear term $Z_t X_t$. It only allows to calculate asymptotic solutions for small intensities α. However, an increasing of n and α detroys the convergence of the infinite determinant, associated to (17).

3.2 Stochastic state transformation

As already shown in [6], we avoid the ill-conditioned equations (17) by applying an orthogonalizing procedure which eliminates the eigenbehaviour of the filter process Z_t. For this purpose, we introduce the new state vector Y_t by an stochastic state transformation.

$$Y_t = M(Z_t)X_t, \qquad E(\|M(Z_t)\|^p) < \infty. \tag{18}$$

Herein, $M(Z_t)$ is an unknown 3x3 transformation matrix defined on the Ornstein-Uhlenbeck process Z_t. The matrix $M(Z_t)$ is assumed to be non-singular and bounded in the stochastic sense, e.g. by the p-th expectation of a matrix norm, as noted in (18). Since Z_t is stationary and gaussian, X_t and Y_t have the same stability properties. The transformation (18) leads to

$$dY_t = \Lambda Y_t dt + \sigma M_z M^{-1} Y_t dW_t, \tag{19}$$

$$M(A + \alpha Z_t R) - \omega_g M_z Z_t + \tfrac{1}{2}\sigma^2 M_{zz} = \Lambda M. \tag{20}$$

In (19), the unknown matrix $\mathbf{\Delta}$ is determined by the equation (20).

We postulate now that the transformation $\mathbf{M}(Z_t)$ yields a time-invariant drift term such that the 3x3 matrix $\mathbf{\Delta}$ has constant elements independent on Z_t. Taking the expectation in (19) the diffusion term is vanishing. Consequently, we obtain a closed moments equation.

$$\dot{E}(\mathbf{Y}_t) = \mathbf{\Delta}\, E(\mathbf{Y}_t), \qquad E(\mathbf{Y}_t) = \exp(\lambda t)\, \mathbf{c}. \tag{21}$$

Its stability can easily be checked by the exponential function, noted above. In case that the matrices $\mathbf{A}$ and $\mathbf{R}$ are commutative, the equation (20) can be solved by

$$\mathbf{M}(Z_t) = \exp(\frac{\alpha}{\omega_g}\mathbf{R}\, Z_t), \qquad (\, \mathbf{A}\,\mathbf{R} = \mathbf{R}\,\mathbf{A}\,), \tag{22}$$

$$d\mathbf{Y}_t = [\mathbf{A} + \tfrac{1}{2}(\sigma\alpha/\omega_g)^2\mathbf{R}]\mathbf{Y}_t dt + (\sigma\alpha/\omega_g)\mathbf{R}\mathbf{Y}_t dW_t, \tag{23}$$

yielding the new Ito equation (23) of a linear form in correspondence to the well-known Stratonovich integral. Hence, the equation (20) represents a deterministic eigenvalue problem for the determination of the transformation matrix $\mathbf{M}(Z_t)$ and associated stability matrix $\mathbf{\Delta}$.

Without loss of generality the unknown eigenvalue matrix $\mathbf{\Delta}$ in (20) can be diagonalized by $\mathbf{\Delta} = \mathrm{diag}(\lambda_l)$ for $l = 1,2$ and 3. Therewith, it is possible to rewrite (20) into a vector differential equation

$$\tfrac{1}{2}\sigma^2\mathbf{m}_{zz} - \omega_g z\, \mathbf{m}_z + (\mathbf{A}^T + \alpha z\, \mathbf{R}^T - \lambda\mathbf{I})\mathbf{m} = 0, \qquad (\, \mathbf{A}\,\mathbf{R} \neq \mathbf{R}\,\mathbf{A}\,) \tag{24}$$

for each row vector $\mathbf{m}^{(l)}(z)$ of the matrix $\mathbf{M}(z)$. Herein, we recognize the backward operator of the filter (15). It is the Hermitian operator if we substitute the new variable x to get the following normalized form:

$$\mathbf{m}_{xx} - 2x\, \mathbf{m}_x + \frac{2}{\omega_g}(\mathbf{A}^T + \frac{\alpha\sigma}{\sqrt{\omega_g}}\, x\, \mathbf{R}^T - \lambda\mathbf{I})\, \mathbf{m} = 0, \qquad (\, x = z\sqrt{\omega_g}\,/\sigma\,). \tag{25}$$

Obviously, x is the density variable of a normalized distribution of the parametric excitation by Z_t.

3.3 Solution by Hermite polynomials

The deterministic eigenvalue problem (25) is solved by means of the Hermite polynomials $H_n(x)$.

$$\mathbf{m}(x) = \sum_{n=0}^{\infty} H_n(x)\mathbf{c}_n, \qquad (-\infty < x < \infty), \quad n = 0,1,2,\ldots, \tag{26}$$

$$[\mathbf{A}^T-(\lambda+n\omega_g)\mathbf{I}]\mathbf{c}_n + (\alpha\sigma/\sqrt{\omega_g})\mathbf{R}^T[\tfrac{1}{2}\mathbf{c}_{n-1}+(n+1)\mathbf{c}_{n+1}] = 0. \tag{27}$$

By insertion into (25) we obtain the algebraic equations (27) and therewith an infinite determinant for the calculation of the eigenvalue λ. In comparison with the moments equations (17) we observe a resolving concentration of the secondary diagonals in (27) which implies the convergence of the associated infinite determinant.

To proof this property we evaluate the equation (27) for the matrices $\mathbf{A}$ and $\mathbf{R}$, given in (17), and for the coefficient vector $\mathbf{c}^T = (x_n, y_n, z_n)$. This yields the scalar equation [7]

$$f_n z_n + g_{n-1}z_{n-1} + h_{n+1}z_{n+1} = 0, \quad n = 0,1,2,\ldots \tag{28}$$

$$\left.\begin{aligned} f_n &= [4\omega_1^2+(\lambda+n\omega_g)(\lambda+n\omega_g+4D\omega_1)](\lambda+n\omega_g+2D\omega_1),\\ g_{n-1} &= (4D\omega_1+2\lambda+2n\omega_g-\omega_g)\alpha\sigma/\sqrt{\omega_g},\\ h_{n+1} &= 2(n+1)(4D\omega_1+2\lambda+2n\omega_g+\omega_g)\alpha\sigma/\sqrt{\omega_g}. \end{aligned}\right\} \tag{29}$$

According to H. von Koch [3] the equations (28) are divided by f_n. In this normalized form the associated infinite determinant is convergent if the following absolute sum is finite.

$$\sum_{n=0}^{\infty}\left|\frac{g_n h_{n+1}}{f_n f_{n+1}}\right| \rightarrow \sum_{n=0}^{\infty}\frac{1}{n^3}\,. \tag{30}$$

For increasing n we find a majorant with $1/n^3$ by which the convergence of the infinite determinant is proofed.

Because of its band structure the determinant Δ of (28) can be evaluated by the two-step recursion formula (31).

$$\left.\begin{aligned} \Delta_0 &= f_0, \quad \Delta_1 = f_0 f_1 - g_0 h_1, \quad n = 0,1,2,\ldots,\\ \Delta_{n+2} &= f_{n+2}\Delta_{n+1} - e_{n+2}\Delta_n\,(\alpha\sigma)^2/\omega_g,\\ e_{n+2} &= 2(n+2)(4D\omega_1+2\lambda+2n\omega_g+3\omega_g)^2. \end{aligned}\right\} \tag{31}$$

Starting with the first two determinants Δ_0 and Δ_1 we calculate the next one Δ_2 simply by multiplying with f_2 and e_2. Subsequently, we go on up to a certain approximation order $n \geq 1$. For given parameters α, σ, D and ω_1 we may calculate those eigenvalues $\lambda = \alpha \mp i\beta$ for which the real part and the imaginary part of the determinant are vanishing simultaneously. However, in the stability investigation of second order moments, this two-parametric evaluation of the determinant can be reduced as follows.

3.4 Stability evaluation

From [8] we know that the critical eigenvalue $Re(\lambda) = \alpha = 0$ implies a vanishing imaginary part $Im(\lambda) = \beta = 0$; i.e. we observe only the monotone instability in the time behaviour of the second moments.

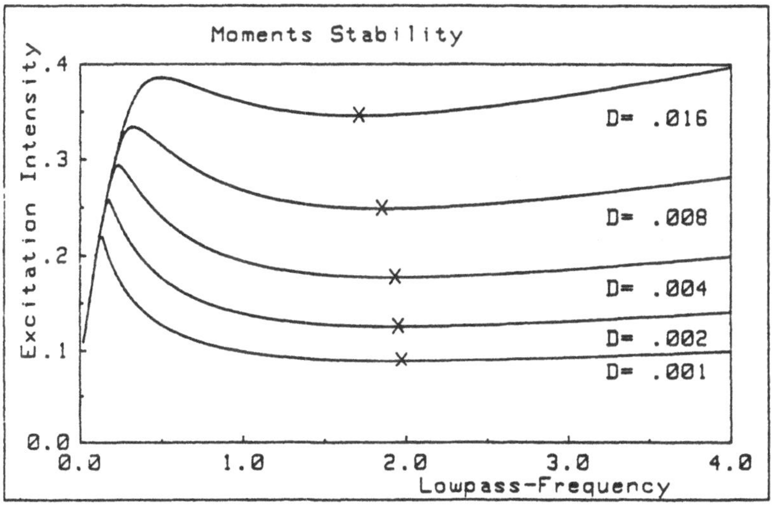

Fig. 2: Second order moments stability of the parametric oscillator under filtered noise excitation

Hence, the eigenvalue problem (31) can be reduced to the real-valued, one-parametric evaluation of $\Delta(\alpha\sigma, \lambda=0) = 0$. We find the critical intensity simply by increasing the positive product $(\alpha\sigma)^2$ up to the first zero of the determinant (31).

For physical interpretations, finally, it is suitable to discuss the following two different evaluations:

$$\left.\begin{array}{lll} \text{a) Normed zero spectrum value:} & (\sigma/\omega_g)^2 = 1, \\ \text{b) Normed excitation variance:} & \sigma^2/2\omega_g = 1. \end{array}\right\} \qquad (32)$$

The latter is shown in figure 2 where the critical intensity α of a normed excitation variance is plotted over its bandwidth ω_g for several damping values D and for $\omega_1 = 1$. Herein, we observe cross-wise marked minima situated near twice the natural frequency of the system. This corresponds to the deterministic case of harmonic parameter excitations.

4. STABILITY IN THE P-TH MEAN

Almost sure stability and related investigations of the Lyapunov exponents have been started by F. Kozin [9]. Extensions and applications of the multiplicative ergodic theorem are given in [10]. In the following we make an attempt to avoid such extensive simulation techniques by the derivation of deterministic eigenvalue problems for the stability investigation of suitable state vector norms.

4.1 Derivation of the eigenvalue problem

To start with the simplest example we consider the oscillator with multiplicative white noise $\dot{W}_t$ in the restoring term.

$$\ddot{X}_t + 2D\omega_1\dot{X}_t + \omega_1(\omega_1 + \sigma\dot{W}_t)X_t = 0, \qquad E(\dot{W}_t\dot{W}_s) = \delta(t-s). \qquad (33)$$

Analogue to (14), ω_1 is the natural frequency of the oscillator, D is a dimensionless damping measure and σ denotes the intensity of the parametric noise excitation. We introduce the stochastic amplitude A_t and the phase process Ψ_t by

$$X_t = A_t\cos\Psi_t, \qquad \dot{X}_t = \omega_1 A_t \sin\Psi_t, \qquad (34)$$

$$P_t = A_t^p = \left[\sqrt{X_t^2 + (\dot{X}_t/\omega_1)^2}\,\right]^p, \quad (0 \leq p < \infty) \qquad (35)$$

and subsequently the p-th norm process P_t. The associated Ito differential equations read as follows:

$$d\Psi_t = -[\omega_1 + 2D\omega_1 \sin\Psi_t \cos\Psi_t + \sigma^2 \sin\Psi_t \cos^3\Psi_t]dt - \sigma\cos^2\Psi_t dW_t, \qquad (36)$$

$$dP_t = \tfrac{1}{2}p\sigma^2 P_t \cos^4\Psi_t dt + \tfrac{1}{2}p(p-1)\sigma^2 P_t \sin^2\Psi_t \cos^2\Psi_t dt - - p2D\omega_1 P_t \sin^2\Psi_t dt - p\sigma P_t \sin\Psi_t \cos\Psi_t dW_t. \qquad (37)$$

Both equations are linear in P_t and nonlinear in Ψ_t. The phase process Ψ_t is decoupled from the stochastic norm P_t. For $p \Rightarrow \infty$ the norm (35) represents the supremum or Tschebyscheff norm, for $p = 2$ it is the gaussian norm and for $p \Rightarrow 0$ we obtain the Lyapunov exponent of the zeroth norm. According to [11], this limiting moment determines the almost sure stability of the system.

Similar to the state transformation (18) in chapter 3.2 we introduce a new norm process S_t by the scalar function $T(\Psi_t)$ defined on the stochastic phase process Ψ_t.

$$S_t = T(\Psi_t)\, P_t, \qquad \Rightarrow \qquad \dot{E}(S_t) = \lambda\, E(S_t), \qquad (38)$$

$$dS_t = \lambda S_t dt + g(\Psi_t, S_t)dW_t, \qquad \sigma_1^2 = \sigma^2/\omega_1. \qquad (39)$$

Therewith, we obtain the new Ito differential equation (39). Moreover, the transformation $T(\Psi_t)$ is defined on the stochastic phase process in such a way that the new drift term λS_t in (39) is independent on the phase Ψ_t. This postulation results into the following equation.

$$\tfrac{1}{2}\sigma_1^2\cos^4\psi\; T_{\psi\psi}(\psi) - [1+2D\sin\psi\cos\psi+\sigma_1^2(1-p)\sin\psi\cos^3\psi]\; T_\psi(\psi) + + [\tfrac{1}{2}p(p-1)\sigma_1^2\sin^2\psi\cos^2\psi+\tfrac{1}{2}p\sigma_1^2\cos^4\psi-p2D\sin^2\psi]\; T(\psi) = \lambda\; T(\psi). \qquad (40)$$

Obviously, the time-invariant condition (40) represents a deterministic second order eigenvalue problem for the determination of the unknown transformation function $T(\psi)$ and associated eigenvalue λ. For $\sigma_1 = 0$ e.g., the equation (40) degenerates to

$$(1+D\sin 2\psi)\; T_\psi(\psi) + (\lambda + 2D\sin^2\psi)\; T(\psi) = 0, \qquad (41)$$

$$T(\psi) = C\; (1+D\sin 2\psi)^{p/2}, \qquad (\lambda = -pD) \qquad (42)$$

and possesses the periodic solution (42) for the eigenvalue $\lambda = -pD$. It is nonsingular and real-valued for $0 \leqslant D \leqslant 1$.

4.2 Orthogonal expansion of the problem

We find an asymptotic solution [12] of the eigenvalue problem (40) by a double-periodic, complex-valued Fourier expansion.

$$T(\psi) = \sum_{n=-\infty}^{\infty} \exp(i2n\psi)\, z_n, \qquad n = 0, \mp 1, \mp 2, \ldots, \tag{43}$$

$$c_{n-2}^{n} z_{n-2} + b_{n-1}^{n} z_{n-1} + a_n z_n + b_{n+1}^{p} z_{n+1} + c_{n+2}^{p} z_{n+2} = 0. \tag{44}$$

Inserting it into (40) and comparing all coefficients we obtain the algebraic equation (44). For the stability investigation it can be written in a real-valued form with the following infinite determinant.

$$\Delta(\sigma) = \begin{vmatrix} a_o & 2b_1 & 0 & 2c_2 & 0 & 0 & 0 & 0 & \cdots \\ b_o & a_1+c_1 & 2 & b_2 & 0 & c_3 & 0 & 0 & \\ 0 & -2 & a_1-c_1 & 0 & b_2 & 0 & c_3 & 0 & \\ c_o & b_1 & 0 & a_2 & 4 & b_3 & 0 & c_4 & \\ 0 & 0 & b_1 & -4 & a_2 & 0 & b_3 & 0 & \cdots \\ \vdots & & & & & & \vdots & & \end{vmatrix}. \tag{45}$$

The determinant possesses the coefficients

$$a_n = (p^2+2p-12n^2)\sigma_1^2/16 - pD - \lambda, \qquad c_n = [p(2-p)-4n^2+4n(1-p)]\sigma_1^2/32,$$
$$b_n = (p+2n)D/2 + [p-4n^2+2n(1-p)]\sigma_1^2/8, \quad \sigma_1^2 = \sigma^2/\omega_1, \quad n = 0,1,\ldots \tag{46}$$

It is noted that the convergence conditions of H. von Koch are not applicable in (45). This is a consequence of the second order operator in (40) which becomes singular at the periodic boundaries $\psi = \mp\pi/2$.

In spite of this bad condition of the determinant (45) we are able to evaluate it for the calculation of the critical variance σ^2 and a simultaneously vanishing of the real part and the imaginary part of the eigenvalue λ. Increasing the approximation order n of (45) we observe convergence properties in a wide parameter range. The corresponding numerical results are shown in figure 3 for n = 0, 1, 2, 3 and 4, marked by different line types. They are evaluations of (45) cut off at m = 2n+1 elements. The applied norm powers are p = 0, 1, 2, and 4. For p = 2 we obtain the boundary of the mean square stability and for p = 0 the almost sure stability condition. Naturally, there are now many questions related to the convergence of (45). Are there better expansions of the eigenvalue problem (40)? Is there any influence of the norm chosen in (40) on the stability boundaries if other linear combinations of the squared state processes are applied?

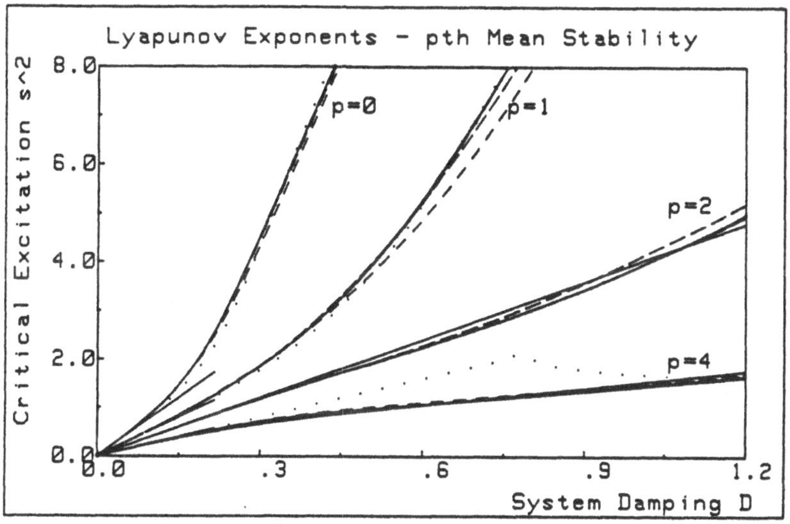

Fig. 3: P-th norm stability condition of the parametric oscillator under white noise

REFERENCES

1. Coddington, E.A.; Levinson, N.: Theory of Ordinary Differential Equations, McGraw-Hill, New York 1955.
2. Whittaker, E.T.; Watson, G.N.: A Course of Modern Analysis, University Press, Cambridge 1958.
3. Koch von, H.: Sur les determinants infinis et les equations

differentielles lineaires, Acta Math. 16 (1892/93), 217-295.

4. Klotter, K.; Kotowski, G.: Über die Stabilität der Lösungen Hillscher Differentialgleichungen mit drei unabhängigen Parametern, ZAMM 23 (1943) 3, 149-155.
5. Bogdanoff, J.L.; Kozin, F.: Moments of the output of linear random systems, J. Acoust. Soc. Amer., 34 (8), 1962, 1063-1066.
6. Wedig, W., Stochastische Schwingungen - Simulation, Schätzung und Stabilität, ZAMM 67 (4), 1987, T34-T42.
7. Bräutigam, H.: Beiträge zur Momentenstabilität parametererregter Schwingungssysteme, Diss. Universität Karlsruhe 1982.
8. Nevelson, M.B.; Khasminskii, R.Z.: Stability of stochastic systems (in russ.), Problemy Peredachi Informatsii 2, No 3, 1966, 76-91.
9. Kozin, F.; Sugimoto, S.: Decision criteria for stbility of stochastic systems from observed data, in: Stochastic Problems in Dynamics (ed. by B.L.Clarkson), Pitman, London 1977, 8-33.
10. Arnold, L.; Kliemann, W.: Qualitative theory of stochastic systems, in: Probabilistic Analysis and Related Topics, Vol. 3, Academic Press, New York, 1981.
11. Kozin. F.; Sugimoto, S: Relations between sample and moment stability for linear stochastic differential equations, in: Proc. Conference Stochastic Diff. Eq. (ed. by D. Mason), Academic Press, 1977, 145-162.
12. Wedig, W.: Berechnung des p-ten Lyapunov Exponenten über funktionalanalytische Methoden, GAMM-Tagung 87, Stuttgart, to appear in ZAMM 88.

PARAMETER IDENTIFICATION
OF ROAD SPECTRA AND NONLINEAR OSCILLATORS

Walter V. Wedig
Institute of Technical Mechanics
University of Karlsruhe
D-7500 Karlsruhe, Kaiserstr.12, FRG

ABSTRACT

For on-line simulations it is important to apply complete filters for the generation of spectrum-compatible random processes. The spectra may be given by corresponding evaluations of actual measurements or by nonlinear models of dynamic systems under white noise. In both cases, we are able to derive the denominator and numerator parameters of the filter transfer functions ensuring asymptotic stability and physical existence of the complete filter.

1. FORMULATION OF THE PROBLEM

The paper is dealing with the identification of given power spectra $\bar{S}(\omega)$ which can be defined by a nonlinear modelling of dynamic systems under white noise or by a piecewise analytic frequency distribution derived from actual measurements of stationary random processes and its modelling in the spectral range. In both cases, we are interested to identify a linear time-invariant system under white noise for the generation of spectrum-compatible random processes. Since exact solutions of power spectra are not obtainable for nonlinear systems under white noise, a number of approximation techniques have been developed [1]. Some of these techniques are perturbation methods [2],[3], equivalent linearization techniques [4],[5], cumulant closure [6],[7] and stochastic averaging methods [8],[9]. In the following text, we propose to go back to infinite moment and correlation equations by applying orthogonal expansions via Hermite polynomials. First numerical results show monotone convergence properties in correspondence to the numerical method of steepest descent. The obtained spectra results have the rational form of complete filters.

Consequently, the dynamic modelling of stationary nonlinear processes has to be performed by complete dynamic systems, in general. Restricting our interest to one-dimensional problems such filters are described by single differential equations of the following form:

$$\sum_{i=0}^{n} T_i X_t^{(i)} = \dot{W}_t, \qquad Y_t = \sum_{i=0}^{n-1} S_i X_t^{(i)}, \tag{1}$$

Herein, X_t is the state process of the filter, $X_t{}^{(i)}$ are time derivatives, W_t is the normed Wiener process and Y_t is the simulation process of interest. The time-invariant parameters of the filter are T_i and S_i. It is noted that the time-discrete versions of (1) are so-called ARMA-models whereby the autoregressive part (AR) corresponds to the left part of (1) and the moving average (MA) follows from the right equation in (1). In the frequency range, the complete filter (1) is determined by the denominator polynomial $T(j\omega)$ and the numerator polynomial $S(j\omega)$.

$$T(j\omega) = \sum_{i=0}^{n} (j\omega)^i T_i, \qquad S(j\omega) = \sum_{i=0}^{n-1} (j\omega)^i S_i. \tag{2}$$

The zeros of $T(j\omega)$ decide the stability behaviour of the filter. The poles of $S(j\omega)$ determine its physical existence. Hence, the parameters T_i and S_i of both polynomials (2) have to be calculated in such a way that complex-valued results are avoided. Furthermore, the eigenvalues of the denominator polynomial have to be situated in the negativ half plane. Finally, we are interested in best approximations in the sense of the L_2- Norm satisfying the following minimal condition:

$$\int_{-\infty}^{+\infty}\left[\frac{S(j\omega)\ S(-j\omega)}{T(j\omega)\ T(-j\omega)} - \bar{S}(\omega)\right]^2 d\omega = \text{Min.!} \tag{3}$$

As already mentioned, $\bar{S}(\omega)$ is the given power spectrum of a stationary process. Its squared deviation from the spectrum of the filter (1) becomes minimal in the entire frequency range $-\infty < \omega < +\infty$ for suitably chosen parameters T_i and S_i. Obviously, the minimal problem (3) is highly nonlinear. It is non-convex and therefore not solvelable, in general. To avoid this difficulty, we are looking for a suboptimal processing with unique and complete solutions which may be used as starting values for an iterative solution of (3).

2. NONLINEAR POWER SPECTRA

We start with the first identification problem that a given power spectrum is defined by a nonlinear stochastic system under white noise. To explain the method, it is sufficient to consider a nonlinear first order system.

$$\dot{X}_t + \omega_g X_t + \gamma\omega_g X_t^3 = \sigma\dot{W}_t, \qquad E(\dot{W}_t\dot{W}_s) = \delta(t-s). \tag{4}$$

Herein, X_t is the nonlinear state process, $\dot{W}_t$ is white noise with the intensity σ, ω_g is the limiting frequency and γ is the parameter of the cubic term in (4). In the following, it is shown that the analysis of such nonlinear systems leads automatically to transfer functions of the form (2) from which we are able to determine the filter parameters T_i and S_i in a simple and direct way.

2.1 Stationary higher order moments

The analysis of nonlinear power spectra has to be prepared by the calculation of higher order moments. For this purpose, we may apply the Fokker-Planck technique [10] in deriving stationary density distributions of X_t and finding the associated moments by numerical integrations. A better approach [11] starts with the increment of even powers of the process X_t

$$d(X_t^{2n+2}) = 2(n+1)X_t^{2n+1}dX_t + (n+1)(2n+1)X_t^{2n}\sigma^2 dt, \qquad (n = 0,1,..) \quad (5)$$

from which we obtain the corresponding moment equations simply by inserting (4) and taking the expectation.

$$\left.\begin{aligned} \dot{m}_{2n+2} &= 2(n+1)\omega_g[-m_{2n+2} - \gamma m_{2n+4} + \sigma_1^2(2n+1)m_{2n}], \\ m_{2n} &= E(X_t^{2n}), \quad n = 0,1,2,\ldots, \quad \sigma_1^2 = \sigma^2/(2\omega_g). \end{aligned}\right\} \quad (6)$$

In the stationary case, the time derivatives $\dot{E}(.)$ are vanishing. Consequently, the higher order moments of X_t are determined by the matrix equation, as follows:

$$\begin{bmatrix} 1 & \gamma & 0 & 0 \ldots \\ -3\sigma_1^2 & 1 & \gamma & 0 \\ 0 & -5\sigma_1^2 & 1 & \gamma \ldots \\ \vdots & & \vdots & \end{bmatrix} \begin{bmatrix} m_2 \\ m_4 \\ m_6 \\ \vdots \end{bmatrix} = \sigma_1^2 \begin{bmatrix} 1 \\ 0 \\ 0 \\ \vdots \end{bmatrix}. \quad (7)$$

We calculate the solution of (7) by means of Cramer's rule and evaluate the associated determinants via recurrence formulas.

For the mean square $E(X_t^2) = m_2$ e.g., this is performed by

$$\left.\begin{aligned} & m_2^{(p)} = E^{(p)}(X_t^2) = \sigma_1^2\, \Delta_p/d_p, & & \gamma_1 = \gamma\, \sigma_1^2\,, \\ & d_{p+2} = d_{p+1} + (2p+5)\gamma_1\, d_p, \quad d_o = d_1 = 1, & & p = 0,1,2,\ldots, \\ & \Delta_{p+2} = \Delta_{p+1} + (2p+5)\gamma_1\, \Delta_p, \quad \Delta_o = 1, & & \Delta_1 = 1 + 3\gamma_1. \end{aligned}\right\} \quad (8)$$

Herein, p denotes the approximation order by cutting off the equation (7) at a certain row or column number p starting with the initial determinant values d_o, d_1, Δ_o and Δ_1 for the numerator d_p and the denominater Δ_p, respectively. In figure 1, we show a numerical evaluation of (8) by plotting the square mean of X_t over the nonlinear parameter γ for an increasing approximation order p. Obviously, we obtained an upper bound for even numbers p, a lower bound for odd p and therewith an inclusion of the strong solution which is valid for all parameter values $0 \leq \gamma < +\infty$.

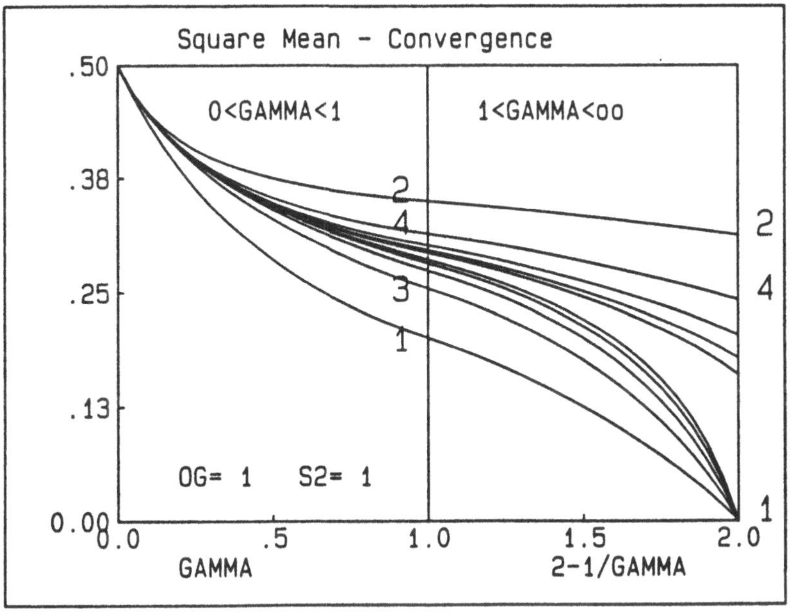

Fig. 1: Upper and lower bounds of square means in dependence on the approximation order p

2.2 Nonlinear correlations and spectra

The method, explained above, can easily be extended to the derivation of nonlinear correlation functions. For this purpose, we derive the increment of an odd power of X_t and multiply it by X_s for $t \geq s$.

$$X_s dX_t^{2n+1} = -(2n+1)\omega_g X_s X_t^{2n+1} dt + (2n+1)\sigma X_s X_t^{2n} dW_t -$$
$$- (2n+1)\gamma\omega_g X_s X_t^{2n+3} dt + n(2n+1)\sigma^2 X_s X_t^{2n-1} dt. \qquad (9)$$

Taking the expectation in (9), the diffusion term is vanishing and the rest gives a correlation differential equation of the form

$$\left.\begin{aligned} &R'_{2n+2}(\tau) + (2n+1)[\omega_g R_{2n+2}(\tau) + \gamma\omega_g R_{2n+4}(\tau) - n\sigma^2 R_{2n}(\tau)] = 0, \\ &R_{2n+2}(\tau) = E(X_s X_{s+\tau}^{2n+1}), \quad n = 0,1,2,\dots \quad t = s + \tau \quad (\tau \geq 0), \end{aligned}\right\} \tag{10}$$

in restricting our interest to the stationary case. Hereby, the time variable s is fixed and τ denotes a positive time shifting. Dashes are time derivations with respect to τ. The differential equations (10) are solved by means of the Laplace transforms $\mathcal{L}\{R_{2n}(\tau)\} = X_{2n}(s)$ in dependence on the Laplace variable $s = j\omega$. Under the initial conditions $R_{2n}(0) = E(X_s^{2n}) = m_{2n}$ we obtain a matrix equation as follows:

$$\begin{bmatrix} s+\omega_g & \gamma\omega_g & 0 & 0 & \dots \\ -3\sigma^2 & s+3\omega_g & 3\gamma\omega_g & 0 & \\ 0 & -10\sigma^2 & s+5\omega_g & 5\gamma\omega_g & \dots \\ \vdots & & & \vdots & \end{bmatrix} \begin{bmatrix} X_2(s) \\ X_4(s) \\ X_6(s) \\ \vdots \end{bmatrix} = \begin{bmatrix} m_2 \\ m_4 \\ m_6 \\ \vdots \end{bmatrix}. \tag{11}$$

It is noted that the equation (11) has the same structure as (7). It can therefore be solved by similar determinant recursions, as before.

Because of the symmetry of $R_{2n}(\tau)$, the solution of (11) can be supplemented by $R_{2n}(-\tau)$. In the frequency domain $s=j\omega$, this implies the addition of conjugate complex Laplace transforms by which we finally obtain the associated power spectrum of interest.

$$S_2(\omega) = \int_{-\infty}^{+\infty} E(X_s X_{s+\tau}) \exp(-j\omega\tau)\, d\tau = X_2(j\omega) + X_2(-j\omega). \tag{12}$$

For a first analytic evaluation of (12), we apply the first two equations in (11) and the first three ones in (7). The corresponding results are

$$m_2^{(2)} = \sigma_1^2 \frac{1+5\gamma_1}{1+8\gamma_1}, \qquad \gamma m_4^{(2)} = \sigma_1^2 \frac{3\gamma_1}{1+8\gamma_1}, \qquad \gamma_1 = \gamma \frac{\sigma^2}{2\omega_g},$$

$$S_2^{(1)}(\omega) = \frac{\sigma^2}{1+8\gamma_1} \frac{9\omega_g^2(1+6\gamma_1+8\gamma_1^2) + (1+8\gamma_1)\omega^2}{9(1+2\gamma_1)^2\omega_g^4 + 2(5-6\gamma_1)\omega_g^2\omega^2 + \omega^4} \geq 0. \tag{13}$$

The calculated spectrum (13) is non-negative definite for all parameters $\gamma \geqslant 0$ and for all spectral frequencies $-\infty < \omega < +\infty$. For $\gamma \equiv 0$, the spectrum goes over to the well-known linear form. Finally, it is noted that the result (13) agrees identically with the power spectrum of the complete filter, given in (1) and (2).

$$S(\omega) = \frac{S(j\omega)\ S(-j\omega)}{T(j\omega)\ T(-j\omega)} = \frac{S_o^2 + S_1^2\,\omega^2}{T_o^2 + (T_1^2 - 2T_oT_2)\omega^2 + T_2^2\omega^4}\ . \tag{14}$$

Comparing all coefficients of the denominator and numerator polynomials in (13) and in (14), we find

$$\left.\begin{array}{lll} T_2 = 1, & T_1 = 4\,\omega_g, & T_o = 3(1+2\gamma_1)\,\omega_g^2, \\ S_1 = \sigma, & S_o = 3\sigma\omega_g \sqrt{(1+6\gamma_1+8\gamma_1^2)/(1+8\gamma_1)}\ , & \end{array}\right\} \tag{15}$$

and therewith physically existent S_i parameters and asymptotically stable eigenvalues of the denominator polynomial. If better approximations are needed, the analysis can be extended to complete filters of higher order. In figure 2 we find such higher order evaluations of

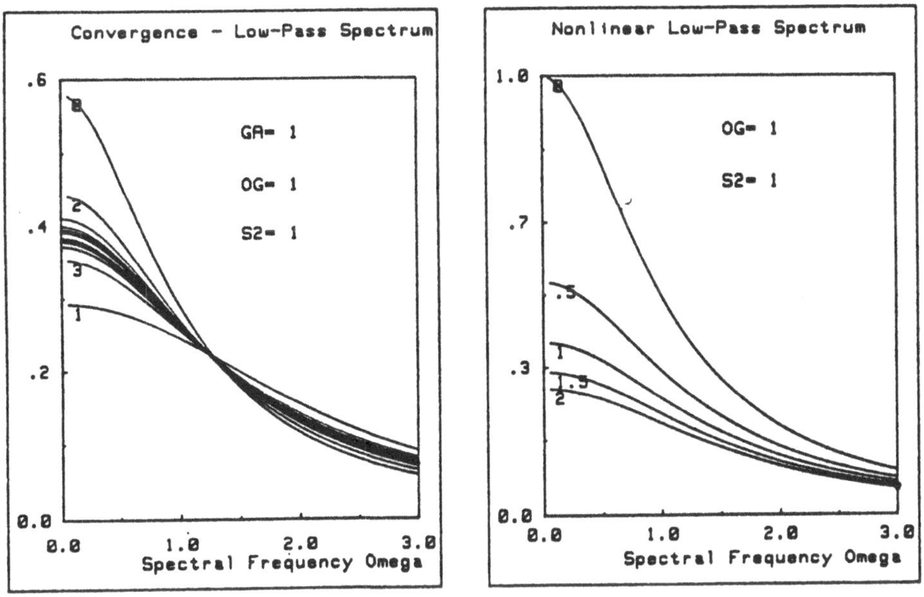

Fig. 2: Nonlinear low-pass spectra - influence of approximation order and of nonlinearity

(11) and (12). On the left side of the figure, the nonlinear parameter is fixed with $\gamma = 1$ in order to show the convergence velocity with increasing approximations $n = 0,1,2,..,10$. Hereby, $n = 0$ denotes the linear case. On the right side, we see the influence of the nonlinear parameter γ which reduces the power spectrum in the lower frequency range and removes it to the higher system frequencies ω_g, $3\omega_g$, $5\omega_g$,... as indicated in the power spectrum expansion (11).

3. STOCHASTIC DUFFING OSCILLATOR

More insight into nonlinear effects of stochastic dynamic systems is given by the Duffing oscillator under white noise.

$$\left.\begin{array}{ll} \ddot{X}_t + 2D\omega_1\dot{X}_t + \omega_1^2 X_t + \gamma\omega_1^2 X_t^3 = \sigma \dot{W}_t, & D,\ \omega_1,\ \gamma > 0, \\ X_{1,t} = X_t, \quad X_{2,t} = \dot{X}_t, & E(\dot{W}_t\dot{W}_s) = \delta(t-s). \end{array}\right\} \qquad (16)$$

Since there are now two state processes, the displacement $X_{1,t}$ and the velocity $X_{2,t}$, the correlation analysis, explained before, has to be extended, correspondingly.

3.1 Two-dimensional moments

For the stochastic Duffing oscillator (16) we investigate the increments of even sum powers of the two state processes.

$$\left.\begin{array}{l} d(X_{1,t}^{2n+2-k}X_{2,t}^k) = (2n+2-k)X_{1,t}^{2n+1-k}X_{2,t}^k dX_{1,t} + \\ + kX_{1,t}^{2n+2-k}X_{2,t}^{k-1}dX_{2,t} + \tfrac{1}{2}\sigma^2 k(k-1)X_{1,t}^{2n+2-k}X_{2,t}^{k-2}dt, \end{array}\right\} \begin{array}{l} n = 0,1,2,... \\ k = 0,..2n+2. \end{array} \qquad (17)$$

Inserting the increments $dX_{1,t}$ and $dX_{2,t}$, defined by (16), and taking the expectation we obtain the following moment equations.

$$\left.\begin{array}{l} \dot{m}_{2n+2-k,k} = (2n+2-k)m_{2n+1-k,k+1} + \tfrac{1}{2}\sigma^2 k(k-1)m_{2n+2-k,k-2} - \\ - 2D\omega_1 k m_{2n+2-k,k} - k\omega_1^2 m_{2n+3-k,k-1} - k\gamma\omega_1^2 m_{2n+5-k,k-1}, \\ E(X_{1,t}^{2n+2-k}X_{2,t}^k) = m_{2n+2-k,k}, \quad n = 0,1,2,..,\ k = 0,1,..2n+2. \end{array}\right\} \qquad (18)$$

In the stationary case, the time derivatives are vanishing. Consequently, the time-invariant moments are determined by the right-hand side of (18).

$$\left[\begin{array}{ccc|ccccc|c} 0 & -2 & 0 & 0 & 0 & 0 & 0 & 0 & \cdots \\ \omega_1^2 & 2D\omega_1 & -1 & \gamma\omega_1^2 & 0 & 0 & 0 & 0 & \\ 0 & 2\omega_1^2 & 4D\omega_1 & 0 & 2\gamma\omega_1^2 & 0 & 0 & 0 & \\ \hline 0 & 0 & 0 & 0 & -4 & 0 & 0 & 0 & \\ 0 & 0 & 0 & \omega_1^2 & 2D\omega_1 & -3 & 0 & 0 & \\ -\sigma^2 & 0 & 0 & 0 & 2\omega_1^2 & 4D\omega_1 & -2 & 0 & \\ 0 & -3\sigma^2 & 0 & 0 & 0 & 3\omega_1^2 & 6D\omega_1 & -1 & \\ 0 & 0 & -6\sigma^2 & 0 & 0 & 0 & 4\omega_1^2 & 8D\omega_1 & \cdots \\ \hline \vdots & & & & & & & \vdots & \end{array}\right] \left[\begin{array}{c} m_{2,o} \\ m_{1,1} \\ m_{o,2} \\ \hline m_{4,o} \\ m_{3,1} \\ m_{2,2} \\ m_{1,3} \\ m_{o,4} \\ \hline \vdots \end{array}\right] = \left[\begin{array}{c} 0 \\ 0 \\ \sigma^2 \\ \hline 0 \\ 0 \\ 0 \\ 0 \\ 0 \\ \hline \vdots \end{array}\right] . \tag{19}$$

These moment equations do not lead to convergent solutions if we evaluate them by the same procedure applied in the scalar case. Already, the first approximation performed by means of the four submatrices noted in (19) leads to

$$m_{2,o} = \sigma_1^2 \, \frac{1+3D^2-9\gamma\sigma_1^2/4}{1+3D^2+3\gamma\sigma_1^2(1+12D^2)/4} \,, \qquad \sigma_1^2 = \frac{\sigma^2}{4D\omega_1^3} \,, \tag{20}$$

and therewith to negative square means for an increasing nonlinear parameter γ. As shown in figure 4, the power or normal moment equations (19) can only be applied in the sense of an asymptotic method restricted to a small parameter $\gamma \ll 1$.

However, the evaluation of (19) becomes convergent if we make use of the a-priori knowledge that in the stationary case, both state processes $X_{1,t}$ and $X_{2,t}$ are statistically independent and that $X_{2,t}$ is normally distributed with the variance $E(X_{2,t}^2) = \sigma^2/(4D\omega_1)$. This follows from the exact solution of the associated Fokker-Planck equation which is derivable if there exists a potential function [12]. Therewith, the equation (19) or (18) can be reduced to the investigation of the stationary displacement process $X_{1,t}$ itself, simply by setting $k = 1$ and $\dot{E}(.) = 0$ in (18).

$$E(X_{1,t}^{2n+2-k}X_{2,t}^{k}) = E(X_{1,t}^{2n+2-k})E(X_{2,t}^{k}) = m_{2n+2-k}n_k,$$

$$-(2n+1)\sigma_1^2\, m_{2n} + m_{2n+2} + \gamma\, m_{2n+4} = 0, \quad n = 0,1,2,\ldots \qquad (21)$$

It is obvious that (21) coincides identically with the moment equations (6) of the scalar problem. Consequently, we can apply the same solution technique in obtaining the same results, already shown in the figure 1.

3.2 Correlations of the Duffing oscillator

If we restrict our interest to asymptotic properties, the analyis above can be extended to the investigation of stationary correlation functions of the Duffing oscillator (16). They are determined by the following differential equations for positive time shifting $\tau \geq 0$.

$$R'_{2n+1-k,k}(\tau) = (2n+1-k)R_{2n-k,k+1}(\tau) - k2D\omega_1R_{2n+1-k,k}(\tau) - \qquad (22)$$
$$- k\omega_1^2R_{2n+2-k,k-1}(\tau) - k\gamma\omega_1^2R_{2n+4-k,k-1}(\tau) + \tfrac{1}{2}\sigma_1^2k(k-1)R_{2n+1-k,k-2}(\tau).$$

The equations (22) are derived via the increment $d(X_{1,t}^{2n+1-k}X_{2,t}^{k})$ multiplied by $X_{1,s}$ for $t = s + \tau$. Taking the expectation we obtain the correlation function valid for positive difference times $\tau \geq 0$.

$$R_{2n+1-k,k}(\tau) = E(X_{1,s}X_{1,t}^{2n+1-k}X_{2,t}^{k}), \quad \begin{matrix} n = 0,1,2,3,\ldots\ldots \\ k = 0,1,\ldots\ 2n+1. \end{matrix} \qquad (23)$$

Subsequently, we apply the Laplace transformation to (22). Under the initial conditions

$$R_{2n+1-k,k}(0) = E(X_{1,s}^{2n+2-k}X_{2,s}^{k}) = m_{2n+2-k}\, n_k,$$

$$\mathfrak{L}\{R'_{2n+1-k,k}(\tau)\} = s\, X_{2n+1-k,k}(s) - m_{2n+2-k}\, n_k \qquad (24)$$

the time derivative in (22) goes over to the algebraic expression, noted above. Finally, we find the power spectra of the nonlinear oscillator simply by adding the conjugate complex Laplace transforms of X(s).

$$S_{2n+1-k,k}(\omega) = X_{2n+1-k,k}(j\omega) + X_{2n+1-k,k}(-j\omega), \qquad s = j\omega \qquad (25)$$

Herein, the Laplace variable s is replaced by the circle frequency ω.

It is suitable to write the associated equations in matrix form.

$$\left.\begin{aligned}&(\mathbf{A} + s\mathbf{I})\mathbf{X}(s) = \mathbf{m},\\ &\mathbf{S}(\omega) = \mathbf{X}(j\omega)+\mathbf{X}(-j\omega), \qquad \mathbf{Q}(\omega) = j\,[\mathbf{X}(j\omega)-\mathbf{X}(-j\omega)].\end{aligned}\right\} \tag{26}$$

Herein, $\mathbf{m}$ is the moment vector and $\mathbf{X}(s)$ contains all Laplace transforms of the nonlinear correlation functions.

$$\begin{aligned}&\mathbf{m}^T = (m_2,\ 0 \mid m_4,\ 0,\ m_2n_2,\ 0 \mid m_6,\ 0,\ m_4n_2,\ 0,\ m_2n_4,\ \ldots),\\ &\mathbf{X}^T(s) = (X_{1,o},\ X_{o,1} \mid X_{3,o},\ X_{2,1},\ X_{1,2},\ X_{o,3} \mid X_{5,o},\ \ldots\ldots).\end{aligned} \tag{27}$$

Accordingly, the correlation matrix $\mathbf{A}$ has the following form:

$$\mathbf{A} = \left[\begin{array}{cc|cccc|ccc} 0 & -1 & 0 & 0 & 0 & 0 & 0 & 0 & 0\ \ldots \\ \omega_1^2 & 2D\omega_1 & \gamma\omega_1^2 & 0 & 0 & 0 & 0 & 0 & 0 \\ \hline 0 & 0 & 0 & -3 & 0 & 0 & 0 & 0 & 0 \\ 0 & 0 & \omega_1^2 & 2D\omega_1 & -2 & 0 & \gamma\omega_1^2 & 0 & 0 \\ -\sigma^2 & 0 & 0 & 2\omega_1^2 & 4D\omega_1 & -1 & 0 & 2\gamma\omega_1^2 & 0 \\ 0 & -3\sigma^2 & 0 & 0 & 3\omega_1^2 & 6D\omega_1 & 0 & 0 & 3\gamma\omega_1^2 \\ \hline 0 & 0 & 0 & 0 & 0 & 0 & 0 & -5 & 0 \\ 0 & 0 & 0 & 0 & 0 & 0 & \omega_1^2 & 2D\omega_1 & -4 \\ 0 & 0 & -\sigma^2 & 0 & 0 & 0 & 0 & 2\omega_1^2 & 4D\omega_1\ldots \\ \vdots & & & & & & & \vdots & \end{array}\right]. \tag{28}$$

Furthermore, $\mathbf{S}(\omega)$ denotes the vector of all real-valued power spectra and $\mathbf{Q}(\omega)$ contains the associated imaginary parts. Therewith, we obtain the final vector equations

$$[\mathbf{A}^2 + \omega^2\mathbf{I}]\ \mathbf{S}(\omega) = 2\ \mathbf{A}\ \mathbf{m}, \qquad [\mathbf{A}^2 + \omega^2\mathbf{I}]\ \mathbf{Q}(\omega) = 2\omega\ \mathbf{m} \tag{29}$$

for both, the real-valued auto-spectra and the imaginary-valued cross-spectra.

As already mentioned, we restrict our interest to asymptotic results in order to show first effects in the frequency distribution of the Duffing oscillator under white noise. For this purpose, we apply the initial conditions of the second approximation

$$\left.\begin{aligned} &m_2 = \sigma_1^2 \frac{1+5\gamma_1}{1+8\gamma_1}, \qquad \gamma m_4 = \sigma_1^2 \frac{3\gamma_1}{1+8\gamma_1}, \qquad \gamma_1 = \gamma\sigma_1^2 = \frac{\gamma\sigma^2}{4D\omega_1^3}, \\ &X_{1,o}(s) = \frac{\sigma_1^2}{1+8\gamma_1} \sum_{i=o}^{5} Z_i s^i / (\sum_{i=o}^{6} T_i s^i), \qquad S_{1,o} = X_{1,o} + X_{1,o}^* . \end{aligned}\right\} \quad (30)$$

and evaluate the correlation matrix **A** up to the first four submatrices, noted in (28). This yields a linear complete filter of sixth order with the following parameters:

$$\left.\begin{aligned} &T_6 = 1, \qquad T_5 = 14D\omega_1, \qquad T_2 = (19+236D^2+96D^4)\omega_1^4, \\ &T_4 = (11+68D^2)\omega_1^2, \qquad T_1 = 6D\omega_1^5[13+16(\gamma_1+2D^2)], \\ &T_3 = 4D\omega_1^3(23+34D^2), \qquad T_o = 9\omega_1^6[1+8D^2(1+2\gamma_1)]. \end{aligned}\right\} \quad (31)$$

$$\left.\begin{aligned} &Z_5 = 1+5\gamma_1, \qquad Z_1 = 3\omega_1^4[3+6\gamma_1-10\gamma_1^2+4D^2(16+8D^2+69\gamma_1+40D^2\gamma_1)], \\ &Z_4 = 14D\omega_1(1+5\gamma_1), \qquad Z_o = 18D\omega_1^5[1+2\gamma_1+10\gamma_1^2+8D^2(1+4\gamma_1)], \\ &Z_3 = \omega_1^2[10+47\gamma_1+68D^2(1+5\gamma_1)], \quad Z_2 = 4D\omega_1^3[20+91\gamma_1+34D^2(1+5\gamma_1)]. \end{aligned}\right\} \quad (32)$$

The denominator polynomial (31) is asymptotically stable. However, the

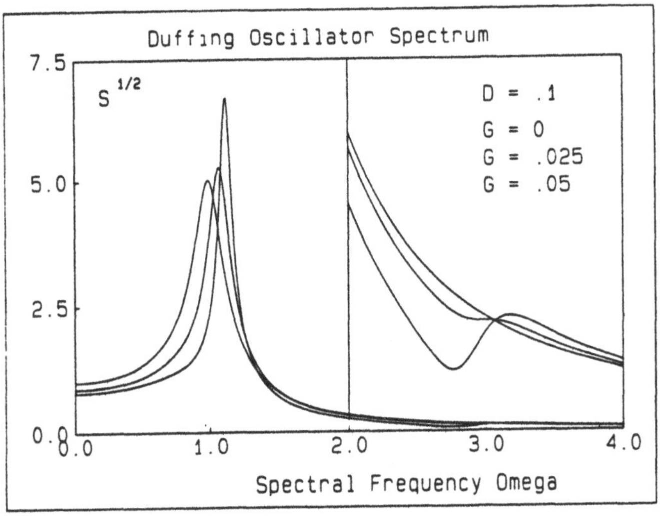

Fig. 3: Power spectrum of the Duffing oscillator
- nonlinear resonances near ω_1 and $3\omega_1$

numerator polynomial leads to negative-valued power spectra for increasing parameter γ. Therefore, in a first numerical evaluation of $S_{1,0}(\omega)$, we neglect the quadratic terms $O(\gamma_1^2)$ in Z_0 and Z_1. As shown in figure 3, the nonlinear parameter γ shifts the main resonance near ω_1 to higher frequencies and generates a further resonance at $\omega = 3\omega_1$. Naturally, these are well expected results of the stochastic Duffing oscillator.

3.3 Hermite moments

As already mentioned, the complete moment equation (19) does not lead to convergent results. Since the spectral matrix (28) has an analogue structure, we expect the same divergence if the spectral analysis, explained above, is continued to higher approximations. For this reason, we have to restart the moment analysis by means of an orthogonalizing procedure which avoids the non-convergent evaluation, performed before. The orthogonalizing is based on Hermite polynomials, already applied in [13] for the investigation of moment stability in nonlinear parametric systems. These orthogonal functions are defined by gaussian distributions. Consequently, they possess the main advantage that all stationary Hermite moments of linear stochastic systems are vanishing identically. This results in vanishing drift terms in applying a corresponding transformation of the state processes.

We start the Hermite analysis with the introduction of the normalized state processes U_t, V_t and derive the associated Ito equations.

$$\left.\begin{array}{lll} U_t = X_{1,t}\dfrac{1}{\sigma}\sqrt{2D\omega_1^3}, & dU_t = \omega_1 V_t dt, & \gamma_1 = \gamma\sigma^2/4D\omega_1^3, \\ V_t = X_{2,t}\dfrac{1}{\sigma}\sqrt{2D\omega_1}, & dV_t = \sqrt{2D\omega_1}\, dW_t - \omega_1(2DV_t+U_t+2\gamma_1 U_t^3)dt. \end{array}\right\} \quad (33)$$

Subsequently, we introduce the Hermite polynomials $H_k(U_t)$, $H_l(V_t)$ and calculate the increment of the product of both polynomials.

$$d[H_k(U_t)H_l(V_t)] = 2kH_{k-1}H_l dU_t + 2lH_kH_{l-1}dV_t + 4D\omega_1 l(l-1)H_kH_{l-2}dt. \quad (34)$$

After the insertion of the Ito equations (33) into the increment (34), we apply the Hermite recursion formulas in order to substitute powers of U_t and V_t by corresponding Hermite polynomials. Finally, we take the expectation in (34) and go over to the stationary case $\dot{E}(.) = 0$ in arriving at

$$\left.\begin{array}{l} H_{k,l} = E[H_k(U_t)H_l(V_t)], \qquad k+l = 2,4,6,\ldots \qquad \gamma_1 = \gamma\sigma^2/4D\omega_1^3, \\ 2DlH_{k,l} + l[1+3(k+l)\gamma_1]H_{k+1,l-1} - kH_{k-1,l+1} + \\ + \tfrac{1}{2}l\gamma_1 H_{k+3,l-1} + 6\gamma_1 lk^2 H_{k-1,l-1} + 4\gamma_1 lk(k-1)(k-2)H_{k-3,l-1} = 0. \end{array}\right\} \quad (35)$$

The equations (35) determine the stationary values of the Hermite moments $H_{k,l}$. With $H_{o,o} = 1$, they have the following matrix form:

$$\left[\begin{array}{ccc|ccccc|c} 0 & -2 & 0 & 0 & 0 & 0 & 0 & 0 & \cdot\cdot \\ 1+6\gamma_1 & 2D & -1 & \tfrac{1}{2}\gamma_1 & 0 & 0 & 0 & 0 & \\ 0 & 2+6\gamma_1 & 4D & 0 & \gamma_1 & 0 & 0 & 0 & \\ \hline 0 & 0 & 0 & 0 & -4 & 0 & 0 & 0 & \\ 54\gamma_1 & 0 & 0 & 1+12\gamma_1 & 2D & -3 & 0 & 0 & \\ 0 & 48\gamma_1 & 0 & 0 & 2+18\gamma_1 & 4D & -2 & 0 & \\ 0 & 0 & 18\gamma_1 & 0 & 0 & 3+18\gamma_1 & 6D & -1 & \\ 0 & 0 & 0 & 0 & 0 & 0 & 4+12\gamma_1 & 8D & \cdot\cdot \\ \hline \vdots & & & & & & & \vdots & \end{array}\right] \left[\begin{array}{c} H_{2,o} \\ H_{1,1} \\ H_{o,2} \\ \hline H_{4,o} \\ H_{3,1} \\ H_{2,2} \\ H_{1,3} \\ H_{o,4} \\ \hline \vdots \end{array}\right] = \left[\begin{array}{c} 0 \\ -6\gamma_1 \\ 0 \\ \hline 0 \\ -24\gamma_1 \\ 0 \\ 0 \\ 0 \\ \hline \vdots \end{array}\right]. \quad (36)$$

In comparison with the power or normal moments (19), we observe a non-linear correction term already in the first submatrix of (36). Evaluating this first approximation we obtain the results $H_{1,1} = H_{o,2} = 0$ and

$$H_{2,o} = E[H_2(U_t^2)] = \frac{-6\gamma_1}{1+6\gamma_1}, \qquad \Rightarrow \qquad E(X_{1,t}^2) = \frac{\sigma^2}{4D\omega_1^2}\,\frac{1+3\gamma_1}{1+6\gamma_1}. \qquad (37)$$

Hence, the stationary square mean of the Duffing process $X_{1,t}$ is positive definite for all parameter values $0 \leq \gamma < \infty$. Moreover, the first approximation (37) coincides with the corresponding result in

(13) if both are expanded for small parameters $\gamma_1 \ll 1$.

In figure 4, we show a numerical evaluation of these Hermite moments for increasing approximation order n = 1,2,..,7 applying the complete equations (35). We observe a fast convergence velocity in the parameter range $0 \leqslant \gamma \leqslant 2$. In opposite to this, the analogue evaluation of the complete equations (19) of the power or normal moments diverges for increasing nonlinear parameter γ. It is important to note that the moment equation (19) is only valid in the asymptotic sense for small values $\gamma \ll 1$ and for limited numbers of submatrices or limited approximation order.

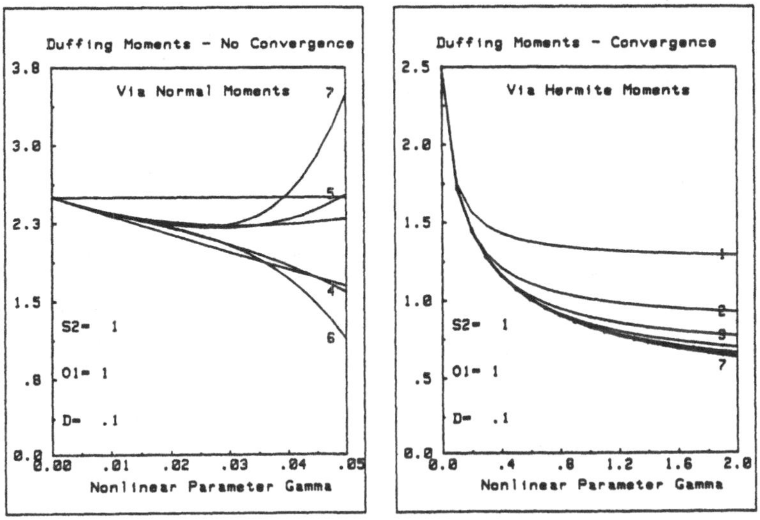

Fig. 4: Convergence properties of Hermite moments in comparison with those of normal moments

4. LINEAR IDENTIFICATION OF ROAD SPECTRA

In the following we pass over to the problem of spectrum identification for those cases where no mechanical models are available but only spectral measurements which are given by piecewise or pointwise valuable frequency distributions of stationary random processes. Such measurements are shown in figure 5. They represent power spectra of typical road surfaces [14] plotted over the way frequencies in a double logarithm scale. Close by, we see a correspondingly constructed power spectrum skechted by interrupted lines and its approximation by a complete filter of the forms (1) und (2). Such a modelling is important for road-vehicle systems where we are interested in on-line simulations of statistically representative road surfaces.

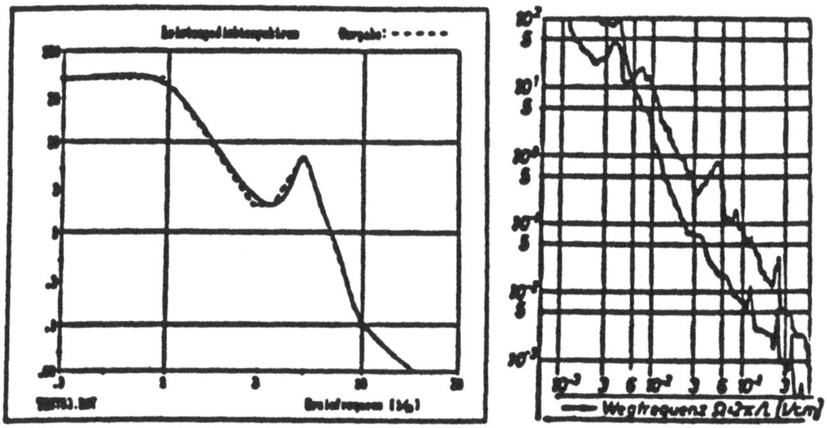

Fig. 5: Measured power spectra of typical road surfaces and their identification by complete filters

Accordingly, a given road spectrum $\bar{S}(\omega)$ is defined by piecewise analytic functions $\bar{S}_i(\omega)$ which are linear in a double logarithm scaling.

$$\left.\begin{aligned} &\bar{S}_i(\omega) = X_i\ (\omega/\omega_i)^{a_i}, \qquad a_i = \frac{\lg(X_i/X_{i-1})}{\lg(\omega_i/\omega_{i-1})}, \qquad i = 1,2,\dots k, \\ &\bar{S}_1(\omega) = X_1, \quad (0 \le \omega \le \omega_1), \qquad \bar{S}_{k+1}(\omega) = X_k(\omega/\omega_k)^{-2}, \quad (\omega_k \le \omega \le \infty). \end{aligned}\right\} \qquad (38)$$

This poygonal spectrum is determined by the corner frequencies ω_i and the associated spectral values X_i. Since road spectra are not measurable in the low and high frequency ranges, we are allowed to assume special slope parameters a_i for both ranges. It is suitable to fix the first parameter with $a_1 = 0$ and the last one with $a_{k+1} = -2$. The latter choice implies the physical existence of approximated spectra. Finally, we apply the Fourier transformation

$$\bar{R}(\tau) = \frac{1}{\pi} \int_0^{+\infty} \bar{S}(\omega) \cos(\omega\tau)\, d\omega \tag{39}$$

in order to get the correspondingly given correlation function $\bar{R}(\tau)$ for an equivalent description of the random road trace.

4.1 Suboptimal correlation parameters

As already mentioned in chapter 1, we are looking for an suboptimal processing to calculate the unknown parameters T_i and S_i of the complete filter (1) or (2) in such a way that the given spectrum $\bar{S}(\omega)$ and correlation $\bar{R}(\tau)$ are approximated in the L_2-sense. For this purpose, we go over to the correlation differential equation associated to (1).

$$\sum_{i=0}^{n} T_i R_x^{(i)}(\tau) = 0, \qquad (\ \tau \geqslant 0\ ), \qquad\qquad R_x(\tau) = E(X_t X_{t+\tau}). \tag{40}$$

It is derived from (1) by replacing the time variable t by $s=t+\tau$ and multiplying (1) by X_t. Taking the expectation, we obtain the homogeneous differential equation (40) where (i) denotes time derivatives with respect to a positive correlation time τ. It is worth to note that (40) represents a causal dynamic system, which is independent on the filter parameters S_i. Solving (40) under the initial conditions $\bar{R}^{(i)}(0)$, we know $R_x(\tau)$ and are then able to calculate the correlation function $R_y(\tau)$ in which we are interested for the approximation of the given correlation $\bar{R}(\tau)$. According to (1), $R_y(\tau)$ is a linear combination of all filter correlations $R_x^{(i)}(\tau)$.

$$R_y(\tau) = E(Y_t Y_{t+\tau}) = \sum_{k,l}^{n-1} S_k S_l (-1)^l\, R_x^{(k+l)}(\tau), \quad (\ \tau > 0\ ). \tag{41}$$

Consequently, both correlations $R_x(\tau)$ and $R_y(\tau)$ satisfy the same differential equation (40). Because of its homogeneity, both are completely differentiable and integrable with respect to positive times.

Therefore, we are allowed to insert the given correlation function $\bar{R}(\tau)$ into the equation (40) in order to calculate adapted filter parameters T_i. Taking the least square in the time range $0 \leq \tau < \infty$, we obtain the following minimal problem [13]:

$$\int_0^\infty \Big[\sum_{i=0}^{n} T_i\, \bar{R}^{(i)}(\tau)\Big]^2 d\tau = \text{Min.!} \qquad (\text{e.g. } T_o = 1\ ), \tag{42}$$

$$\sum_{j=0}^{n} \int_0^\infty \bar{R}^{(i)}(\tau)\, \bar{R}^{(j)}(\tau) d\tau\, T_j = 0, \qquad (\ i = 0,1,\dots n). \tag{43}$$

It is solved by differentiating with respect to T_i yielding the linear algebraic system (43) for the determination of the parameters T_i. The equations (43) are inhomogeneous if one parameter is fixed e.g. with the first one $T_o = 1$. Since the given correlation function $\bar{R}(\tau)$ is asymptotically stable and the applied system (40) is causal, the denominator parameters T_i calculated from (43) are asymptotically stable, as well, i. e. all eigenvalues of the T-polynomial have negative real parts.

The evaluation of the integrals in (43) is performed by splitting up the given correlation function $\bar{R}(\tau)$ into two parts in correpondence to the frequency ranges $0 \leq \omega \leq \omega_k$ and $\omega_k \leq \omega \leq \infty$.

$$\bar{R}(\tau) = \bar{R}_s(\tau) + \bar{R}_p(\tau), \qquad \bar{R}_p(\tau) = \frac{1}{\pi} X_k \omega_k^2 \int_0^{+\infty} \frac{1}{\omega^2} \cos(\omega\tau) d\tau. \tag{44}$$

According to the integral representation of sine integrals [15], the last correlation $\bar{R}_p(\tau)$ can be expressed by the form

$$\bar{R}_p(\tau) = \frac{1}{\pi} X_k \omega_k \int_0^{\pi/2} \exp(-\omega_k \tau \cos x) \cos(x + \omega_k \tau \sin x)\, dx. \tag{45}$$

The identity of (44) and (45) follows by one differentiation of both representations. Obviously, in the form (45), the given correlation $\bar{R}_p(\tau)$ is now completely differentiable. It has the initial values

$$\left.\begin{aligned} &\bar{R}_p^{(1)}(0) = -\, X_k \omega_k^2/2, && \bar{R}_p^{(2n+3)}(0) = 0, \\ &\bar{R}_p^{(2n)}(0) = (-1)^{n+1}\, X_k\, \omega_k^{2n+1}/[\pi(2n-1)], && n = 0,1,2,\dots\ . \end{aligned}\right\} \tag{46}$$

The associated squared correlation integrals are calculated to

$$\int_0^\infty [\bar{R}_p(\tau)]^2 d\tau = \omega_k X_k/(6\pi), \qquad \int_0^\infty [\bar{R}_p^{(1)}(\tau)]^2 d\tau = \omega_k^3\, X_k^2/(2\pi), \tag{47}$$

$$\int_0^\infty [\bar{R}_p^{(n+1)}(\tau)]^2 d\tau = X_k^2\ \omega_k^{2n+3}/[2\pi(2n-1)], \qquad n = 1,2,3,\dots . \quad (47)$$

All other integrals of $\bar{R}(\tau)$ and corresponding initial conditions are calculable applying the Parseval equation and partial integration rules.

4.2 Spectral moments of the filter

Knowing the filter parameters T_i from the above calculation, we are now able to solve the differential equation (40) under the given initial conditions in order to determine an approximating correlation function $R(\tau)$. The solution is performed by means of the Laplace transformation $X(s)$ of $R(\tau)$ which leads to a rational function $Z(s)/T(s)$ in dependence on the Laplace variable $s = j\omega$.

$$\mathfrak{L}\{R(\tau)\} = X(s) = \frac{Z(s)}{T(s)}, \qquad R^{(i)}(0) = \bar{R}^{(i)}(0), \quad i = 0,1,..n-1. \quad (48)$$

This is a homogeneous and causal solution valid for $\tau \geqslant 0$. For the negative time range, we find corresponding results. Since both are vanishing outside their definition ranges, they can simply be added in order to get the complete non-causal correlation function. In the frequency range, this implies the addition of the conjugate complex transfer functions $X(j\omega)$ and $X(-j\omega)$ which leads to a real-valued approximation spectrum $S(\omega)$ of the given road spectrum $\bar{S}(\omega)$.

$$S(\omega) = \frac{Z(j\omega)}{T(j\omega)} + \frac{Z(-j\omega)}{T(-j\omega)} = \frac{S(j\omega)\ S(-j\omega)}{T(j\omega)\ T(-j\omega)}, \quad (49)$$

$$Z(j\omega)T(-j\omega) + Z(-j\omega)T(j\omega) = S(j\omega)S(-j\omega). \quad (50)$$

On the left side of (49), we find the frequency polynomials $Z(s)$ and $T(s)$ calculated from the homogeneous problem. On the riht side there are the unknown filter parameters S_i of the inhomogeneous part of the filter (1). Consequently, they are calculable by equating (50). We perform this by the determination of the zeros of the polynomials $Z(j\omega)$ and $T(j\omega)$. Then, the equation (50) is separable into a linear form from which the numerator parameters S_i are determined by a simple comparison of all coefficients.

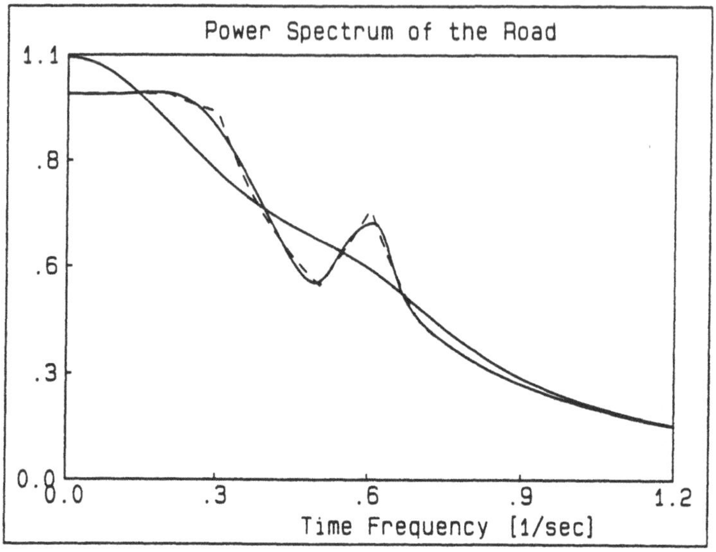

Fig.6: A typical road spectrum in linear scaling and its approximation by filters of order 3 and 7

In figure 6 we show a numerical example of the identification procedure, described above. The given road spectrum is denoted by interrupted lines in a linear scaling. It has the following corner frequency values:

$$\omega_1 = 0.2, \quad \omega_1 = 0.3, \quad \omega_2 = 0.5, \quad \omega_4 = 0.6, \quad \omega_5 = 0.7,$$
$$X_1 = 1, \quad X_2 = 0.95, \quad X_2 = 0.5, \quad X_4 = 0.7, \quad X_5 = 0.4.$$

The approximating spectra are denoted by drawn out lines. They belong to identified filters of the order n=3 and n=7. In the latter case, the filter parameters T_i and S_i are calculated to

$$T_o = 0, \quad T_1 = 8.186, \quad T_2 = 33.896, \quad T_3 = 92.119, \quad T_4 = 182.028,$$
$$T_5 = 283.530, \quad T_6 = 256.457, \quad T_7 = 248.221, \quad S_o = 0.999, \quad S_1 = 6.592,$$
$$S_2 = 22.311, \quad S_3 = 48.121, \quad S_4 = 96.021, \quad S_5 = 80.042, \quad S_6 = 109.892.$$

The denominator polynomial is asymptotically stable. The numerator parameters S_i are real-valued, i.e. the identified filter is physically existent. The results in figure 6 show a high convergence rate with increasing approximation order n.

Finally, it should be mentioned that the initial conditions (48) represent the spectral moments of the given power spectrum which are identically satisfied by the applied identification procedure.

$$\bar{R}(\tau) = \frac{1}{2\pi}\int_{-\infty}^{+\infty} \bar{S}(\omega)\exp(j\omega\tau)d\omega, \qquad \bar{R}^{(i)}(0) = \frac{1}{2\pi}\int_{-\infty}^{+\infty} (j\omega)^i\bar{S}(\omega)d\omega. \quad (51)$$

Therefore, we observe best approximations of the third order filter in the high frequency range and less good results in the lower one. The identification leads to non-real parameters S_i if the approximated spectrum becomes negative. However, this is only possible if we change the special frequency decaying $a_{k+1} = -2$ outside the approximation range. Naturally, an overspill effect may also occur if spectral jumps are to be identified. If the spectral value at the jump is too low, the approximated spectrum becomes negative which is a consequence of the L_2-norm. Hence, the total identification procedure remembers to the well-known solution technique of boundary value problems. In that sense, the spectral moments (51) are boundary conditions which have to be identically satisfied by suitably chosen S_i parameters. Only, the T_i parameters are free. They are applied to optimize the dynamic behaviour of the identified correlation function $R(\tau)$.

4.3 Orthogonalizing identification

The applied identification procedure can be improved in several steps. Instead of the differentiations in (40) or in (51), we can introduce integrations of the form:

$$R^{(-1)}(\tau) = \int_{+\infty}^{\tau} R(\sigma)d\sigma, \qquad R^{[-(k+1)]}(\tau) = \frac{1}{k!}\int_{+\infty}^{\tau} (\tau-\sigma)^k R(\sigma)d\sigma. \quad (52)$$

Hence, we get integro-differential correlation equations for (40)

$$\sum_{i=-n}^{m} T_i\, R^{(i)}(\tau) = 0, \qquad (\ \tau \geqslant 0, \quad m,n > 0\ ) \quad (53)$$

if we use both, the differentiated correlation functions (i>0) as well as the integrated ones ($i<0$). Of course, we can apply to (53) the same identification procedure as shown in the forgoing chapter.

For example, let us consider a given power spectrum which is constantly distributed in the low frequency range and has a parabolic decay in the upper range.

$$\left.\begin{aligned} &\bar{S}(\omega) = X_u \quad \text{for } |\omega| \leq \omega_k, \qquad \bar{S}(\omega) = X_o(\omega_k/\omega)^2 \quad \text{for } |\omega| \geq \omega_k, \\ &\bar{R}(\tau) = \frac{1}{\pi}\int_0^{\omega_k} X_u \cos\omega\tau \, d\omega + \frac{1}{\pi}\int_{\omega_k}^{\infty} X_o(\omega_k/\omega)^2 \cos\omega\tau \, d\omega, \quad (-\infty<\tau<+\infty). \end{aligned}\right\} \quad (54)$$

Obviously, the associated correlation function $\bar{R}(\tau)$ is valid in the entire time range. This is the reason why $\bar{R}(\tau)$ is not integrable or differentiable. However, similar as shown in (45), it is possible to derive the equivalent representation [15]

$$\begin{aligned} \bar{R}(\tau) = \frac{1}{\pi}\,\omega_k\,[X_u \int_0^{\pi/2} &\exp(-\omega_k\tau\cos x)\,\cos(x-\omega_k\tau\sin x)\,dx + \\ + X_o \int_0^{\pi/2} &\exp(-\omega_k\tau\cos x)\,\cos(x+\omega_k\tau\sin x)\,dx] \end{aligned} \quad (55)$$

which is completely integrable and differentiable for $\tau \geq 0$.

$$\left.\begin{aligned} &\bar{R}^{(n)}(0) = \frac{1}{\pi}(-1)^n \omega_k^{n+1}[\frac{X_u}{n+1}\sin(n+1)\frac{\pi}{2} + \frac{X_o}{n-1}\sin(n-1)\frac{\pi}{2}], \\ &\int_0^{+\infty}[\bar{R}^{(n)}(\tau)]^2 d\tau = \frac{1}{2\pi}\,\omega_k^{2n+1}\left\{\frac{X_u^2}{|2n+1|} + \frac{X_o^2}{|2n-3|} - \frac{X_u X_o}{2n-1} * \right. \\ &\qquad \left. * \,[\mathrm{sign}(2n+1) + \mathrm{sign}(2n-3)]\right\}, \quad n = 0, \mp 1, \mp 2, \mp 3, \ldots \end{aligned}\right\} \quad (56)$$

All other integrals needed for the least square procedure (43) are calculated by the recurrence formula

$$\int_0^{\infty}\bar{R}^{(n)}(\tau)\bar{R}^{(n+m)}(\tau)d\tau = -\bar{R}^{(n)}(0)\bar{R}^{(n+m-1)}(0) - \int_0^{\infty}\bar{R}^{(n+1)}(\tau)\bar{R}^{(n+m-1)}(\tau)d\tau.$$

This follows by partial integration.

The next figure shows two typical examples, the first one has a steady frequency distribution. The second one has a spectral jump at the corner point ω_k. Both spectra are identified by complete filters of the order $n = 3$, 5 and 7. For the correlation equation (53), we chose $m = n$, i. e. the same number of differentiations and integrations of the given correlation function. This special choice possesses the advantage that the associated initial conditions or spectral moments are weighting the high frquencies and the low distributions, simultaneously. Therefore, the filter spectra, evaluated in figure 7, are exact in both frequency ranges in the sense of a Taylor series and of an asymptotic expansion, respectively. Approximations are only observable in the middle frequency range with typical overspill effects for the jump example [13].

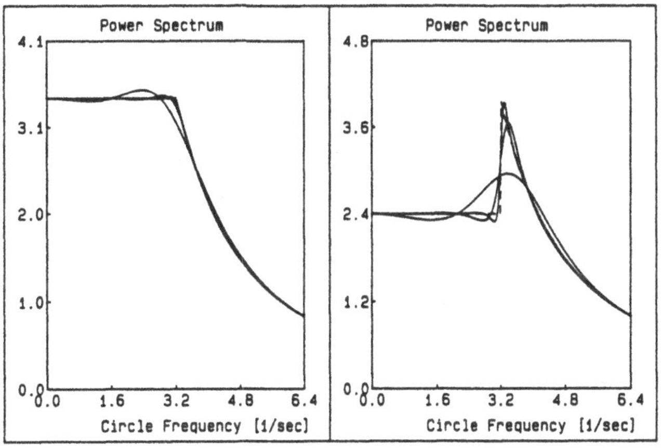

Fig. 7: Power spectra with two different distributions
- identified filters of the order n = 3, 5 and 7

We come now to a last improvement. It avoids the fixation of the inhomogeneous parameter T_o. Each other choice leads to different parameters T_i and to different approximations of the spectra, as well. For this purpose, finally, we rewrite the minimal problem (42) into the following form:

$$\sum_{i,j=-n}^{m}\sum T_i T_j \int_o^{\infty} \bar{R}^{(i)}(\tau)\bar{R}^{(j)}(\tau)d\tau = \sum_{i,j=-n}^{m}\sum t_i t_j \; \alpha_{ij} \; = \text{Min.!} \tag{57}$$

Herein, the T_i parameters are normalized by

$$T_i = \frac{t_i}{\sqrt{\int_o^{\infty}[\bar{R}^{(i)}(\tau)]^2 d\tau}} \; , \; \alpha_{ij} = \frac{\int_o^{\infty} \bar{R}^{(i)}(\tau)\bar{R}^{(j)}(\tau)d\tau}{\sqrt{\int_o^{\infty}[\bar{R}^{(i)}(\tau)]^2 d\tau \int_o^{\infty}[\bar{R}^{(j)}(\tau)]^2 d\tau}} \tag{58}$$

so that the correlation integrals in (57) go over to the coefficients α_{ij}, noted in (58). It is obvious that the minimal formulation (57) represents a quadratic form which has the best minimal solution if we calculate the eigenvalues of the matrix (α_{ij}).

$$\sum_{i,j=-n}^{m}\sum t_i t_j \alpha_{ij} = \text{Min.!} \quad \Rightarrow \quad \sum_{i=-n}^{m} (\alpha_{ij} - \lambda\, \delta_{ij})\, t_j = 0. \tag{59}$$

The eigenvalue λ of interest is positve definite and decreases for increasing approximation order. Consequently, an identical solution to the given spectrum is reached if the first positive λ value is vanishing. In this case, further approximations lead to double eigenvalues, then to triple eigenvalues and so on. These properties can easily be checked in studying the simple problem of a given low-pass spectrum.

In conclusion, it should be remarked that the application of the orthogonalizing method combined with an optimal choice of the numbers n and m in (59) is only sensitive if the approximation has to be stopped at a low filter order n+m. Then, the calculated suboptimal parameters T_i and S_i can be used as starting values for a nonlinear investigation of the original minimal problem, introduced in the very beginning.

$$\int_{-\infty}^{+\infty}\left\{ \frac{S(j\omega)S(-j\omega)}{T(j\omega)T(-j\omega)} - \bar{S}(\omega)\right\}^2 d\omega = \text{Min.!} \tag{60}$$

Naturally, the spectral moments, mentioned above, are not satisfied by this nonlinear formulation. Moreover, the obtained results are better, but they do not guarantee the asymptotic stability of the filter nor its physical existence.

5. CONCLUSION

The contribution is dealing with the filter identification of given stationary random processes which are defined by a nonlinear stochastic model or by piecewise analytic spectral distributions. In both cases, we apply linear identification methods based on correlation equations and corresponding moments.

The infinite moment equations of nonlinear stochastic systems have the same structure as the associated correlation equations in the Laplace domain. Their analysis can considerably be improved by applying orthogonalizing procedures based on Hermite polynomials. They lead to convergent results if the linear part of the dynamic system is asymptotically stable.

In case that a dynamic model is not available, the filter identification is performed by optimizing the homogeneous correlation operator. The inhomogeneous parameters of the filter follow from the spectral moments of the given spectra. The identification is once more improved by an orthogonalizing methods which leads to best eigenvalues and to main directions in the quadratic forms of the operator parameters associated to the complete filter.

REFERENCES

1. Caughey,T.K.: Nonlinear theory of random vibrations,- in: Advances in Applied Mechanics, 11, Academic Press, 1971, pp. 209-253.
2. Crandall, S.H.: Perturbation techniques for random vibrations of nonlinear systems. J. Acoustical Soc. Am., 35, 1963, pp.1700-1705.
3. Wedig, W.: Regions of instability for a linear system with random parametric excitation, in: Lecture Notes in Math., 294, Springer-Verlag, New York, 1972, pp. 160-172.
4. Booton, R.C.: The analysis of nonlinear control systems with random inputs. IRE Trans. Circuit Theory, 1, 1954, pp. 32-34.
5. Caughey, T.K.: Equivalent linearization technique. J. Acoustical Soc. Am., 35, 1963, pp. 1706-1711.
6. Bogdanoff, J.L. and Kozin, F.: Moments of the output of linear random systems. J. Acoustical Soc. Am., 34, 1962, pp. 1063-1066.
7. Wu, W.F. and Lin, Y.K.: Cumulant-neglect closure for non-linear

oscillators under random parametric and external excitations. Int. J. Non-linear Mech., 19, 1984, pp.349-362.

8. Stratonovich, R.L.: Topics in the Theory of Random Noise, 1, Gordon and Breach, New York, 1963.
9. Ariaratnam, S.T.: Dynamic stability of a column under random loading, in: Pro. Int. Conf. on Dynamic Stability of Structures, Pergamon, New York, 1967, pp. 267.
10. Caughey, T.K.: Derivation and application of the Fokker-Planck equation, J. Acoust. Soc. Am., 35, 1963, p. 1683.
11. Ito, K.: On a formula concerning stochastic differentials. Nagoya Math. J. Jap., 3, 1951, p.55.
12. Parkus, H.: Random Processes in Mechanical Sciences, CISN Courses and Lectures, No 9, Springer-Verlag, Wien, 1969.
13. Wedig, W.: Stochastische Schwingungen - Simulation, Schätzung und Stabilität, ZAMM 67, 4, 1987, T34-T42.
14. Braun, H.: Spektrale Dichte von Fahrbahnunebenheiten. Tagungsband der VDI-Tagung Akustik und Schwingungstechnik, VDI-Verlag, Düsseldorf, 1970, pp. 539-546.
15. Magnus, W., Oberhettinger, F., Soni, R.P.: Formulas and Theorems for the Special Functions of Mathematical Physics, Springer-Verlag, Berlin, 1966, pp.347-348.

MODELING, ANALYSIS AND ESTIMATION OF VEHICLE SYSTEMS

W. Schiehlen
University of Stuttgart, Stuttgart, FRG

ABSTRACT

Ground vehicles are subject to random vibrations in the vertical direction due to stochastic guideway irregularities. The vertical motions of vehicles affect the ride comfort of passengers and goods as well as the ride safety. The ride comfort can be evaluated from the accelerations of the human body while the criterion on ride safety follows from the dynamic wheel loading. A thorough dynamical analysis of vehicle vibrations requires a mathematical modeling of the guideway roughness, the vehicle itself and the sensation of the human being. The resulting stochastic differential equations may be analysed by spectral density or covariance methods, respectively. However, it is often cumbersome and difficult to obtain reliable information on the parameters of the total dynamical system. Therefore, parameter estimation methods have to be applied in vehicle dynamics for improved results.

1. INTRODUCTION

The stochastic analysis of vehicle vibrations and their consequences to the ride characteristics requires an overall investigation of guideway roughness, vehicle dynamics and human response to vibration exposure. Moreover, the parameters of the guideway and vehicle equations of motion have to be estimated from measurements resulting in the inverse problem with respect to the analysis.

Many aspects of these problems have been treated in numerous publications, see Robson [1], Dinca [2] and Hedrick [3]. Two fundamental methods are available for the investigation of vehicle vibrations: the spectral density analysis in the frequency domain and the covariance analysis in the time domain. In these lectures both methods will be discussed and compared from a general point of view. A first comparison for the most simple vehicle model has been presented in [4]. The theoretical background for complex vehicles traveling with constant speed can be found in [5]. First results of the parameter estimation using time domain methods are due to Kozin and Kozin [6], Wedig [7] and Ref. [8]. Applications to vehicle systems have been presented by Kallenbach [9] and Schäfer [10].

2. MODELING OF VEHICLE SYSTEMS

The primary models of vehicle systems refer to guideway roughness and the vehicle components represented by the elements of multibody systems.

2.1 Excitation by Guideway Roughness

The guideway roughness results with respect to the four wheels of a two-axle vehicle in four random processes summarized in a 4x1-vector excitation process

$$\xi(s) = [\xi_{fr}(s) \quad \xi_{fl}(s) \quad \xi_{rr}(s) \quad \xi_{rl}(s)]^T \tag{1}$$

where s is the space coordinate of the longitudinal motion of the vehicle, Fig.1. Since the 2x1-vector process $\xi_r(s)$ of the rear axle is only delayed by the axle distance Δs, the 2x1-vector process $\xi_f(s)$ of the front axle is discussed in more detail.

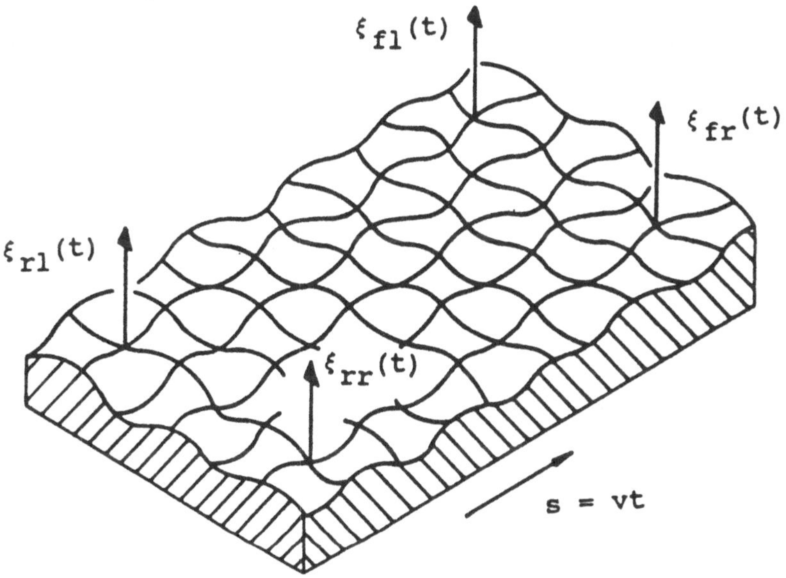

Fig. 1 Guideway roughness and random excitation process

Numerous measurements have shown that the guideway roughness can be characterized by a Gaussian, ergodic and stationary process with zero mean value:

$$m_\xi = E\{\xi(s)\} = 0 \quad , \tag{2}$$

$$R_\xi(\bar{s}) = E\{\xi(s)\, \xi^T(s-\bar{s})\} \quad , \tag{3}$$

$$S_\xi(\Omega) = \frac{1}{2\pi} \int_{-\infty}^{+\infty} R_\xi(\bar{s})\, e^{-i\Omega\bar{s}}\, d\bar{s} \quad , \tag{4}$$

$$P_\xi = R_\xi(0) = \int_{-\infty}^{+\infty} S_\xi(\Omega) d\Omega \quad , \tag{5}$$

where m_ξ is the mean value, $R_\xi(\bar{s})$ the correlation matrix, $S_\xi(\Omega)$ the spectral density matrix and P_ξ the covariance matrix including the variances or the squared standard deviations, respectively, as diagonal elements. Usually, the stochastic properties of guideway processes are found by measurements and they are presented as spectral densities. For vibration analysis the measurements are approximated in the frequency domain by polynomials or in the space domain by shape filters, respectively. A first order model of a scalar process reads for example as

$$S_\xi(\Omega) = \frac{Q_s}{2\pi} \frac{1}{\Omega_0^2 + \Omega^2} \tag{6}$$

or

$$\frac{d\xi}{ds} = -\Omega_0\, \xi(s) + w(s) \tag{7}$$

where Ω_0 is a constant filter coefficient and Q_s is the intensity of a white noise process $w(s)$ with $R_w(\bar{s}) = Q_s \delta(\bar{s})$.

In vehicle dynamics, the right and left trace of a spatial guideway represent stochastically dependent processes. Thus, the spectral density matrix follows as

$$S_{\xi f}(\Omega) = \begin{bmatrix} S_r(\Omega) & S_{rl}(\Omega) \\ S_{rl}(\Omega) & S_l(\Omega) \end{bmatrix} \quad . \tag{8}$$

On the other hand, for a spatial guideway the medium trace ξ_M and the trace difference $2\xi_D$ are found to be independent processes. With the 2x1-vector of the front axle

$$u_f(s) = [\xi_{fM}(s) \quad \xi_{fD}(s)] \tag{9}$$

the kinematical relations, Fig. 2, can be expressed as

$$\xi_f(s) = H\, u_f(s) \quad , \quad H = \begin{bmatrix} 1 & 1 \\ 1 & -1 \end{bmatrix} \quad . \tag{10}$$

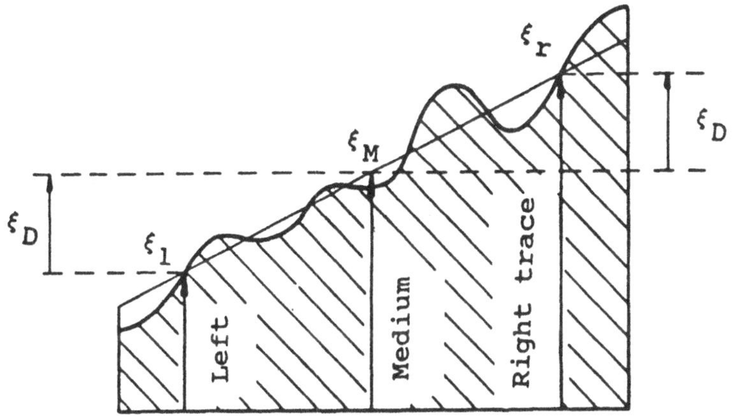

Fig. 2 Cross-section of guideway at front or rear axle respect.

Then, it yields for the spectral density matrix

$$S_{uf}(\Omega) = \begin{bmatrix} S_M(\Omega) & 0 \\ 0 & S_D(\Omega) \end{bmatrix} \tag{11}$$

and the corresponding first order shape filter equations reads as

$$\frac{du_f}{ds} = F_s\, u_f(s) + G_s\, w(s) \tag{12}$$

where F_s and G_s are constant diagonal 2x2-matrices and w(s) means a stochastically independent 2x1-white noise process with a diagonal 2x2-intensity matrix. Even with a simple first order shape filter as used in (12), guideway irregularities can be modeled very well.

For the rear axle the shape filter approach (10), (12) can also be applied, if the space delay Δs is introduced in the white noise excitation process. Then it remains

$$\xi(s) = [\xi_f(s) \quad \xi_r(s)] = [\xi_f(s) \quad \xi_f(s-\Delta s)] \ . \tag{13}$$

This is also true for an arbitrary number of axles.

Random vehicle vibrations take place in the time domain. Therefore the guideway roughness models have to be transformed from the space to the time domain. For a constant vehicle speed v one obtains

$$s = vt \quad , \quad \omega = v\Omega \quad . \tag{14}$$

Then it follows for (8), e.g.,

$$S_{\xi f}(\omega) = \frac{1}{v}\, S_{\xi f}(\Omega) \tag{15}$$

and the shape filter (12) is transformed to

$$\frac{du_f}{dt} = F_t\, u_f(t) + G_t\, w(t) \tag{16}$$

with the white noise process w(t) characterized by $R_w(\bar{t}) = Q\delta(\bar{t})$ and $F_t = vF_s$, $G_t = vG_s$, $Q = Q_t = \frac{1}{v} Q_s$. Furthermore, in the rear axle equations the space distance Δs has to be replaced by a time interval Δt acccording to (14).

Some special cases are now easily obtained. For $F_t = 0$ it is obvious from (16) that the excitation process is white velocity noise, a very handy approximation. With $\xi_D(t) \equiv 0$ it follows from (10) that the traces are parallel, $\xi_{fr}(t) = \xi_{fl}(t)$, $\xi_{rr}(t) = \xi_{rl}(t)$.

If a variable vehicle speed $v(t) = \text{const}$ is considered, then the description (15) in the frequency domain is no longer consistent. However, the time domain representation (16) remains consistent even if the coefficients become time-variant and the stochastic processes prove to be nonstationary.

2.2 Mathematical Models of Vehicles

The choice of the mathematical model for a vehicle depends on the technical problem under consideration. There are three models in mechanics available for different geometry and stiffness properties, Table 1. The final decision for one or more of these models can be made with respect to the technical problem.

The equations of motion read for nonlinear ordinary multibody systems as

$$M(y)\,\ddot{y}(t) + k(y,\dot{y},t) = q(y,\dot{y},t) \quad , \tag{17}$$

for linear finite element systems as

$$M\,\ddot{y}(t) + D\,\dot{y}(t) + K\,y(t) = h(t) \tag{18}$$

and for linear continuous systems in modal representation as

$$\ddot{y}(t) + 2\,\mathrm{diag}\,(\delta_j)\,\dot{y}(t) + \mathrm{diag}\,(\omega^2_j)\,y(t) = f(t) \tag{19}$$

where $y(t)$ is the corresponding position vector and $q(t)$, $h(t)$ and $f(t)$ are excitation vectors. The equations are completed by the coefficient matrices M, D, K representing inertia, damping and stiffness.

Model	Geometry	Stiffness
Multibody System (MBS)	Complex	Inhomogen
Finite Element System (FES)	Complex	Homogen
Continuous System (COS)	Simple	Homogen

Table 1. Mathematical models for vehicles

For the dynamical analysis of vehicle vibrations, with respect to ride comfort and ride safety, vehicles are mainly modeled as multibody systems. This means that all parts of the vehicle are considered as rigid bodies with inertia interconnected by rigid bearings. The bodies are subject to additional applied forces and torques by supports, springs, dampers and servomotors without inertia, Fig. 3. The method of multibody systems is very well developed, a recent state of the art is given by Haug [12] and in Ref. [13].

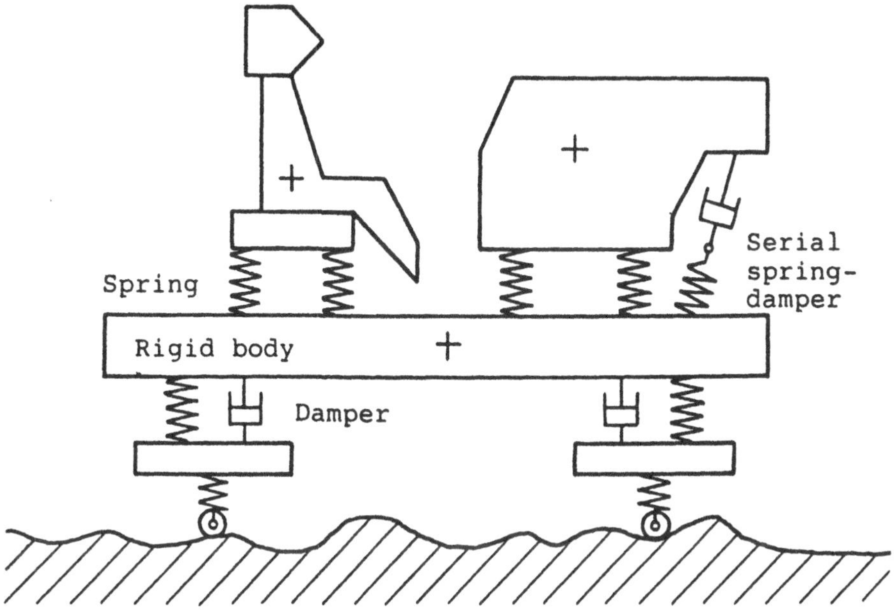

Fig. 3 Vehicle model as multibody system

A vehicle modeled as multibody system is uniquely characterized by f generalized coordinates according to the f degrees of freedom. The nonlinear equations of motion can be written as a fx1-vector differential equation in the more detailed form

$$M(y)\,\ddot{y}(t) + k(y,\dot{y}) = q(y,\dot{y},s,t) \tag{20}$$

where M is the fxf-inertia matrix, k and q are fx1-vectors of coriolis and generalized forces, respectively, y(t) is the fx1-position vector of the generalized coordinates and s(t) is a ρx1-vector summarizing state variables of additional forces. Equation (20) has to be supplemented by a ρx1-vector differential equation

$$\dot{s} = \dot{s}(y,\dot{y},s,t) \tag{21}$$

The generalized forces can be interpreted as follows:

i) position-dependent forces $q_1(y)$ due to nonlinear springs and weight,

ii) velocity-dependent forces $q_2(y,\dot{y})$ due to nonlinear damping and friction,

iii) general forces $q_3(y,s)$ generated, e.g., by a serial spring - damper configuration,

iv) time-dependent forces $q_4(t)$ due to stochastic guide- way excitations.

Equations (20) and (21) can be summarized in a nx1-vector differential equation

$$\dot{x} = a(x,t) \tag{22}$$

using the nx1-state vector

$$x(t) = [y^T(t) \quad \dot{y}^T(t) \quad s(t)^T]^T \tag{23}$$

and a nonlinear nx1-vector function $a(x,t)$.

Often the kinematics of a vehicle can be linearized while nonlinear force characteristics have to be regarded. Then, the state equation (22) can be rewritten as

$$\dot{x}(t) = A\, x(t) + f(x(t)) + B\, \xi(t) \quad . \tag{24}$$

Here, A is the nxn-system matrix, $f(x)$ an odd nonlinear nx1-vector function without linear terms, B represents the nx4-excitation matrix and $\xi(t)$ the stochastic 4x1-vector process. With respect to (10) and (13) the excitation term in (24) can also be written as

$$B\, \xi(t) = B_f\, H\, u(t) + B_r\, H\, u(t-\Delta t) \quad . \tag{25}$$

This relation can also be extended to an arbitrary number of axles.

The equations of motion (20) of complex multibody systems may be found by paper and pencil or by computer-aided analysis. In Ref. [14] the equations of motion are shown obtained by the symbolical formalism NEWEUL .

On the one hand, (24) and (25) represent a stochastically excited state equation and the response $x(t)$ can be found by stochastic analysis. On the other hand, the matrices A, B may depend on s unknown parameters,

$$A(p) = \sum_{j=1}^{s} p_j A_j \quad ,$$

$$B(p) = \sum_{j=1}^{s} p_j B_j \quad , \tag{26}$$

where p is the sx1-parameter vector. Then, from the measurement of response $x(t)$ the unknown parameters can be estimated using (24) and (25), again.

3. STOCHASTIC ANALYSIS OF VEHICLE SYSTEMS

Mechanical systems under stochastic excitation are represented by filter and state equations subject to white noise. Moreover, the response of the mechanical system may affect additionally creature and environment. Then, additional dynamical phenomena has to be considered using more differential equations. This will be shown for vehicle systems.

3.1 Human Sensation of Random Vibrations

The guideway roughness generates mechanical random vibrations of the vehicle acting on the human body. Therefore, the stochastic analysis has to be applied not only to the vehicle but also to the passenger. The evaluation of the human sensation depends on the state and excitation vector processes which are assumed Gaussian, ergodic and stationary.

The objectively measurable mechanical vibrations affecting the guideway and the vehicle can be evaluated by scalar vibration variables

$$w_k(t) = c_k^T \dot{x}(t) + d_k^T \xi(t) \quad , \quad k = 1, 2, \ldots \tag{27}$$

where c_k and d_k are the corresponding weighting vectors. Typical vibration variables in vehicle dynamics are the car body acceleration and dynamical wheel load variation. By definition, the scalar vibration variables $w_k(t)$ are also Gaussian, ergodic and stationary processes characterized by the mean value

$$m_{wk} = c_k^T m_x + d_k^T m_\xi = 0 \tag{28}$$

and the variances

$$\sigma_{wk}^2 = c_k^T P_x c_k + 2c_k^T P_{x\xi} d_k + d_k^T P_\xi d_k \quad . \tag{29}$$

The covariance matrices P are defined by (5) with respect to the processes $x(t)$ and $\xi(t)$. The covariance matrix $P_{x\xi}$ represents the coupling of these processes. If the excitation follows from white noise guideway roughness, $P_\xi \to \infty$, then only vibration variables with $d_k = 0$ are admissible. This may be a restriction for the dynamical wheel load computation in the ride safety analysis.

The human sensation of mechanical vibrations differs from the objectively measurable vibrations. Numerous physiological investigations have shown that the subjective sensation is proportional to the acceleration and depends on the dynamics of the human body. Furthermore, it is assumed that the human sensation can be characterized by scalar perception variables $\bar{w}_k(t)$ depending on the position of the passenger and on the objectively measurable vibration variables $w_k(t)$. The relation between the perception variables $\bar{w}_k(t)$ and the vibration variables $w_k(t)$ can be represented by shape filters using frequency response methods or differential equations in the time domain. However, the perception variables $\bar{w}_k(t)$ are also stochastic processes to be characterized by the corresponding variance $\sigma^2_{\bar{w}k}$. In the national standard [15] the human statement to vibration exposure is characterized by the number K of the perception proportional to the variance of the corresponding perception variable $\bar{w}_k(t)$.

With respect to vehicle vibrations only the vertical acceleration $a(t)$ will be used. Then, the perception follows as

$$K = \sigma^2_{\bar{a}} = \alpha^2 \int_{-\infty}^{+\infty} |f_a(\omega)|^2 \, S_a(\omega) \, d\omega \tag{30}$$

where $\alpha = 20 \; s^2/m$ is a constant, $f_a(\omega)$ is the frequency response of the vertical sensation and $S_a(\omega)$ is the spectral density of the vertical acceleration $a(t)$. The frequency response $f_a(\omega)$ is given by the international standard [16] and shown in Fig. 4. In the time domain the frequency response is replaced by a second order shape filter

$$\bar{a}(t) = \bar{h}^T \bar{v}(t) \quad , \quad \dot{\bar{v}}(t) = \bar{F} \bar{v}(t) + \bar{g} \, a(t) \quad , \tag{31}$$

where $\bar{v}(t)$ is the 2x1-filter state vector, $\bar{F}$ the 2x2-filter matrix, $\bar{g}$ the 2x1-input vector and $\bar{h}$ the 2x1-output vector. The following numbers may be used

$$\bar{F} = \begin{bmatrix} 0 & 1 \\ -1200 \; 1/s^2 & -50 \; 1/s \end{bmatrix}, \quad \bar{g} = \begin{bmatrix} 0 \\ 1 \end{bmatrix}, \quad \bar{h} = \begin{bmatrix} 500 \; 1/s^2 \\ 50 \; 1/s \end{bmatrix} \tag{32}$$

resulting in an approximated frequency response also shown in Fig. 5. Now, the perception reads as

$$K = \sigma_{\bar{a}}^2 = \alpha^2 \bar{h}^T P_{\bar{v}} \bar{h} \tag{33}$$

where $P_{\bar{v}}$ is the 2x2-covariance matrix of the shape filter process $\bar{v}(t)$.

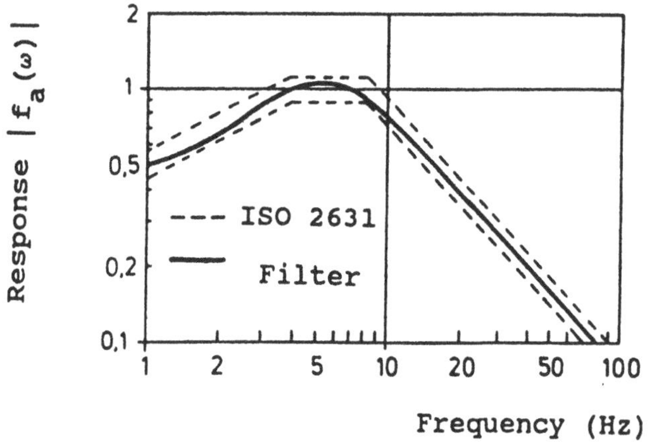

Fig. 4 Frequency response of human sensation

The final results (30) and (33) are given in the frequency domain and the time domain, respectively. It turns out that an infinite integral has to be evaluated in the frequency domain while in the time domain only an algebraic matrix operation is required.

3.2 Stochastic Analysis of Total System

The modeling of guideway roughness, vehicle dynamics and human sensation results in a strongly coupled total system under stochastic excitation. Since the excitation and sensation shape filters are linear subsystems, the dynamical equations of motion will be statistically linearized. Then, the state equation (24) can be replaced by

$$\dot{x}(t) = [A + N(P_x)]\, x(t) + B\, \xi(t) \tag{34}$$

where N is a nxn-matrix describing the nonlinearities as a function of the covariance matrix P_x of the system.

3.2.1 Spectral Density Analysis

The spectral density analysis begins with the guideway roughness (15) and the state equation (34) resulting in the spectral density of the state vector process

$$S_x(\omega;P_x) = \hat{F}(\omega;P_x)\; B\; S_\xi(\omega)\; B^T\; \hat{F}^T(-\omega;P_x) \tag{35}$$

where

$$\hat{F}(\omega;P_x) = (i\omega E - A - N(P_x))^{-1} \tag{36}$$

is the nxn-frequency response matrix of (34) and

$$S_\xi(\omega) = \text{diag}\,\{S_{\xi f}(\omega) \quad S_{\xi f}(\omega)\; e^{-i\omega\Delta t}\} \tag{37}$$

represents the 4x4-matrix of the guideway roughness according to (15). Furthermore, the nxn-covariance matrix

$$P_x = \int_{-\infty}^{+\infty} S_x(\omega;P_x)\, d\omega \tag{38}$$

is required. It turns out, that the relations (35) and (38) are strongly nonlinear; they can be solved only by a combined numerical integration and iteration procedure where the infinite frequency domain in (38) has to be restricted and divided in a finite number of steps. Thus, numerical errors are preprogrammed.

In a next step, the spectral density of the vibration variables (27) has to be computed,

$$S_{wk}(\omega) = (c_k^T \hat{F} B + d_k^T) S_\xi (B^T \hat{F}^T c_k + d_k) \tag{39}$$

Finally, the perception K follows from (30) by another numerical integration with the above mentioned errors and problems.

The spectral density analysis is very well-known and often used in engineering problems. However, if characteristic numbers like the perception are needed, or nonlinearities are at hand, respectively, the spectral density analysis suffers from numerical problems due to infinite integrals in the frequency domain.

3.2.2 Covariance Analysis

The covariance analysis begins with the representation of the total system by an extended state equation

$$\dot{\hat{x}}(t) = \hat{A}(\hat{P}) \hat{x}(t) + \hat{B}_f w(t) + \hat{B}_r w(t-\Delta t) \tag{40}$$

where

$$\hat{x}(t) = [u_f(t) \quad u_r(t) \quad x(t) \quad \bar{v}(t)]^T \tag{41}$$

is the extended (6+n)-state vector and $\hat{A}$, $\hat{B}_f$, $\hat{B}_r$ are corresponding matrices following from (16), (34), (25) and (30). Then, an extended (6+n)x(6+n)-covariance matrix $\hat{P} = P_{\hat{x}}$ can be found from the algebraic Ljapunov matrix equation

$$\hat{A}(\hat{P}) \hat{P} + \hat{P} \hat{A}(\hat{P}) + \hat{B}_f Q \hat{B}_f^T + \hat{B}_r Q \hat{B}_r^T$$

$$+ \hat{\Phi}(\Delta t) \hat{B}_r Q \hat{B}_f^T + \hat{B}_f Q \hat{B}_r^T \hat{\Phi}^T(\Delta t) = 0 \tag{42}$$

where the fundamental matrix

$$\hat{\Phi}(\Delta t) = e^{\hat{A}\Delta t} = \hat{E} + \hat{A}\Delta t + \frac{1}{2}\hat{A}^2(\Delta t)^2 + \ldots \tag{43}$$

may be obtained by series expansion. The Ljapunov equation (38) results immediately in the covariance matrix, without any numerical integration. However, the Ljapunov equation (42) is also strongly nonlinear; it can be solved iteratively as pointed out in [5].

In a second step the variances of the vibration variables (27) are obtained by the algebraical matrix operations (29). Further, the unknown matrix in (32) is immediately available as submatrix,

$$\hat{P} = \left[\begin{array}{c|c} & \\ \hline & P_{\bar{v}} \end{array}\right] \quad . \tag{44}$$

For the numerical solution of the linear part of the Ljapunov equation (42) the method of Smith [17] may be applied, since the matrix A is asymptotically stable by definition.

The stochastic analysis of mechanical systems may be performed by the spectral density analysis or the covariance analysis, respectively. The spectral density analysis is very well-known and widely applied since the spectral density of the output process can be easily obtained for linear time-invariant systems. If mechanical systems are primarily rated by the output variances, it is more accurate and economic to compute immediately the covariance matrix of the total system. The theoretical and numerical elements of the covariance analysis are available for the application to all kinds of mechanical systems.

4. ESTIMATION OF VEHICLE SYSTEMS

For the modeling of ground vehicles the method of multibody systems is well qualified. In contrary to the method of finite element systems and of continuous systems, multibody systems include also large nonlinear motions. The modeling by rigid bodies and discrete constraint and force elements fits well to the engineer's imagination. However, it is often cumbersome and difficult to estimate the parameters of the elements due to inertia, stiffness and damping properties. Traditionally, the parameters are determined by simple calculations using the information at the drawings or by special experiments after disassembling the vehicle. In every case, the parameters have major errors. Thus, there is a strong motion to apply parameter estimation methods also to vehicle dynamics.

Estimation methods have been used by Reichelt [18] for the parameter identification of the equations of motion of an automobile's lateral motion. It was found that the spectral analysis method is more accurate than the method of least squares in the time domain. The parameters of complex models for the vertical motion of automobiles were estimated in Ref. [19] with good success using the covariance analysis method in the time domain.

This approach has been recently applied by Schäfer [10] for large scale experiments with vehicles. In this lecture the parameter estimation for models of vertical vehicle vibrations are treated. In particular, estimation is considered as the inverse problem of analysis.

4.1 Vehicle Model and Analysis

In the test department the excitation of the four wheels of an automobile can be chosen independently as well as small excitation amplitudes can be adjusted. Thus, the 4x1-excitation process is simply given by the shape filter

$$\dot{u}(t) = F\,u\,(t) + g\,w\,(t) \tag{45}$$

and the nx1-state equation reads as

$$\dot{x}(t) = A(p)\,x(t) + B(p)\,u(t) \tag{46}$$

according to (16), (24) and (26), the nonlinear vector function f(x) is neglected. Shape filter and vehicle system are assumed to be asymptotically stable, $R_e\,\lambda_i(F) < 0$, $R_e\,\lambda_j(A) < 0$, $i = 1(1)4$, $j = 1(1)\,n$. Then, the system response x(t) is given as a stationary process with vanishing mean value $m_x = 0$ and a nxn-covariance matrix $C_{xx}(\tau)$ where τ means the scalar correlation time. As an extension of (42), the following Ljapunov matrix equations are obtained with $B_f = B$, $B_r = 0$, $\Delta t = 0$, written for the original matrices:

$$A\,C_{xx} + C_{xx}\,A^T + B\,C_{xu}^T + C_{xu}\,B^T = 0\,, \tag{47}$$

$$A\,C_{xu} + C_{xu}^T\,A^T + B\,C_{uu}^T + C_{uu}\,B^T = 0\,, \tag{48}$$

$$F\,C_{uu} + C_{uu}\,F^T + gg^T e^{F^T \tau} = 0\,, \tag{49}$$

where the nx4-covariance matrix $C_{xu}(t)$ and the 4x4-covariance matrix $C_{uu}(t)$ are introduced.

4.2 Covariance Methods for Estimation

The covariance methods are characterized by stochastic excitation processes and by a reduction of the influence of measurements errors due to an additional evaluation filter. The covariance methods have been mentioned by Wedig [7] and in Ref. [8]. Recently, Kallenbach [9] thoroughly established these methods.

The measuring system is shown in Fig. 5. The evaluation filter is characterized by its state equation

$$\dot{z}(t) = H\, z(t) + J\, u(t) + L\, x(t) \tag{50}$$

where $z(t)$ is the lx1-filter vector and H, J, L are constant matrices of appropriate dimension.

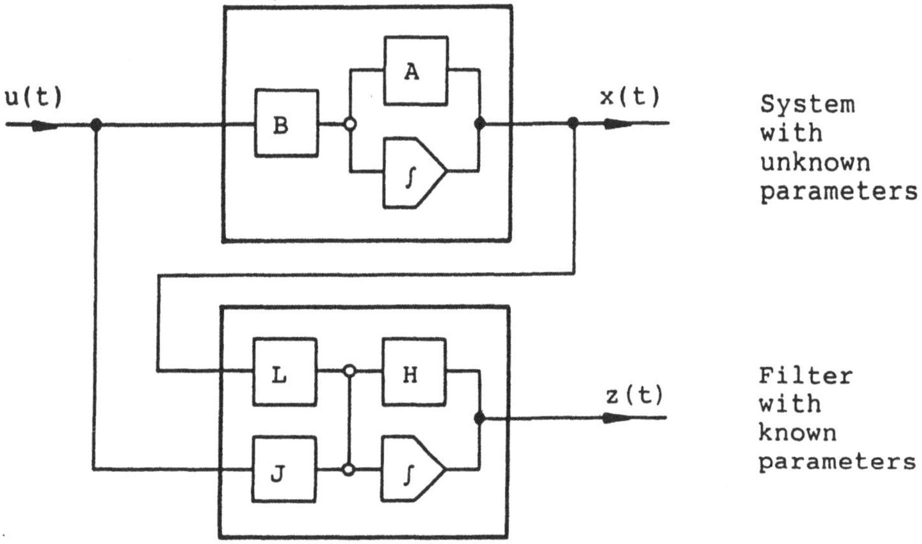

Fig. 5 Measuring system according to covariance analysis

Regarding (46) and (50), the Ljapunov matrix equation (48) has to be extended, and it remains

$$A(p)C_{xz}(\tau)+B(p)C_{uz}(\tau)+C_{xz}(\tau)H^T+C_{xu}(\tau)J^T+C_{xx}(\tau)L^T=0. \tag{51}$$

For the estimation of the unknown parameters summarized in the sx1-vector p, the covariance matrices in (51) have to be found by measurements. For this purpose, the data of the vector processes are sampled at equidistant time instants and processed according to the definition of covariance as

$$C_{xz}(\tau) = \frac{1}{N-1} \sum_{i=1}^{N} \left[x(t_i+\tau)-m_x\right] \left[z(t_i)-m_z\right]^T \tag{52}$$

where N is the number of sample instants and m the mean value of the corresponding process.

For a sufficient number of different correlation times τ_k, k = 1(1)m, a linear overdetermined system of algebraic equations is obtained from (26) and (51) which can be rewritten as

$$Wp = r . \tag{53}$$

The least square solution reads as

$$p = (W^T W)^{-1} W^T r \tag{54}$$

where the matrix W is assumed to be regular with respect to its columns. The lmnxs-matrix W and the lmnxl-vector r depend not only on the estimated variances but also on the filter matrices H, J, L which can be arbitarily chosen. In this respect the covariance methods are similar to the well-known instrumental variable methods.

The numerical algorithms for the computation of (52) and (54) require special treatment for minimal errors, see Kallenbach [9].

4.3 Experimental Results

The parameters of the front axle suspension of an automobile have been estimated by Schäfer [10] using covariance methods. Under the assumption of a parallel excitation of both front wheels, the system has two degrees of freedom, Fig. 6.

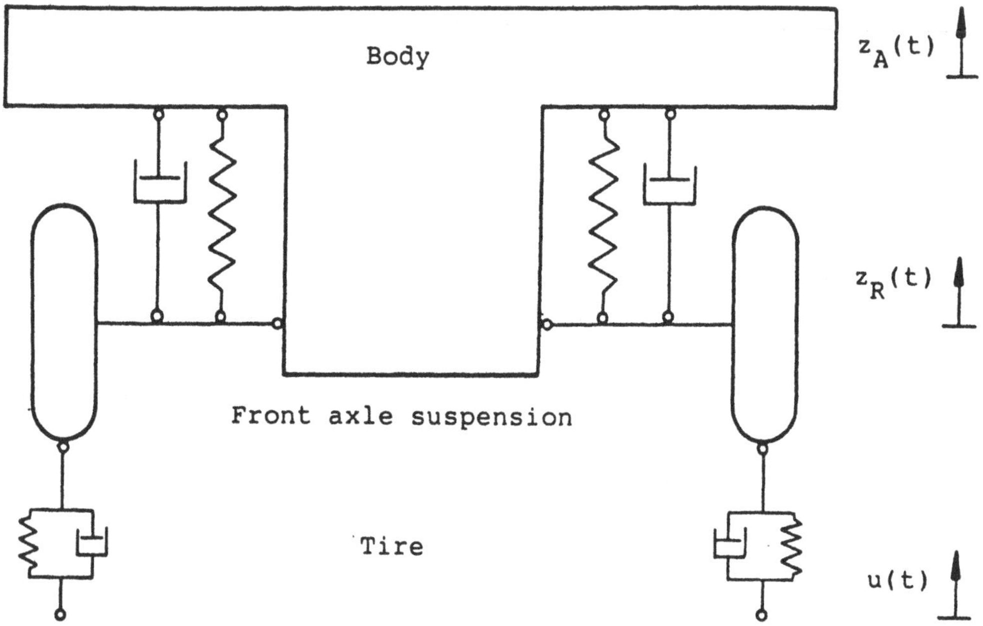

Fig. 6 Front axle suspension

The equations of motion are given as

$$\begin{bmatrix} m_A & 0 \\ 0 & m_R \end{bmatrix} \begin{bmatrix} \ddot{z}_A \\ \ddot{z}_R \end{bmatrix} + \begin{bmatrix} d_A+d_H & -d_A \\ -d_A & d_A+d_R \end{bmatrix} \begin{bmatrix} \dot{z}_A \\ \dot{z}_R \end{bmatrix}$$

$$+ \begin{bmatrix} c_A & -c_A \\ -c_A & c_A+c_R \end{bmatrix} \begin{bmatrix} z_A \\ z_R \end{bmatrix} = \begin{bmatrix} 0 & 0 \\ c_R & d_R \end{bmatrix} \begin{bmatrix} u \\ \dot{u} \end{bmatrix} \tag{55}$$

The index A characterizes the parameter of the body, the index R represents the front axle related parameters. The parameter d_H is due to the friction between the rear axle and the rear wheels appearing only during measurements in the test department. Fig. 7 shows the results of the simulation and of the experiment. The vehicle model and the estimated parameters approach the reality very well.

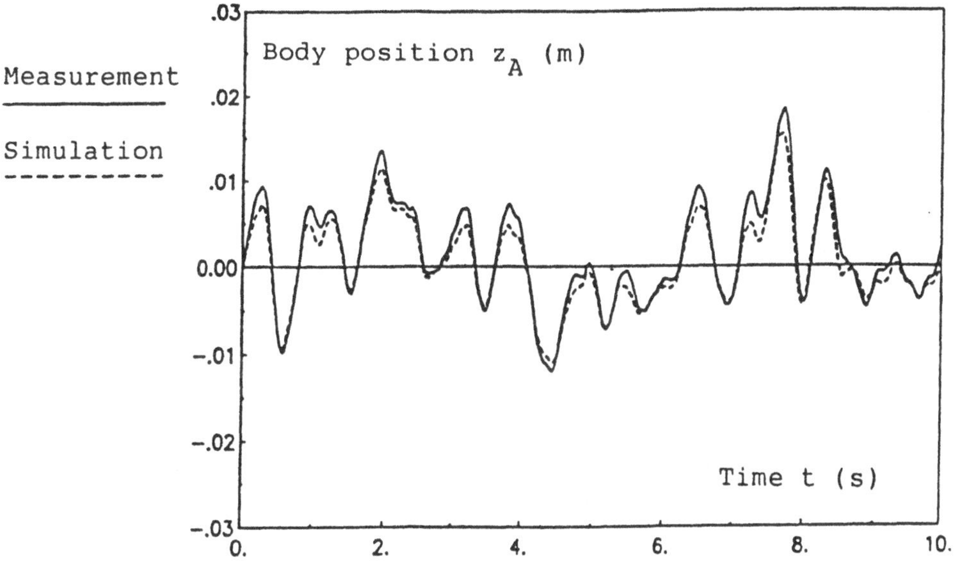

Fig. 7 Comparison of measurement and simulation

4.4 Future Trends

In vehicle system engineering the parameter estimation is of increasing importance. Detailed models with a large number of degrees of freedom result in many parameters to be identified by measurements in the test department. The excitation by stochastic processes is similar to real world of random vehicle vibrations. The covariance methods are well qualified for the needs in vehicle dynamics. Further research activities are necessary to include nonlinear phenomena and to identify the parameters of the stochastic excitation processes.

REFERENCES

[1] Robson, J.D.: Random Vibrations. Amsterdam/...: Elesvier Publ. Comp., 1984.

[2] Dinca, F. and Teodosiu: Nonlinear and Random Vibrations. New York: Academic Press, 1973.

[3] Arslan, V.A. and Hedrick, J.K.: Nonlinear Dynamic Response of a Locomotive via a Nongaussian Statistical Linearization Prodecure. In: The Dynamics of Vehicles. Wickens, A.H. (ed.) Lisse: Swets and Zeitlinger, 1982, S. 521-534.

[4] Müller, P.C. and Schiehlen, W.O.: Linear Vibrations. Dordrecht: Martinus Nijhoff Publ., 1985.

[5] Müller, P.C.; Popp, K. und Schiehlen, W.O.: Berechnungsverfahren für stochastische Fahrzeugschwingungen. Ing.Arch. 49 (1980), S. 235-254.

[6] Kozin, F. and Kozin, C.H.: A Moment Technique for System Parameter Identification. Shock and Vibration Bull. 8 (1968) S. 119-131.

[7] Wedig, W.: Stochastic Identification of Stiffness and Damping Matrices. In: CISM Course on Structural Identification and Parameter Estimation. Lecture Notes. Wien: Springer-Verlag, 1983.

[8] Weber, H.I. and Schiehlen, W.O.: A Filter Technique for for Parameter Identification. Mech. Res. Com. 10 (1983), S. 259-265.

[9] Kallenbach, R.: Kovarianzmethoden zur Parameteridentifikation zeitkontinuierlicher Systeme. Fortschr. Ber. VDI- Düsseldorf: VDI-Verlag, 1987.

[10] Schäfer, P.: Parameteridentifikation von Fahrzeugmodellen mit nichtinearen Kraftgesetzen unter Praxisbedingungen. Diplomarbeit DIPL-15, Institut B für Mechanik, Universität Stuttgart, 1987.

[11] Schiehlen, W.: Technische Dynamik. Stuttgart: Teubner, 1985.

[12] Haug, E.J. (ed.): Computer Aided Analysis and Optimization of Mechanical System Dynamics. Berlin/...: Springer-Verlag 1984.

[13] Bianchi, G. and Schiehlen, W. (eds): Dynamics of Multibody Systems. Berlin/...: Springer-Verlag 1986.

[14] Schiehlen, W.: Modelling of Complex Vehicle Systems. In: The Dynamics of Vehicles. Hedrick, J.K. (ed.) Lisse: Swets and Zeitlinger, 1984, S. 548-563.

[15] VDI-Richtlinie 2057. Beurteilung der Einwirkung mechanischer Schwingungen auf den Menschen. Düsseldorf: Verein Dt. Ing., 1975-1979.

[16] International Standard ISO 2631. Guide for the Evaluation of Human Exposure to Whole-Body Vibrations. Int. Org. Standardization 1974.

[17] Smith, P.A.: Matrix Equation XA+BX=C. SIAM J. Appl. Math. 16 (1968), S. 198-201.

[18] Reichelt, W.: Identifikationsmethoden für die Fahrzeugdynamik. Automobilt. Z. 86 (1984), S. 391-397.

[19] Schiehlen, W. and Kallenbach, R.: Modeling and Identification of Linear Multibody Systems. In: Interdynamics '85. Heimann, B.; Friedrich, H. (eds.). Karl-Marx-Stadt: Akademie der Wiss. der DDR, 1986, Part 2, S. 219-227.

STOCHASTIC MODELLING AND ANALYSIS OF FATIGUE

K. Sobczyk
Polish Academy of Sciences, Warsaw, Poland

ABSTRACT

The text contains a verified version of the lectures presented in CISM on the recent advances in stochastic modelling and analysis of fatigue fracture of materials.

In the first (introductory) part the description of basic concepts, mechanisms and empirical hypotheses is briefly presented (in order to introduce a reader into the subject). The next sections contain the discussion of the most notable recent achievements in stochastic modelling of fatigue crack growth in real materials. The so called evolutionary probabilistic models, the stochastic differential equation model (obtained via randomization of empirical equations) and random cumulative jump models recently proposed are shortly characterized.

I. FATIGUE: BASIC CONCEPTS, MECHANISMS, EMPIRICAL HYPOTHESES

1. PHYSICAL AND EMPIRICAL BACKGROUND

Fatigue of materials has been recognized as an important cause of failure of engineering structures.
In general, fatigue can be defined as a phenomenon which takes places in components and structures subjected to time - varying external actions and which manifests itself in deterioration of the material resistance to carry intendend loading.

In most applications fatigue damage results from the joint actions of the time-variable applied stress and external enviromental factors. If one wish to account for the effects of vibrations on materials and structures one of the first thoughts which comes to mind is that of mechanical fatigue.

The fatigue phenomenon is today deemed to originate in local field, in the material or, in other words, in sliding of atomic layers. This sliding is caused by a combination of dislocations and local stress concentration. It is assumed that each slip, no matter how small, is connected with a small deterioration of the material structure. Under variable stress conditions there is migration of dislocations and there is localized plastic deformation. Microscopic cracks are created which can grow and joint together to produce major cracks. <u>Nucleation</u> and <u>growth</u> of cracks are commonly regarded as basic cause of fatigue processes and ultimate fatigue failure.

The fatigue life of a structural element can be subdivided into some periods as shown below (cf.[13])

complete life time

nucleation	microcrack growth	macrocrack growth	final failure

10 A ⟶ 1μm ⟶ 10μm ⟶100μm ⟶ 1mm ⟶

nucleation period — crack growth period

The begining and the end of each period is not easily defined (exepct for the last one).

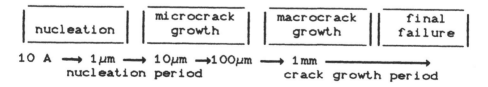

Microscopical investigations shows that nucleation period starts early in the fatigue life. It may be of importance to ask: when does a microcrack become a macrocrack? Various definitionas can be given (cf.[13])

a) a crack is a macrocrack if it is large enough to be seen by the naked eye;

b) a crack is a macrocrack if it has sufficient lenght to ensure that local conditions responsible for crack nucleation do not longer affect crack growth;

c) a crack is a macrocrack as soon as fenomenological fracture mechanics is applicable; i.e. as soon as the stress intensity factor has a real meaning for describing its growth.

Of course, none of above definitions are precise. They give however, good feeling on the content of word "a crack" as it is used in fatigue mechanics.

The number of fatigue cracks presented or nucleated in a specimen is usually very large. The experiments show, however, that even though there is a large number of cracks distributed in a specimen, always a "dominant crack" can be distinguished which is primarily responsible for the ultimate damage. So, the length of such a dominant crack is usually taken as a measure of fatigue.

Physical mechanism of fatigue crack extension on an atomic level is actually anknown, but on the larger scale some important experimental observations have been made. On the macroscopic level - which is of our concern - the feature of the crack extension are:

(i) cracks usually grow in a plane perpendicular to the main principal stress (at least as long as the crack rate is low);

(ii) crack growth is a repetitive process, which is concluded from the observation of striations on the fracture surface, where each striation has been shown to represent the crack advance for one cycle of the load;

(iii) crack growth take place mainly during a tensile (positive) part of the loading cycle; a compressive stress (negative) does not effect significantly crack growth;

(iv) crack growth is stimulated by many factors of different physical origin; in general, crack growth is a function of:

1) stress variability range: $\Delta S = S_{max} - S_{min}$

2) frequency of cyclic londing: ω

3) stress ratio: $R = \frac{S_{min}}{S_{max}}$

4) internal (residual) stress: S_r

5) material proporties (e.g. yield limit) C

6) crack length L

7) temperature θ
8) enviroment ζ (e.g. humid air, salt water).

The crack growth rate is usually denoted as $\frac{dL}{dN}$ and is understood as the crack extension ΔL of a crack (of length L) in one cycle. It has been found that a quantity which plays a main role in describing crack growth is the stress intensity factor (K).

2. STRESS INTENSITY FACTOR

The stress distribution in a cracked element can be calculated by use of the theory of elasticity (assuming linear elastic behaviour of the element). The simplest problem is in the case of an infinite sheet loaded by a tensile stress S. The solution for the stresses in the sheet are still complex, and can not be given as explicit functions. However, if we are interested in stresses in neighbourhood of the crack tip only (i.e. $r \ll L$, where L - crack length) the formulae for the stress components become relatively simple (asymptotic solution for the near field stress):

$$\sigma_{xx}\;,\;\sigma_{xy}\;,\;\sigma_{yy}$$

$$\sigma_{ij} = \frac{K}{\sqrt{2\pi r}}\, f_{ij}(\varphi)$$

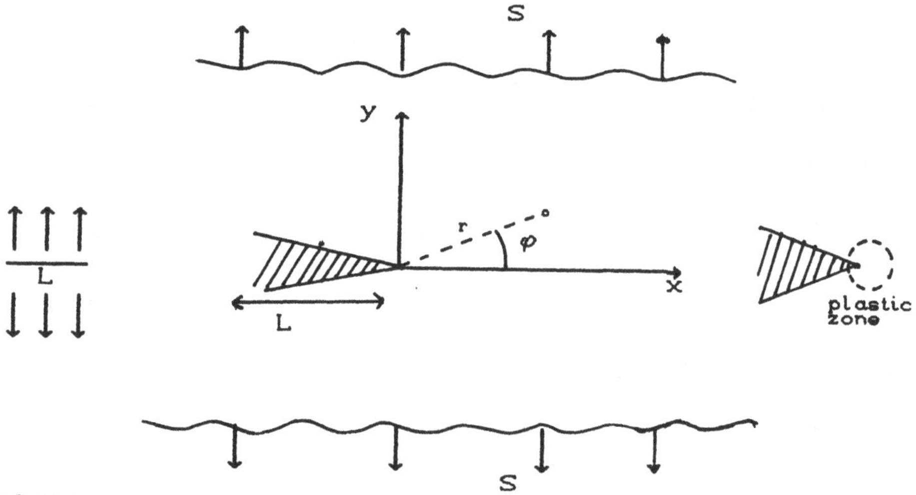

where

$$K = BS\sqrt{\pi L}$$

is called the <u>stress intensity factor</u>;
S - represents far-field stress resulting from the applied load, B = B(L) is a factor which accounts for the shape of the component and the crack geometry (in the case of an infinite sheet B = 1).

It is seen that the stress field has singularity at r=0 ($\sigma_{ij} \longrightarrow \infty$ as $r \longrightarrow 0$). Of course, infinite elastic stresses are not posiible and ductile materials will therefore show some plasticity at the tip of the crack. The size and shape of the plastic zone can be derived when the yield criterion is known.

For the case of purely plastic zone imbedded in an elastic matrix, the radius r_p of the roughly circlular plastic zone is

$$r_p = \frac{1}{2\pi}\left[\frac{K}{\sigma_y}\right]^2, \quad \text{plane stress}$$

$$r_p = \frac{1}{6\pi}\left[\frac{K}{\sigma_y}\right]^2, \quad \text{plane strain}$$

where σ_y is the yield stress. The plastic zone plays an important role in studying various phenomena associated with fatigue crack growth.

From the definition: $K = BS\sqrt{\pi L}$ it follows that a cyclic variation of S will cause a similar cyclic variation of K . Hence, the stres intensity in the crack tip will be characterized by K_{max} and K_{min} .

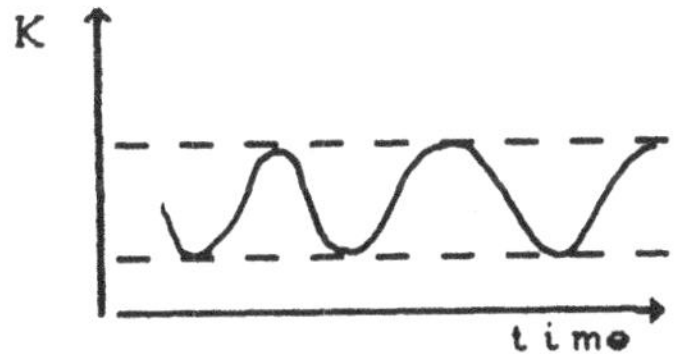

$$K_{max} = BS_{max}\sqrt{\pi L}$$

$$K_{min} = BS_{min}\sqrt{\pi L}$$

$$R = \frac{S_{min}}{S_{max}} = \frac{K_{min}}{K_{max}}$$

<u>Remark.</u> The stress intensity factor K depends on the boundary conditions (when the geometry of the body and the character of loading are given). In the literature there exist expressions for K obtained for some important geometrical configurations. Also, solving the appropriate thermoelastic problem, it is possible to determine K for thermal loading. If one wish to account for the internal stresses (generated in material during fabrication process) the stress intensity factor K should be supplemented by

K_r - the stress intensity factor associated with residual (internal) stress.

3. EMPIRICAL CRACK GROWTH EQUATIONS

Experimental investigations of fatigue crack growth have shown that the most suitable parameter for characterizing the growth rate is the stress intensity factor K. So, this factor is commonly used to describe the macroscopic crack growth. Of course, the success of application of the stress intensity factor to the analysis of growth data depends on solution for K (solution of the appropriate boundary value problems) for the complicated geometry for real structural elements. It is usually belived that such a solution can be obtained by analytical methods or by the finite element techniques.

The mathematical hypotheses concerning fatigue crack growth rate, deduced from experimental data are commonly called - the fatigue crack growth equations (or laws). The equations existing today in the literature have been identified for constant-amplitude cyclic loading. The genaral form of the developed equations is

$$\frac{dL}{dN} = f(\Delta K) \tag{1.1}$$

The most simple and generally used is the following Paris-Erdogan law

$$\frac{dL}{dN} = C\,(\Delta K)^n \tag{1.2}$$

where C and n are regarded as material constants.

It turns out, however, that constant C depends on many factors and its numerical values deviate greatly from experiment to experiment. As a matter of fact, C is not a material constant (as was initially assumed by Paris), but it depends on the mean stress R, frequency, temperature, material inhomogeneity etc. Especially, the stress ratio R has been recognized to have significant influence on fatigue crack growth. So, instead of (1.2) the equation of fatigue crack growth should be written as

$$\frac{dL}{dN} = f(R, \Delta K) \tag{1.3}$$

The experiments indicate that the effect of mean stress R under the tensile stress results in increasing the fatigue crack growth rate. So, above equation is written in more specified form

$$\frac{dL}{dN} = c\ g(R)(\Delta K)^n \tag{1.4}$$

A number of laws proposed in the literature falls into general form (1.4). For example, a relation proposed by Brock and Schijve (1963) is

$$\frac{dL}{dn} = c\left[\frac{1}{1-R}\right]^2 (\Delta K)^3 \tag{1.5}$$

Newman at al. (1972) suggested the following laws:

$$\frac{dL}{dN} = c(1 + \alpha R)^q (\Delta K)^q \quad ; \tag{1.6}$$

experiments with the aluminium welds and some steels showed that $\alpha = \frac{3}{4}$ and $3 \le q \le 7$.

Another empirical equation which is regarded as one of the most notable is the Forman equation

$$\frac{dL}{dN} = \frac{C(\Delta K)^n}{(1-R)K_c - \Delta K} \tag{1.7}$$

where K_c is a fracture toughness (a critical level of K coressponding to unstable fracture).

Considering the relation between $\frac{dL}{dN}$ and ΔK there are two limitations to ΔK:

a) if ΔK becomes too large a static failure will occur immediately, which implies that K_{max} exceeded to fracture toughness K_c which is equivalent to: ΔK exceeding $(1-R)K_c$;

b) the other limitation comes from the fact that crack will not grow if ΔK is too low; there is a threshold value ΔK_o implying that crack growth requires $\Delta K > \Delta K_o$.

Analytically, these limitations can be written as:

$$\frac{dL}{dN} \longrightarrow 0 \text{ as } \Delta K - \Delta K_o \longrightarrow 0$$

$$\frac{dL}{dN} \longrightarrow \infty \text{ as } K_c - K_{max} \longrightarrow 0$$

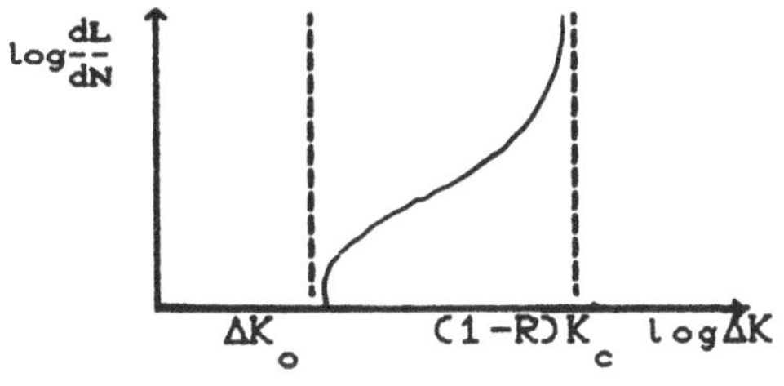

To include these limitations into crack growth equations one can write the Forman equation as

$$\frac{dL}{dN} = \frac{C(\Delta K)^n}{(1-R)(K_c - K_{max})} \quad \text{or} \quad \frac{dL}{dN} = \frac{C(\Delta K - \Delta K_c)^n}{(1-R)(K_c - K_{max})} \qquad (1.8)$$

It was found in some experiments that ΔK_o depends on the mean stress R and the relation proposed is

$$\Delta K_o = A(1 - R)^{\beta}$$

where A is constant and $\beta = 0.5 - 1.0$.

Forman equation gives good results (compared with experiments) for the crack growth rate in steel and aluminium alloys. However, when the data of tests on viscoelastic materials were analyzed using this equation, no success was achieved.

Based on the outcome of an extensive test programme covering a wide range of the stress intensity factor at various loading frequences the following relationship has been proposed by Radon J.C., Ard G., Culver E., (1974) for the polymer-like materials

$$\frac{dL}{dN} = \beta\lambda^n ,$$

$$\lambda = (K^2_{max} - K^2_{min}) = 2\Delta K\, K_{max} . \qquad (1.9)$$

Let us come back to the Paris-Erdogan equation (1.2)

$$\frac{dL}{dN} = C(\Delta K)^n$$

which is the simplest and the most often used.
Based on a comparison with experimental data the value of n is usually predicted to be in the interval between 2 and 4. C is numerical factor dependent on loading conditions, the mean level of stress R , the material properties and the characteristics of enviroment.

It is clear that in the case of homogeneous cyclic loading the relationship between number of cycles N and time t is simple, namely $N = \omega t$ and

$$\frac{dL}{dt} = \frac{dL}{dN}\frac{dN}{dt} ; \quad \frac{dL}{dt} = \omega \frac{dL}{dN} ,$$

ω - constant frequency
In the case of infinite elastic sheet loaded (at infinity) by a tensile stress S , the Paris-Erdogan equation takes

the form

$$\frac{dL}{dt} = C\,\omega\,\pi^{1/2}(\Delta S)^n\,L^{n/2} \qquad (1.10)$$

$$= C_1(\Delta S)^n\,L^{n/2} \quad , \quad C_1 = \omega\,\pi^{1/2}\,C$$

The solution satysfying the initial condition: $L(0) = L_o$ is

$$L(t) = \begin{cases} L_o \exp\,[C_1(\Delta S)^2 t] & , \quad n = 2 \\ \left[L_o^{\frac{2-n}{2}} + \frac{2-n}{2}\,C_1(\Delta S)^n t\right]^{\frac{2}{2-n}} & , \quad n \neq 2 \end{cases} \qquad (1.11)$$

It is seen that for $n > 2$ there exists a finite instant of time $t = \tau$ in which the solution explodes to infinity. From the solution given above one easily obtains the time (or the number of load cycles) in which crack reaches a crictical value L^*

$$t(L^*) = \begin{cases} \frac{1}{C_1(\Delta S)^2}\ln\frac{L^*}{L_o} & n = 2 \\ \frac{2}{2-n}\;\frac{1}{C_1(\Delta S)^n}\left(L^{*\frac{2-n}{2}} - L_o^{\frac{2-n}{2}}\right) & n \neq 2 \end{cases} \qquad (1.12)$$

It is worth noting that at failure

$$t(L^*)(\Delta S)^n = \text{constant} \qquad (1.13)$$

The solution (1.11) is completely determined by the five parameters: L_o , $C_1 = (C, \omega)$, n , ΔS .

4. FATIGUE CRACK GROWTH UNDER VARIABLE-AMPLITUDE LOADING

The traditional fatigue crack growth equations - including these presented above - are based on fixed stress level fatigue experiments (constant amplitude (homogeneous) cyclic loading) and they do not take into account so called interaction effects due to irregularity of loading, including random loading.

Various types of variable - amplitude loading occur in reality or are generated in laboratories (to increase fundamental understanding of the crack growth mechanism). These loading can be classified as follows: step loading, flight-simulation loading (sequence of loads in flights, distribution function of load cycle etc.), programmed block loads, overloads (single overload, repeated overloads etc.), random loading.

Of particular interest is the retardation in fatigue crack growth rate which normally follows a high overload. As pointed out by many researches this retarded growth can have a significant influence on the fatigue life of a structure.

An important question is: how to predict fatigue crack growth for variable - amplitude loading?

There is no general standarized method, but a number of procedures have been put forward which have some predictive ability. Since the stress intensity factor range ΔK plays a key role in the case of constant amlitude loading it is usually belived that this quantity should also occur in hypotheses concerned with fatigue crack growth under variable amplitude loading.

Among the methods of prediction of crack growth under variable amlitude loading let us mention the following (cf. [13]) ones.

1. <u>Cycle-by-cycle (non-interactive) prediction</u>

$$L_N = L_o + \sum_{i=1}^{N} \Delta L_i \tag{1.14}$$

where L_o is the initial value (for $N = 0$) and $\Delta L_i = \frac{dL}{dN}$ for ΔK and R in cycle i as under constant - amplitude loading.

The deficiency of such a scheme is that ΔL_i are independent of the history of the preceding crack growth.

2. <u>Concept of equivalent K</u>

The equivalent K - concept was proposed mainly for randomly irregular loading. The basic assumption is that an equivalent ΔK can be found, which under constant-amplitude (CA) loading will give the same crack rate as the random loading; that is

$$\left\{ \frac{dL}{dN} = f(\Delta K_{eq}) \begin{matrix} \text{random} \\ \text{loading} \end{matrix} \right\} = \left\{ \frac{dL}{dN} = f(\Delta K) \begin{matrix} \text{const.} \\ \text{ampl.} \\ \text{loading} \end{matrix} \right\} \tag{1.15}$$

Most ofen the root mean squere (rms) value of random loading was proposed for this purpose, i.e. $\Delta K_{eq} = \Delta K_{rms}$ and a relationship to constant - amplitude data is abandoned. In this case the fatigue crack growth equation is postulated in the form

$$\frac{dL}{dN} = f(\Delta K_{rms}) \quad , \; \Delta K_{rms} = C\, S_{rms} \sqrt{\pi L} \tag{1.16}$$

3. Crack closure model of Elber

This is an empirically based model which uses an effective stress range concept. The effective stress range is defined as

$$\Delta S_{eff} = S_{max} - S_c \tag{1.17}$$

where S_c is the closure stress (determined experimentally).

If we put

$$C_f = \frac{S_c}{S_{max}} = \frac{K_c}{K_{max}}$$

than

$$\Delta S_{eff} = S_{max} (1 - C_f)$$

and, if the Paris-Erdogan equation is used, we have

$$\frac{dL}{dN} = C (\Delta K_{eff})^n = C [K_{max} - K_c]^n$$
$$= C [(S_{max} - S_c) \sqrt{\pi L}]^n \tag{1.18}$$

If S_c is defined as a function of the previous load history, than the above equation would predict crack growth interaction effects. This model is often used to predict retarded crack growth caused by overloads.

4. Wheeler model

In order to account for delayed growth due to overload, Wheeler proposed the following equation

$$\frac{dL}{dN} = C_p f(\Delta K) \tag{1.19}$$

where C_p - retarding factor ($C_p \le 1$) is expressed in terms of plastic zone sizes created in the crack tip.

There are also some other particular proposals to describe crack growth in complicated loading situations (e.g. Willenborg model, residual force model etc.). None of these empirical models is accepted to be superior to the others. It is important to formulate more uniform approach to modelling and analysis of fatigue crack growth. Most likely, this aim can be realized with use of the probabilistic reasoning.

II. FATIGUE RELIABILITY

5. SCATTER IN FATIGUE DATA

Because of a very complex nature of fatigue phenomenon as well as the presence of various uncertainties in basic factors provoking fatigue, it is no doubt nowedays that a fatigue of real materials should be regarded as a random phenomenon.

When data from fatigue experiments (under constant - amplitude loading) are studied (cf.[9]) it is easily concluded that the sample functions of crack length (versus number of cycles) are all different and each sample is rather irregular although some smoothing was performed where the curves were drawn

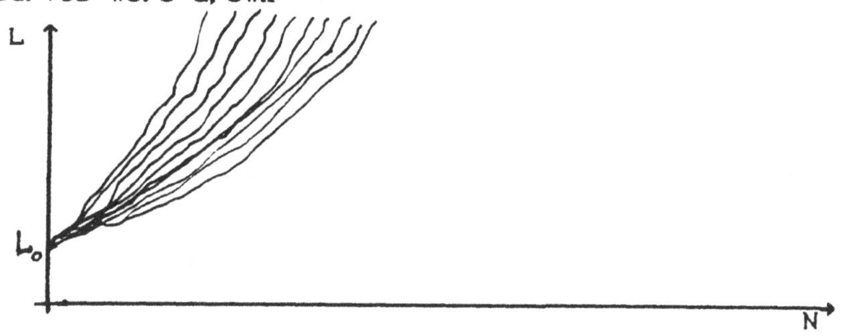

From: Virkler et al.
Trans. ASME J, Eng. Mater. Technol. ,101, 148-153, 1979.

The experiments show that the fatigue life of real mechnical components is characteristically random. A randomness of the fatigue process is completely evident if a structure is subjected to time varying random loading.

6. FATIGUE RELIABILITY; LIFE TIME DISTRIBUTIONS

The statistical nature of data associated with fatigue tests requires a probability-based analysis of the safety of structures subjected to time - varying actions. From engineering point of view, for a structure required to perform safely during its entire service life, it is necessary to have an appropriate probability distribution of ultimate fatigue failure. Such a need leads to the problem which are analogues to these in general reliability theory where estimation of the life time distribution of a system considered is one of main concern.

However, in classical reliability theory (originating in analysis of radioelectrical systems) the system is modelled as an interconnection of statistically independent elements each having a relatively simple reliability characteristic (usually, the probability of functioning or not is assigned). Thus, topological arguments suffice to establish the mode of calculation for the system reliability in terms of its component reliabilities.

In the case of mechanical structures such a model is most often impractical because the individual component functions are not statistically independent and they are determined by the mechanical properties of the material and crucially depend on various external actions. These additional complications (especially charcteristic for random vibration problems) strongly influence system reliability and must be accounted for methodically within the extended - physical reliability theory.
If a deterioration process is due to fatigue we come to the problems of fatigue reliability.

Let us denote a random-time of a structural element by θ. The probability distribution of θ will be denotes by $F(\vartheta) = P(\theta > \vartheta)$. The survival function $\bar{F}(\vartheta)$ and probability density $f(\vartheta)$ are defined as:

$$\bar{F}(\vartheta) = 1 - F(\vartheta) = P(\theta \leq \vartheta),$$

$$F(\vartheta) = \int_{-\infty}^{\vartheta} f(x)dx$$

Several families of probability distributions have been suggested for the random variable θ, or - as it is often stated - for the random variable $N(s)$, the number of load cycles to fatigue failure of material specimen at stress level s. The most important of these are:

a) log-normal distribution,
b) extreme - value distribution,
c) Weibull distribution,
d) Gamma distribution.

Ad. a) The long-normal distribution is defined as:

$$f(\vartheta) = \frac{1}{\vartheta\,\sigma\sqrt{2\pi}} \exp\left[-\frac{1}{2\sigma^2}(\log\vartheta - \mu)^2\right] \tag{2.1}$$

μ, σ - real constants; $\sigma > 0$, $-\infty < \mu < \infty$.
The mean and variancee are:

$$\text{mean} = e^{\mu + \frac{\sigma^2}{2}} \quad ; \quad \text{variance} = e^{2\mu+\sigma^2}\left[e^{\sigma^2} - 1\right]$$

Ad. b) A random variable θ is defined to have an <u>extreme value distribution</u> (or the type I asymptotic distribution of extreme values - according to Gumbel) with parameters ν and β ($\beta > 0$, $-\infty < \nu < \infty$) if

$$F(\theta) = 1 - \exp\left[- e^{-\frac{\theta-\nu}{\beta}}\right],$$

$$f(\theta) = \exp\left[- \frac{\theta-\nu}{\beta} - e^{-\frac{\theta-\nu}{\beta}}\right]. \tag{2.2}$$

The mean and variance are:

$$\text{mean} = \nu + 0.5772\,\beta \quad , \quad \text{variance} = \frac{\pi}{\sqrt{6}}\beta$$

Ad. c) A random variable θ which takes values greater than some number ε, is defined to have the <u>Weibull distribution</u> with parameters ν and k (and with threshold ε) if

$$F(\theta) = 1 - \exp\left[-\left(\frac{\theta-\varepsilon}{\nu}\right)^k\right] \qquad \theta \geq \varepsilon$$

or (2.3)

$$f(\theta) = \frac{k(\theta-\varepsilon)^{k-1}}{\nu^k}\exp\left[-\left(\frac{\theta-\varepsilon}{\nu}\right)^k\right], \quad \theta \geq \varepsilon$$

The mean and variance are:

$$\text{mean} = \varepsilon + \nu\,\Gamma\left(1 + \frac{1}{k}\right),$$

$$\text{variance} = \nu^2\left[\Gamma\left(1 + \frac{2}{k}\right) - \Gamma^2\left(1 + \frac{1}{k}\right)\right]$$

where $\Gamma(r)$ - is the Gamma function

$$\Gamma(x) = \int_0^\infty x^{r-1}e^{-x}dx \quad ; \quad \Gamma(r) = (n-1)! \qquad n \text{ -positive integer}$$

If $\varepsilon = 0$ we obtain a common, two - parameter Weibull distribution.
Ad. d) A random variable θ which takes only positive values is defined to have a Gamma distribution with parameters r and λ ($r = 1, 2, \ldots ; \lambda>0$) if

$$f(\theta) = \frac{\lambda}{\Gamma(r)}(\lambda r)^{r-1} e^{-\lambda\theta} \tag{2.4}$$

The mean and variance are:

$$\text{mean} = \frac{r}{\lambda}, \quad \text{variance} = \frac{r}{\lambda^2}$$

For $r = 1$ we obtain the exponential distribution for which

$$f(\theta) = \lambda e^{-\lambda\theta}, \quad \theta>0 \tag{2.5}$$

It is worth noting that two-parameter distributions fit experimental data on random life time quite well for high stress levels generated in material. When stress levels are lower the appropriate three-parameter distributions (with threshold) are often introduced.

A general and mathematically natural approach to formulation and analysis of the problem of fatigue life distribution is as follows.
The accumulation of fatigue in real materials subjected to realistic actions is, as a function of time t, a certain stochastic process $X(t,\gamma)$ defined for $t\in(t_o,\infty)$, $\gamma \in \Gamma$ where $(\Gamma, \mathcal{F}, P)$ is the probability space that is, the triple consisting of a space of elementary events Γ, a σ-field $\mathcal{F}$ of subsets of Γ and probability P defined on $\mathcal{F}$.

Fatigue failure occurs at such time, $t = \theta(\gamma)$ being a random variable, for which $X(t,\gamma)$ crosses, for the first time, a fixed critical level x^*. Therefore, the fatigue life time distribution is the first passage time distribution defined as

$$\theta = \sup \left\{t: X(\tau,\gamma) < x^* \right\}, \quad t_o \le \tau \le t \tag{2.6}$$

If there is a basis for assuming that the main reason of random dispersion of life-time is randomness in the initial quality of a material element then random process $X(t,\gamma)$ can be represented as

$$X(t,\gamma) = \varphi\left[t;\ A_1(\gamma), \dots,\ A_n(\gamma)\right]$$

where φ is a deterministic function of time and random variables $A_1, \dots, A_n$. The simplest model of this kind is : $X(t,\gamma) = A(\gamma)t$ and it gives simple linear random accumulation.

If randomness in life-time is an inherent feature of fatigue accumulation then a natural model is

$$X(t,\gamma) = \int_{t_0}^{t} \eta(\tau,\gamma)d\tau \qquad (2.7)$$

where $\eta(\tau,\gamma)$ is a non-negative stochastic process characterizing the intensity of fatigue accumulation. This process should be constructed using the knowledge on mechanics and physics of fatigue. In next Sections we shall describe more sophisticated constructions of process $X(t,\gamma)$.

7. FATIGUE FAILURE RATES

Another approach to characterization of fatigue reliability of materials is concerned with the concept of fatigue failure rate. Let T be a random life time of a specimen. Let $F(t)$ be the distribution function of T and $f(t)$ its probability density. The non-negative function $\mu(t)$ defined as follows

$$\mu(t) = \frac{f(t)}{1-F(t)} \qquad (2.8)$$

is called the <u>fatigue intensity function</u> or <u>fatigue failure rate</u> (and hazard function - in reliability theory). In words, $\mu(t)dt$ is the conditional probability that the specimen will fail between t and $t+\Delta t$ given that it has survived a time T greater than t, that is

$$P\left\{ T \in (t,t+\Delta t) \mid T > t \right\} = \mu(t)\Delta t + o(\Delta t)$$

For a given $\mu(t)$ the corresponding distribution function can be found easily since (2.8) can be rewritten as

$$\frac{d}{dt} \ln [1-F(t)] = -\mu(t)$$

and therefore

$$1 - F(t) = \bar{F}(t) = \bar{F}(t_o) \exp \{ - \int_{t_o}^{t} \mu(\tau)\, d\tau \}$$

If $\bar{F}(t_o) = 1$, then

$$F(t) = 1 - \exp \{- \int_{t_o}^{t} \mu(\tau)\, d\tau \} \quad ,$$

$$f(t) = F'(t) = \mu(t)\exp \{- \int_{t_o}^{t} \mu(\tau)\, d\tau \} \quad (2.9)$$

It is well known that if $\mu(t) = \lambda = \text{const.}$, $t \geq t_o$, $\lambda > 0$ then the life-time of a specimen is exponentially distributed. If

$$\mu(t) = k\, a^{-k}\, t^{k-1} \quad , \quad t > 0 \quad (2.10)$$

then the life-time has Weibull distribution (2.3). Many other life-time distributions can be derived from various forms of the failure rate $\mu(t)$. However, éxcept the simplest cases these functions have rather complicated forms (see formula (5.8) for Poisson shock model). It is, therefore, of interest to generalize formula (2.9) to the case when $\mu(t)$ is a realization of a stochastic process, that is $\mu(t) = \mu_\gamma(t)$, where $\gamma \in \Gamma$, and Γ is the space of elementary events on which probability is defined.

In the case of random fatigue failure rate we are no longer interested in $P\{T > t \mid \mu_\gamma(\cdot)\}$. The quantity of interest is the unconditional probability

$$P\{T > t\} = \bar{F}(t) = \langle P\{T > t \mid \mu_\gamma(\cdot)\}\rangle_\gamma = \tag{2.11}$$

$$= \langle \exp\{-\int_{t_0}^{t} \mu_\gamma(\tau)\, d\tau\}\rangle_\gamma$$

where $\langle\cdot\rangle$ denotes an average value.
If $\mu_\gamma(t)$ is a stochastic process defined by deterministic function of t with some random parameter, say $A(\gamma)$, then

$$\bar{F}(t|A) = \exp\{-\int_{t_0}^{t} \mu_\gamma(\tau|A)\, d\tau\}$$

and (2.12)

$$\bar{F}(t) = \int_{-\infty}^{\infty} \bar{F}(t|A)\, d\, F_1(A)$$

where $F_1(A)$ is the distribution function of random parameter $A(\gamma)$.
Let us assume that $\mu(t)$ is a generalized Poisson process with $\lambda(t)$ as the expected number of changes of its value per unit time and with $F_t(z)$ representing the distribution of the size of these changes. It can be shown that in this case (cf. [1])

$$P\{T > t\} = \exp\{\int_0^t \int_0^\infty [e^{(\tau-t)v} - 1]\, \lambda(\tau)\, d\, F_\tau(v)\} \tag{2.13}$$

If

$$F_t(z) = \begin{cases} 0 & , \quad z < 1 \\ 1 & , \quad z \geq 1 \end{cases}$$

then

$$P\{T > t\} = \exp\left\{ \int_0^t [e^{\tau - t} - 1] \, \lambda(\tau) \, d\tau \right\} \qquad (2.14)$$

In the simplest case, when additionaly $\lambda(t) = \lambda = \text{const.}$ we obtain

$$P\{T > t\} = \exp\left[\lambda(1 - e^{-t} - t\right] \qquad (2.15)$$

This is a simple expression for fatigue life-time distribution when instantaneous fatigue (or fatigue intensity function) is a homogeneous Poisson process.

III. EVOLUTIONARY PROBABILISTIC MODELS

8. GENERAL IDEA

A general probabilistic approach to the description of fatigue can be obtained if we regard a considered sample of materials together with fatigue process taking place in it as a certain dynamical system whose states (appropriately defined) change in time and are described by a stochastic vector process $\underline{X}(t) = [X_1(t),\ldots,X_n(t)]$; the component processes $X_i(t)$, $i= 1,2,\ldots,n$ charcterize specific features of the failure accumulation and in the most situations we can assume that $0 \le X_i(t) \le x_i^*$. In order to obtain a model for random fatigue growth one should describe the evolution (in time) of probabilistic structure of the process X(t). Infinitesimal characteristics of such an evolution (depending on all basic factors provoking fatigue) should be deduced from the knowledge on fatigue process in real materials.

Denoting the joint probability density function (which describes the probabilistic structure of a "system") of the process $\underline{X}(t)$ by $p(\underline{x},t|x_0,t_0)$ we can write symbolically the evolution equation in the form

$$\frac{\partial p(\underline{x},t)}{\partial t} = L\,[p(\underline{x},t)\,, S(t),\, \zeta(t)] \qquad (3.1)$$

where L can be termed as "fatigue operator" and it needs to be defined; $S(t)$ denotes appropriate parameter (or parameters) charcterizing external loading and $\zeta(t)$ occurs for denoting all other possible quantities important in fatigue problem. In such formulation (which was first indicated by Bolotin) - the modelling problem lies in proper construction of the "fatigue operator L . This, however, is not easy problem.

In order to make the idea described above efficient one has to introduce some assumptions and hypotheses. The basic hypothesis requires the fatigue process to be Markovian. This hypothesis makes it possible to use large variety of mathematical schemes elaborated in the theory of Markov stochastic processes and leads to interesting results. Since the Markov property means that the "future" of the process depends only on its "presence" and is independent of the "past" one could be afraid that all history - dependent effects have to be excluded. Situation is not so bad if we formulate the problem (as we have done above) within the frame of multidimensional Markov processes; ideologically, the history - dependent effects can be accounted for by appropriate selection of the phase space of the process. Obviously, such "multidimensional" modelling makes calculations more intricate. It is worthy noting that in the next section we shall show that Markovian character of fatigue process (characterized by growth of a dominant crack) can be justified for an important class of materials and external actions.

In what follows we shall briefly discuss the evolutionary modelling of fatigue by use of Markov chains and continuous Markov processes.

9. FATIGUE ACCUMULATION AS MARKOV CHAIN

Let us assume that a specimen considered from the point of view of damage accumulation can be in one of n+1 states: E_0, $E_1, \ldots, E_n = E^*$ where E_0 denotes an "ideal" state and E^* - the state characterizing the ultimate damage. Only the forward transitions from one state to another are possible that is

$$E_0 \longrightarrow E_1 \longrightarrow \ldots \longrightarrow E_k \longrightarrow E_{k+1} \longrightarrow \ldots \longrightarrow E_k = E^* \quad (3.2)$$

Let us assume the following probabilistic mechanism of the trasition : if at time t the specimen is in the state E_k the probability of transition $E_k \longrightarrow E_{k+1}$ in the time interval $(t, t+\Delta t)$ is equal to $q_k \Delta t + o(\Delta t)$. The probability of the transition from E_k to a state different from E_{k+1} is $o(\Delta t)$. The probability of no change is $1 - q_k \Delta t + o(\Delta t)$. Let $P_k(t)$ be the probabilty that a specimen at time t is in the state E_k . It is easy seen that

$$P_k(t+\Delta t) = (1-q_k \Delta t) P_k(t) + q_{k-1} P_{k-1}(t) \Delta t + o(\Delta t) \quad (3.3)$$

Transposing the term $P_k(t)$ on the left-hand side, dividing by Δt and passing to the limit (as $\Delta t \longrightarrow 0$) gives the system of differential equations:

$$\frac{dP_o(t)}{dt} = -q_o P_o(t) ,$$
$$\frac{dP_k(t)}{dt} = -q_k P_k(t) + q_{k-1} P_{k-1}(t) , \quad k \ge 1 \tag{3.4}$$

The initial condition says that at time $t = 0$ the system was in the state E_o, that is

$$P_o(0) = 1 \tag{3.5}$$

The above system of equations (3.4) with (3.5) describes the considered process and constitutes a simple but physically sound example of evolution equation (3.1). If, for instance, the fatigue is characterized by the crack length, the states E_k, $k=0,1,\dots,n$ denote the crack tip position measured by real numbers. Since $E_n = E^*$ is the final state of a specimen, the further transitions are impossible, what means that $q_n = 0$. It can be easily recognized that the process described is Markovian; the probability of a change in the interval $(t, t+\Delta t)$ depends only on the state in which the system is at time t. This is the birth process known from the theory of population growth.

The solution of first equation (3.4) is

$$P_o(t) = e^{-q_o t} \tag{3.6}$$

For given q_k one easily obtains the recursion relation:

$$P_k(t) = q_{k-1}\, e^{-q_o t} \int_0^t e^{-q_k \tau}\, P_{k-1}(\tau) d\tau , \quad k=1,2,.. \tag{3.7}$$

The above formula when $k=n$ gives the probability of failure at time t. It should be noticed that though the model is simple the transition intensities q_k are different for different "fatigue states".

If the states are characterized by the integers (what can be easily asumed when damage is measured by the length of a dominant crack) than the linear dependence of q_k on the "fatigue states" seems to be natural since it reflects the fact that the intensities of transition are greater in the advanced state of damage. So, let

$$q_k = kq \quad , \quad k \geq 1 \quad , \quad q > 0 \tag{3.8}$$

In this case we have

$$\frac{dP_k(t)}{dt} = -\,kq\;P_k(t) + q(k-1)\;P_{k-1}(t) \;,\; k \geq 1 \tag{3.9}$$

and one obtains the Yule-Furry distribution:

$$P_k(t) = e^{-qt}\left[1-e^{-qt}\right]^{k-1} \quad , \qquad k \geq 1 \tag{3.10}$$

which is a monotone describing funtion of the damage state k. If transition intensity is the same for all states, i.e. $q_k = q$ than from (3.6) and (3.7) one obtains by induction the Poisson distribution:

$$P_k(t) = \frac{(qt)^k}{k!}\; e^{-qt} \tag{3.11}$$

It can be concluded that in this case life time of a specimen has a gamma distribution. In the model presented above the only quantity needed for calculation of failure probability is the intensity of the failure transition q_k which should be estimated from experimental data (of course, first the fatigue experiments should be planned correctly to be able to estimate such quantity as q_k).

Recently a series of papers by Bogdanoff, Krieger and Kozin has been published (see also a book [4]) in which another probabilistic damage model has been elaborated which is also based on Markov chain theory. The most essential features of this model are as follows.

Let the damage states be denoted by $x = 1,2,\ldots,b$ where state b denotes failure. A basic concept in the model is a duty cycle (DC) which is understood as a repetitive period of operation in the life of a component during which damage can accumulate. In a constant amplitude loading a duty cycle can correspond to a certain number of load cycles; in the case of an aircraft operation each mission can be divided into duty cycles for taxiing, take-off, landing etc. Time (discrete) $t = 0,1,2,\ldots$ is measured in number of DC's. How demage is accumulated within a DC is not a matter of concern. It is assumed that the increment in damage (which takes place only at the end of each DC) depends in a probabilistic manner only on the DC itself and on the value of the damage accumulated at the start of the duty cycle. It is, however, independent of how the damage was accumulated up to the start of the duty cycle. These assumptions are Markovian ones and the damage process is regarded as discrete-space and dicrete-time Markov chain.

As it is usual, this Markov chain is defined by specifying the initial distribution $\underline{P}_o$ and the trasition matrix $\tilde{P} = \{P_{ij}\}$. The initial state of damage is specified by the vector $\{\pi_i\}$ where π_i is the probability of damgae in state i at $t = 0$ i.e.

$$\underline{P}_o = [\pi_1, \pi_2, \dots, \pi_b] \ , \quad \pi_i \geq 0 \ , \ \sum_{i=1}^{b} \pi_i = 1 \tag{3.12}$$

The transition matrix for a duty cycle is $\{P_{ij}\}$ where P_{ij} is the probability that the damage is in state j after the duty cycle provided that damage was in state i at the begining of the duty cycle. The state of damage at time t is given by the vector

$$\underline{P}_t = [P_t(1), P_t(2), \dots, P_t(b)] \ ; \ P_t(i) \geq 0 \ , \ \sum_{i=1}^{b} P_t(i) = 1 \tag{3.13}$$

where $P_t(i)$ is the probability that damage is in state i at time t . The Markov chain theory gives the relation

$$\underline{P}_t = \underline{P}_o \tilde{P}_1 \tilde{P}_2 \dots \tilde{P}_{t-1} \tilde{P}_t \tag{3.14}$$

where $\tilde{P}_j$ is the transition matrix for the j-th duty cycle. The expression (3.14) completely specifies the probability distribution of damage at any time. If the damage cycles are of the same severity, equation (3.14) reduces to

$$\underline{P}_t = \underline{P}_o \ \tilde{P}^{\,t} \tag{3.15}$$

where $\tilde{P}$ is the common transition matrix. Since matrix multiplication is generally not cummutative, it follows from (3.14) that the order of the duty cycles is important for damage accumulation.

The probability distributions of various random variables associated with the damage accumulation process can be easily determined. For example, the time T_b to failure (to reach state b) has the distribution function $F_b(t)$ given by

$$F_b(t) = P\{T_b \le t\} = \underline{P}_t(b) \qquad (3.16)$$

For many additional detailes the interested reader is referred to the original papers. As authors show the model just described has many advantages.

10. MARKOV DIFFUSION MODEL FOR FATIGUE CRACK GROWTH

Though the fatigue crack grows in time intermitently and in reality constitutes a discontinuous random process it is reasonable to approximate this process by a continuous stochastic process. Such modelling of discontinuous phenomena brings interesting results in other fields, such as population biology, physics, chemical kinetics and other. For example, in the study of the motion of particles in a colloidal suspension this approach leads to the diffusion equation that governs the transition density of a typical particle. It is, therefore, of interest to regard also a fatigue crack growth as diffusion - like stochastic process (cf. papers by Oh [12]).

Let us denote the crack tip position at time t (or crack length) by $L(t)$ assuming that it is a continuous stochastic process. Because the crack growth is irreversible $L(t)$ may only take its values in the interval $[l_o, \infty)$ where $l_o = L(t_o)$. Of course, in practice we are interested in the evolution of $L(t)$ on finite interval $[l_o, l_{cr}]$, where l_{cr} denotes a critical crack size. Let $p(l, t; l_o, t_o)$ be a density of the transition probability of the crack tip. It satisfies the conditions

$$p(l, t; l_o, t_o) = 0 \quad \text{for} \quad l \ge l_o$$
$$(3.17)$$
$$p(l, t; l_o, t_o) \ge 0 \quad \text{for} \quad l \ge l_o$$

To be able to put the further analysis into the frame of difusion Markov process theory we have to assume that the growth of the crack is dependent only upon its current crack length. This (Markovian) assumption must be regarded as hypothesis, which calls for experimental verification (in the next section a fracture mechanics - based theoretical justification of this hipothesis for a certain class of problems will be indicated). Heuristically, we can try to motivate the Markovian hypothesis taking into account the basic factors causing crack growth. In the case of deterministic cycles load there are two main factors: the range of stress intensity factor and the strength distribution of the material in the tip region. The first of these two factors is a deterministic quantity when the tip position is given, whereas the second is not altered by the past positions of the crack tip.

Adopting a common in Markov process theory reasoning we come to the following governing evolution equation for the transition density

$$\frac{\partial p}{\partial t} = -\frac{\partial}{\partial l}\left[a(l)p(l,t)\right] + \frac{1}{2}\frac{\partial^2}{\partial l^2}\left[b(l)p(l,t)\right] \qquad (3.18)$$

where $a(l)$ and $b(l)$ are the coefficients characterizing infinitesimal properties of the process (drift and diffusion coefficients, respectively). To be able to use the model, the coefficients $a(l)$, $b(l)$ have to be propely determined. Using the weakest link model for the prediction of fatigue crack growth rate Oh found the following forms of the coefficients

$$a(l) = C_a l^{1+\frac{1}{m}} \quad , \quad b(l) = C_b l^{2+\frac{1}{m}} \qquad (3.19)$$

where C_a and C_b are constants expressed by integrals (which require numerical evaluation).

Since the growth process is considered on semi-infinite interval $[l_o, \infty)$, it is necessary to specify the boundary conditions. This is related to a more general problem of Feller's classification of boundaries associated with one-dimensional diffusion process. Whithout going into details (interested reader is refered to the mathematical literature) we require first, that at $L(t_o) = l_o$ the values of $p(l,t)$ and $\frac{\partial p}{\partial t}$ must be such that

$$\int_{l_o}^{\infty} p(l,t)dl = 1 \qquad (3.20)$$

To write down this condition in terms of the coefficients $a(l)$ and $b(l)$ let us integrate equation (3.18) first with respect to t and then with respect to l. After interchanging the order of integration on the right-hand side we obtain

$$\int_{l_o}^{\infty} p(l,t)dl \Big|_{t=\tau} - \int_{l_o}^{\infty} p(l,t)dl \Big|_{t=o} =$$

$$= \int_{o}^{\tau} \frac{1}{2} \left\{ \frac{\partial}{\partial l} \left[b(l)p \right] - a(l)p \right\}_{l=l_o}^{l=\infty} dt$$

Since (3.20) is required to be satisfied, the left-hand side is identically zero, and we have the condition

$$\frac{1}{2} \frac{\partial}{\partial l} \left[b(l)p \right] - a(l)p = 0 \quad \text{at} \quad l=l_o \qquad (3.21)$$

Since $l_{cr} \in [l_o, \infty)$ is an absorbing "boundary" within semi-infinite interval $[l_o, \infty)$ we have the condition

$$p(l,t) = 0 \quad \text{at} \quad l = l_{cr} \qquad (3.22)$$

The solution of equation (3.18) satisfying the above conditions provides basic information concerning evolution of the probabilistic structure of the fatigue crack growth.

From the transition density $p(l,t;\, l_o,t_o)$ we can obtain quantities which are of interest to fatigue life prediction. For example, the probability distribution function of the crack size at time is given by

$$F_{L(t)}(l,t;\, l_o,t_o) = P\left\{ L(t)<l \,|\, L(t_o)=l_o \right\} =$$

$$= \int_{l_o}^{l} p(x,t;l_o,t_o)dx \qquad (3.23)$$

and the probability that the crack size exceeds a given

value is the complement of (3.23), i.e.

$$F_{L(t)}(l^*,t;l_o,t_o) = P\left\{L(t) \geq l^* \,|\, L(t_o)=l_o\right\} =$$

$$= 1 - \int_{l_o}^{l^*} p(x,t;\ l_o,t_o)dx \tag{3.24}$$

The graph of $F_{L(t)}(l^*,t;l_o,t_o)$ is often referred to as the crack exceedance curve.

Let $T(l^*)$ be the random time when the crack size $L(t)$ reaches a specific (critical) value l^*. Since the event $\left\{T(l^*) < \tau\right\}$ is the same as the event $\left\{L(t_o+\tau) > l^*\right\}$, the probability distribution $F_T(t)$ of $T(l^*)$ is

$$F_T(\tau) = \int_{l_o}^{l^*} p(x,t_o+\tau;\ l_o,t_o)dx = F_{L(t)}(l^*,t_o+\tau;l_o,t_o) \tag{3.25}$$

which is the same as the probability of crack exceedance but treated as function of τ.

Closing this section - devoted to evolutionary probabilistic modelling of fatigue - we wish to notice that the evolution equations (3.4) and (3.18) should be treated as examplifications of general equation (3.1). Some other evolutionary models can also be constructed; for example the model leading to the following equation

$$\frac{\partial p}{\partial t} = \int K(\underline{x},\underline{x}';t,t_o)\ p\ (\underline{x}',t)\ d\underline{x}' \tag{3.26}$$

(where $K(\underline{x},\underline{x}';t,t_o)$ is the transition rate from state $\underline{x}'$) could bring a fruitful generalization of the Markov chain model discussed above.

IV. FATIGUE CRACK GROWTH: STOCHASTIC DIFFENERNTIAL EQUATION MODEL

11. MODIFIED CRACK GROWTH EQUATION FOR RANDOM LOADING

As it has been indicated in Section I, a wide class of empirical crack growth equations can be writen in the form

$$\frac{dL}{dN} = F(\Delta K,\ R) \tag{4.1}$$

where L is the crack length, ΔK - is the stress intensity factor range and R is the stress ratic.

Though the equations of the form (4.1) have been identified from a fixed stress level experiments (constant amplitude cyclic loading), they also have been adopted for predicting crack growth under random loading by use of some "equivalent" constant quantities characterizing (in average) the random applied stress.

Various quantities associated with random stress process have been proposed to play a role of ΔK. The existing studies stronly suggest that ΔK should be replaced by

$$K_{rms} = S_{rms}\sqrt{\pi L} \tag{4.2}$$

where "rms" means "the root mean squere".

Another modification is concerned with the stress ratio R. It is suggested that in the case of random loading one should use a quantity

$$Q = \frac{\langle S \rangle}{S_{rms}} \tag{4.3}$$

A further departure from the constant-amplitude approach is concerned with the concept of cycle. In the case of sinusoidal loading there is a simple relation to time description, since $N = \omega t$, where ω is the frequency. When the load history is randomly varying in time the definition of a cycle is not straight-forward and not unique. In this case

$$N = \eta(t,\gamma)$$

where $\eta(t,\gamma)$ is a point stochastic process. It is reasonable to assume that $\eta(t,\gamma)$ characterizes a number of local maxima of the stress process in the time interval $[t_o,t]$. The quantity $\eta(t,\gamma)$ defined as:

$$\eta(t,\gamma) = \int_{t_o}^{t} n(s.\gamma)\, ds \tag{4.4}$$

determines an "instanatneous" intensity of N per unit time. Of course

$$\langle \eta(t,\gamma) \rangle = \int_{t_o}^{t} \mu(s) ds,$$

$$\mu(t) = \langle n(t,\gamma) \rangle \tag{4.5}$$

where $\mu(t)$ decribes the average number of maximuma of a stress process per unit time. When $S(t,\gamma)$ is a stationary process $\mu(t) = \mu = \text{const}$. Therefore, it is justified to postulate the following relation between the cyclic and temporal descriptions

$$\frac{dL}{dt} = \mu(t) \frac{dL}{dN} \tag{4.6}$$

Finally, the averaged crack growth rate under random loading can be represented in the form of the following differential equation:

$$\frac{dL}{dt} = \mu(t)\; F(K_{rms},\; Q) \tag{4.7}$$

$$L(t_o) = L_o$$

For stationary random loading $\mu(t)$ = const., and it will be included into the symbol of function F.

12. STOCHASTIC DIFFERENTIAL EQUATION MODEL

It turns out that for a certain class of random processes responsible for a scattering of fatigue data the randomized crack growth equation can be analyzed directly by use of the theory of stochastic Ito (or Stratanovich) differential equations. Such an approach has recently been proposed independently by Sobczyk [14] and Lin with Yang [10]. A more methodical extension of this idea is contained in the author's paper [16]. Here we shall describe only the main features of the method.

Taking into account the general form of the fatigue crack growth equation and introducing random fluctuation of uncertainties in the time we come to the following model

$$\frac{dL}{dt} = F(\Delta K_{rms}, Q) X(t,\gamma) \tag{4.8}$$

$$L(t_o) = L_o$$

where $X(t,\gamma)$ is random process representing the combined effect of unknown random factors of external or internal origin, such as environment, temperature, internal stress etc. As a matter of fact, model (4.8) is flexible enough and $X(t,\gamma)$ can also be used to describe better (than ΔK_{rms} and Q solely) the effect of variations of applied stress and stress ratio, which occur multiplicatively (cf. equations (1.5), (1.6)).

Model (4.8) includes two special cases. When $X(t,\gamma)$ is a "constant" random process, i.e. $X(t,\gamma)=X(\gamma)$ where $X(\gamma)$ is a random variable it means that is totally correlated at all times. A detailed discussion of the basic questions associated with randomization of Paris law via introducing random variables is given by Kozin and Bogdanoff [9]. Though equation (4.8) in the case $X(t,\gamma)=X(\gamma)$ is more general than that considered in [9] its mathematical analysis (when random loading is stationary process, i.e. ΔK_{rms} and Q are constant) would be analogues to that in [9] . The other extreme is a case where $X(t,\gamma)$ is totally independent at any two different times (a white noise). This case will be of our concern here.

Let $X(t,\gamma) = m_x + \xi(t,\gamma)$, where $\xi(t,\gamma)$ is a white noise with intensity D , that is

$$\langle \xi(t_1,\gamma) \, \xi(t_2,\gamma) = 2D \, \delta(t_2 - t_1) \tag{4.9}$$

Equation (4.8) takes now a form

$$\frac{dL}{dt} = m_x \, F(K_{rms}, Q) + F(K_{rms}, Q)\xi(t,\gamma) \tag{4.10}$$

$$L(t_o) = l_o$$

In particular, when crack growth equation (1.4) is adopted

$$F = Cg(Q)\,(\Delta K_{rms})^n = C_1 g(Q)\, S^n_{rms} L^{n/2}(t) \qquad (4.11)$$

Let assume that random applied stress $S(t,\gamma)$ is stationary. In this case

$$F = BL^{n/2}, \qquad B = C_1 g(Q)\, S^n_{rms} = \text{const.} \qquad (4.12)$$

Equation (4.10) can be represented as

$$\frac{dL}{dt} = AL^{1+\nu}(t) + BL^{1+\nu}(t)\,\xi(t,\gamma), \qquad A = m_x B, \quad 1+\nu = \frac{n}{2} \qquad (4.13)$$

$$L(t_o) = l_o$$

This equation interpreted as the Stratanowich stochastic differential equation is equivalent to the following stochastic Itô equation

$$dL(t) = f_1(L,t)dt + f_2(L,t)dW(t)$$
$$L(t_o) = l_o \qquad (4.14)$$

where $W(t)$ is the Brownian motion process, and

$$f_1(L,t) = AL^{1+\nu} + (1+\nu)\, DB^2 L^{1+2\nu}$$
$$f_2(L,t) = 2DBL^{1+\nu} \qquad (4.15)$$

The solution of (4.14) is a diffusion Markov process with the following drift and diffusion coefficients:

$$a(L,t) = a(L) = f_1(L), \qquad b(L.t) = b(L) = f_2^{\,2}(L) \qquad (4.14)$$

The Fokker-Planck-Kolmogorov equation for the transition density $p(l, t; l_o, t_o) = p(l, t)$ is

$$\frac{\partial p}{\partial t} = -\frac{\partial}{\partial l}\Big[a(l)p(l,t)\Big] + \frac{1}{2}\frac{\partial^2}{\partial l^2}\Big[b(l,t)p(l,t)\Big]$$

$$p(l,t)\Big|_{t=t_o} = \delta(l-l_o) \qquad (4.17)$$

This is the same equation as (3.18) which in Sec. III was postulated on the basis of hipothesis that fatigue crack size L(t) is a diffusion Markov process. Here, equation (4.17) was obtained as conclusion that a solution of equation (4.14), which is a fracture mechanics - based equation, is a diffusion Markov process. Therefore, deriviation shown above can be regarded as justification of the difusion Markovian assumption in evolutionary probabilistic approach presented in Sec. III:

Introducting the following transformation (cf. [14])

$$Y(t) = L_o^{-\nu} - L^{-\nu}(t) \tag{4.18}$$

we come to the following expression for the solution of equation (4.17) with boundary conditions stated in Sec.III:

$$p(l,t;l_o,t_o) = \frac{\nu}{l^{1+\nu}} \frac{1}{[4\pi\nu^2 D^2 B^2]^{1/2}} \exp\left\{ - \frac{\left[\frac{1}{l_o^{\nu}} - \frac{1}{l^{\nu}} - A\nu(t-t_o) \right]^2}{4\nu^2 D^2 B^2 (t - t_o)} \right\} \tag{4.19}$$

As we have said in Sec. III from the transition density (4.19) one can get the quantities which are of interest in the crack growth life prediction.

It should be noted that expression (4.19) describes the probability structure of a diffusion Markov process which is only an approximation to the physical crack growth process L(t). In fact, expression (4.19) admits also those values of L which are smaller than l_o ; otherwise, the total probability would not be equal to unity. The error which arises from restricting our "mathematical process" L(t) to the physicaly meaningful values $L \geq l_o$ is, however, greatest when t is near t_o and decreases as $t - t_o$ increases. Since we are interested in reasonably long fatigue lives, the error seems to be negligible.

The deficiency of our diffusion stochastic model indicated above is "compensated" by its advantages. As it has been shown, it allow us to obtain explicit expression for probability density function and requires the estimation of only a few parameters from the crack growth rate data. As it was shown by Lin and Yang the theoretical predictions correlate very well with experimental results.

In the author's paper [16] a detailed analysis of stochastic differential equation (4.14) has been provided using the Itô calculus. The basic qualitative features of the process $L(t)$ were stated both, in linear $(\nu = 0)$ and non-linear $(\nu > 0)$ cases. Also, the life-time distribution has been explicitely determined. It turned out to be the inverse Gaussian distribution, that is

$$p_\tau(t) = \frac{a}{\sqrt{2\pi}\, t^{3/2}} \exp\left\{-\frac{1}{2}\frac{(a-bt)^2}{t}\right\} \tag{4.20}$$

where

$$a = \frac{1}{l_o^\nu\, \nu \sqrt{2D}\, f_2^{o2}} > 0 \quad , \quad b = \frac{f_1^o}{f_2^o\sqrt{2D}} = \frac{m_x}{\sqrt{2D}} > 0 \quad ,$$

and

$$f_2^o = c\, \pi^{1+\nu} \mu_o g(Q)\, S_{rms}^n \quad .$$

V. FATIGUE CRACK GROWTH: RANDOM CUMULATIVE JUMP MODELS

13. FATIGUE SHOCK MODELS

It is widely accepted that the fatigue phenomenon takes place via the formation and growth of cracks in material. The experiments show, however, that always exists a dominant crack in a specimen and it grows intermitently and consists of active and dormant periods. On the other hand, it is reasonable to assume that crack grows mainly due to a sequence a peaks of a random stress process or, in other words due to a sequence of shocks occuring randomly in time as events of some point stochastic process. This leads to a shock model considered by Sobczyk in [15].

Let us denote the length of a dominant crack at arbitrary moment by $L(t)$. We characterize this process by the following random sum of random components

$$L(t) = \sum_{i=1}^{N(t)} Y_i(\gamma) \quad \text{if} \quad N(t) > 0 \tag{5.1}$$

where $N(t)$ is a random number of shocks, or jumps in crack length in the interval $(0,t]$; it is assumed that each stress peak contributes an increment in crack's length. Random variables $Y_i(\gamma)$ $i = 1,2,\ldots$ are assumed to be identically distributed, independent, non-negative random variables.

Let random variable T be the life-time of a specimen subjected to shocks counted by Poisson process $N(t)$. If $P_k(\xi)$ is the probability of surviving the first k shocks, $k = 1,2,\ldots,$ (i.e. the probability that amount of damage measured by $L(t)$ after k shocks is less than or equal to ξ) than the life-time distribution is

$$P\left\{T > t\right\} = F_\xi(t) = \sum_{k=0}^{\infty} P_k(\xi) e^{-\lambda t} \frac{(\lambda t)^k}{k!}, \quad t \ge 0 \tag{5.2}$$

where λ is the intensity of the Poisson stream of shocks. Of course

$$1 \geq P_o(\xi) \geq P_1(\xi) \geq \ldots \tag{5.3}$$

Since $1 - \bar{F}_\xi(t) = F_\xi(t)$ is a distribution function, so $P_k(\xi) \longrightarrow 0$ as $k \longrightarrow \infty$. Probability density corresponding to $F_\xi(t)$, that is the probability density of the time to reach a critical value is

$$f_\xi(t) = \lambda e^{-\lambda t} \sum_{k=0}^{\infty} \frac{(\lambda t)^k}{k!} \left[P_k(\xi) - P_{k+1}(\xi)\right] \tag{5.4}$$

where

$$P_k(\xi) = \int_0^\xi P_{k-1}(x)\, g(\xi - x) dx \quad , \quad P_o(\xi) = 1 \tag{5.5}$$

and $g(x)$ is the probability density of the amount of elementary damage (characterized by random variable Y).

If the size of elementary incereament has an exponential distribution, that is

$$g(x) = \beta e^{-\beta x}$$

then

$$f_\xi(t) = \lambda e^{-(\lambda t + \beta\xi)} \sum_{k=0}^{\infty} \frac{(\lambda t)^k}{k!} \frac{(\beta\xi)^k}{k!} \tag{5.6}$$

Hence, the mean and the variance of the first passage time are

$$\langle T \rangle = \frac{1}{\lambda}(1 + \beta\xi)$$

$$\text{var } T + \frac{1}{\lambda^2}(1 + 2\beta\xi) \tag{5.7}$$

Let us notice that we can easily obtain the fatigue failure rate associated with the model considered. Namely, by virtue of (5.2) and (5.4) we have

$$\mu(t) = \lambda \frac{1 - \sum_{k=0}^{\infty} P_{k+1} \frac{(\lambda t)^k}{k!}}{\sum_{k=0}^{\infty} P_k \frac{(\lambda t)^k}{k!}} , \quad t > 0 \qquad (5.8)$$

Let us notice that $\mu(t) \le \lambda$. Furthermore, $\mu(t) = \lambda$ for some $t > 0$ if and only if $P_k = 0$ for all $k>0$. Since (2.9) holds we also have

$$\bar{F}_{\xi}(t) \ge \bar{F}_{\xi}(t_o)\, e^{-\lambda t} \quad , \quad t > t_o \qquad (5.9)$$

If the sequence of shocks, or the sequence of jumps in the crack length is generated by the maxima of random stress process distributed above a certain level s_o then the intensity occuring in (5.2), (5.4) can be expressed in terms of statistical information of this random process. If the level s_o is high enough so that the local maxima are distributed independently according to Poisson distribution on time axis, then λ can be regarded as the expected rate of maxima above s_o.

If the applied random stress process $S(t)$ is at least twice the mean square differentiable, then λ can be taken as

$$\lambda = \mu_+(s_o, t) = - \int_{-\infty}^{0} \left\{ \int_{s_o}^{\infty} \ddot{s}\, f(s,0,\ddot{s};\, t) \right\} d\ddot{s} \qquad (5.10)$$

where $f(s,\dot{s},\ddot{s};\ t)$ is the joint (one-dimensional) probability density of the process $[S(t), \dot{S}(t), \ddot{S}(t)]$. If $S(t)$ is a stationary random process then $\mu_+(s_o, t)$ is independent of t , that is $\lambda = \mu_+(s_o)$. If, moreover, the process $S(t)$ is Gaussian with mean zero, thean

$$\lambda = \mu_+(s_o) = (2\pi)^{-\frac{3}{2}}(\sigma_s, \sigma_{\dot{s}})^{-2} \int_{s_o}^{\infty} \left\{ |\Lambda|^{-\frac{1}{2}} \exp\left[-\frac{1}{2|\Lambda|} \sigma_{\dot{s}}^2 \sigma_{\ddot{s}}^2 s^2 \right] + \right.$$

$$(5.11)$$

$$\left. + \left(\frac{\pi}{2}\right)^{\frac{1}{2}} \left(\frac{\sigma_{\dot{s}}^3 s}{\sigma_s}\right) \exp\left(-\frac{s^2}{2\sigma_s^2}\right) \left[1 + \operatorname{erf}\left(\frac{\sigma_{\dot{s}}^3 s}{\sqrt{2}\, \sigma_s |\Lambda|^{1/2}} \right) \right] \right\} ds$$

where erf $(\cdot)$ is the error function defined by

$$\operatorname{erf}(x) = \frac{2}{\sqrt{x}} \int_0^x \exp(-u^2) du$$

and

$$\Lambda = \begin{bmatrix} \sigma_s^2 & 0 & -\sigma_{\dot{s}}^2 \\ 0 & \sigma_{\dot{s}}^2 & 0 \\ -\sigma_{\dot{s}}^2 & 0 & \sigma_{\ddot{s}}^2 \end{bmatrix}$$

Exact evaluation of the integral in (5.11) is, in general, difficult. However, one can obtain simple expression for μ_+, the expected total number of maxima per unit time. It has the form

$$\mu_+ = \frac{1}{2\pi} \frac{\sigma_{\ddot{s}}}{\sigma_{\dot{s}}} \qquad (5.12)$$

14. FATIGUE CRACK GROWTH WITH RETARDATION; STOCHASTIC CUMULATIVE BIRTH MODEL

As it was pointed out in Sec. I, of particular interest is the decrease in growth rate (crack retardation) which follows a high overload. Retardation in fatigue crack growth due to overload is a highly complex phenomenon and so far no satisfactory model has been developed to account for all the observed behaviour.

Numerous investigations have made effort to treat the retardation phenomenon quantitatively. Since the stress intensity factor range, ΔK , turned out to be very useful in theoretical prediction of fatigue crack growth for constant amplitude loading the modelling of fatigue crack growth under irregular loading (including single and multiple overloads) is usually also based on various modifications of ΔK. However, there is no agreement concerning the ability of satisfactory prediction of real fatigue crack growth by these models,

It is likely that one reason that the existing quantitative descriptions of retarded crack growth are not satisfactory is that these models are deterministic while the fatigue crack growth process was recognized as highly complicated random phenomenon. So, it is essential to look at crack growth processes from a more general point of view taking into account an inherent randomness of the phenomenon. Such approach leads to the representation of the crack growth by a suitable stochastic process. In the paper by Ditlevsen and Sobczyk (1986) the authors have presented a stochastic cumulative model with underlying birth process.

The sample of the material is regarded as a system whose fatigue states are described by the stochastic process $L(t,\gamma)$, where $L(t,\gamma)$ is interpreted as a length of a dominant crack at time t; $\gamma \in \Gamma$ where Γ is a space of elementary events (sample space) on which the probability is defined (for each $\gamma \in \Gamma$, $L(t,\gamma)$ represents a possible sample function of the crack length). Since crack grows mainly due to sequence of events occuring randomly in time the crack size at time t is

$$L(t,\gamma) = L_o + Y_1(\gamma) + \ldots + Y_{N(t,\gamma)}(\gamma) = L_o + \sum_{i=1}^{N(t,\gamma)} Y_i(\gamma) \quad (5.13)$$

where L_o is the initial crack length (sufficient to propagate), $Y_i(\gamma) = \Delta L_i(\gamma)$ are the random partial crack increments and $N9t,\gamma)$ is an integer-valued stochastic process characterizing a number of crack incerements in the interval $[0,t]$. It is assumed that $Y_i(\gamma)$ are independent and identically distributed non-negative random variables with the common distribution $G(\gamma)$. Randomness of the fatigue crack growth process is, therefore, taken into account through the probabilistic mechanism of the transition from one state to another (process $N(t)$) and by the fact that elementary increments are allowed to be random. Since in modelling of fatigue it is essential the growth intensity be state-dependent the process $N(t,\gamma)$ is assumed to be the birth process (originating in the theory of population growth).

In the case of homogenoeous periodic loading the

infinitesimal intensity of $N(t)$ is assumed as

$$\lambda_k = \lambda^o k \quad , \quad k = 1, 2, \ldots \quad , \quad \lambda^o > 0 \tag{5.14}$$

which means that the probability of transition from state k to $k+1$ in the interval $(t, t+\Delta t)$ is proportional to state k. The average crack size grows exponentially with time and the rate of this exponential growth can be related to the basic characteristics of the loading and material properties.

Let us consider a more general case when the crack growth is generated by periodic loading with n overloads of different magnitude which are sufficiently separated to assure that retardation is effectively over after each overload. Although some sequence effects can, in general, occur the influence of the peak overload sequence will be neglected; this sequence effect is ofen considered secondary as compared to the stress magnitude effects (Porter, 1972). In this case the growth intensity is postulated to have the form

$$\lambda_k = \lambda_k(t) = k\,[\lambda^o - \lambda_{oL}(t; t_1, t_2, \ldots, t_n)] \tag{5.15}$$

with

$$\lambda_{oL}(t; t_1, t_2, \ldots, t_n) = \sum_{i=1}^{n} \mu_i(t,\zeta)\, e^{-\alpha_i (t-t_i)} H_i(t-t_i) \tag{5.16}$$

$$H_i(t-t_i) = \begin{cases} 1 \ , \ t_i + \nu_i < t < t_i + \nu_i + \tau_i \\ 0 \ , \ \text{otherwise} \end{cases} \tag{5.17}$$

where $t_{i+1} - t_i > \tau_i$ and $0 < \lambda_{oL} < \lambda^o$.

The factor $\mu_i(t,\zeta)$ suitable bounded, describes the retarding effect of an overload at $t = t_i$ $(i = 1, 2, \ldots, n)$; α_i is a decay parameter associated with i-th overload. a delay ν_i in retardation can likely be assumed to be the same for each overload, i.e. ν_i. The retardation magnitude function $\mu(t,\zeta)$ depends on time t and a collection of relevant variables denoted symbolically by ζ (e.g. the overload ratio, the stress biaxially etc.). The parameter τ_i is the retardationtime associated with i-th overload;

when $t = t_i + \tau_i$ the growth rate comes back to its level just prior to the overload.

A situation which is ofen investigated experimentally is concerned with the blocks of overloads. It has been observed that when the blocks of overloads are applied instead of a single overload a longer retardation is obtained. If the n overload blocks are separated enough it seems to be justified to chracterize this case by retarding intensity (eqs (5.16), (5.17)) where t_i is a characteristic instant for an i-th overload block and $\mu_i(t,\zeta)$ is replaced by $\bar{\mu}_i(t,\zeta) = \beta_i \mu_i(t,\zeta)$ where $\beta_i > 1$ describes the block effect. Also the retardation times τ_i should be modified adequately.

If the time intervals between overloadings are not long enough to assure independence of the retarding effects the situation is much more complicated. In this case the retardations are less obvious as seen at the crack growth curve. In fact, the effects of interaction of overloadings can be either positive (retardation) or negative (accelerated growth). The existing experimental results give no clear indication of how to characterize quantitatively the interaction effects of overloading. Many mechanisms seems to contribute to these effects and it is difficult to recognize and separate them. It is observed that reapplication of a high overload after a previous one early enough before the retardation from the previous overload has become effective, may considerably reduce the crack growth retardation. A possible way of modelling the crack growth under periodic loading with irregularly occuring overloads (within the cumulative birth model decribed above) has also been proposed in Ditlevsen and Sobczyk (1986); the "birth" growth intensity has been represented as

$$\lambda_k(t) = \lambda_{int}(t)\, \lambda_k = \Psi(t)\, X(t,\gamma)\lambda_k \qquad (5.18)$$

where $\lambda_{int}(t)$ is an interaction factor and characterizes an overall effect of irregularity of peak overloads and the load history. Since this effects are random $\lambda_{int}(t)$ is represented according to eq.(5.18) where $X(t,\gamma)$ describes random effect of overload interaction.

Statistics of crack size

The probabilistic descriptiont of the crack size can be obtained by calculating first the characteristic function of process $L(t,\gamma)$. The first two moments are given by the formulae ($\langle\cdot\rangle$ denotes the average value)

$$\langle L(t)\rangle = L + \langle N(t)\rangle \langle Y(\gamma)\rangle$$
$$= L_o + \langle Y\rangle e^{\eta(t)} , \qquad (5.19)$$

$$\mathrm{var}\, L(t) = e^{\eta(t)} \mathrm{var}\, Y + \langle Y\rangle^2 e^{\eta(t)} \left[e^{\eta(t)} - 1\right] \qquad (5.20)$$

where (cf. Ditlevsen, Sobczyk, 1986):

$$\eta(t) = \int_o^t \lambda(s)\, ds \qquad (5.21)$$

and $\lambda(t)$ is the time varying factor in the birth intensity. In the case of homogeneous periodic loading

$$\lambda(t) = \lambda^o = \text{const.} \qquad (5.22)$$

and in the case of multiple (separated) overloads

$$\lambda(t) = \lambda^o - \lambda_{oL}(t; t_1, \ldots, t_n) \qquad (5.23)$$

where $\lambda_{oL}(t; t_1, \ldots, t_n)$ is given by eq. (5.16) and (5.17).

Of course, in case of eq. (5.22) $\eta(t) = \lambda^o t$ and

$$\langle L(t)\rangle = L_o + \langle Y\rangle e^{\lambda^o t} ,$$
$$\mathrm{var}\, L(t) = e^{\lambda^o t}\left\{\mathrm{var}\, Y + \langle Y\rangle^2 (e^{\lambda^o t} - 1)\right\} \qquad (5.24)$$

In the case (5.23)

$$\eta(t) = \lambda^o t - \sum_{i=1}^{n} \mu_i(\zeta) \frac{1}{\alpha_i} \left[e^{-\alpha_i v_i} - e^{-\alpha_i (t-t_i)}\right] H_i(t-t_i) \qquad (5.25)$$

where it has been assumed that $\mu(t,\zeta) = \mu(\zeta)$.

The model described above needs appropriate empirical verification by comparison with data from the experiment in which the crack growth is treated as random cumulative process. The characteristics of the model such as statistics of random elementary increments as well as the properties of the infinitesimal growth intensity λ^o should then be estimated from the data. Since such a methodology is a future matter the parameters of the model should be related to the existing experimental predictions (cf. Ditlevsen, Sobczyk, 1986). In particular, the retardation time occuring in the model (and equal to: $\tau = \frac{1}{\omega} N_{ret}$) can be expressed in terms of parameters occuring in the Wheeler empirical model according to which

$$\left(\frac{dL}{dN}\right)_{oL} = C_p \left(\frac{dL}{dN}\right)_{C.A.} \tag{5.26}$$

where

$$C_p = \left[\frac{r_{pi}}{L_* + r_{po} - L_i} \right]^m = \left[\frac{r_{pi}}{s - L_i} \right]^m , \quad r_{pi} < s - L_i$$

$$= 1 \qquad\qquad , \quad r_{pi} \geq s - L_i \tag{5.27}$$

and r_{pi} = current plastic zone (in the i-th cycle): L_i = current crack size; r_{po} = plastic zone size caused by overload; L_* = crack size at which overload occured; m = empirical (data fitting) constant; $s = L_* + r_{po}$.

The number of cycles, N_{ret}, over which retardation occurs (after each overload) has been estimated from the above formulae. In particular (since $L_i \approx L_*$ (cf. Broek, Smith, 1979) $C_p = \left(\frac{r_i}{r_{po}} \right)^m$) it can be shown that

$$N_{ret} = \frac{\rho^{2m}}{C\alpha\, \sigma_{ys}^2} \; \frac{\rho^2 - 1}{(\sigma_{max,i})^{n-2} (\pi L_*)^{n/2 - 1}} \tag{5.28}$$

where $\rho = \frac{K_{max,o}}{K_{max,i}}$ is the overload ratio, $K_{max,i}$ is the maximum stress intensity in a given cycle, $K_{max,o}$ is the stress intensity at the overload, $\alpha = 2\pi$ for plane stress and $\alpha = 6\pi$ for strain, C and n are constants occuring in the Paris equation adopted for expressing $\left(\frac{dL}{dN}\right)_{C.A}$.

REFERENCES

1. Antelman G., Savage I.R., Characteristic function of stochastic integrals and reliability problems, Naval Res.Lag.Quart., Vol.12, No 3, 1965.

2. Bogdanoff J.L., A new cumulative damage model, P.I, J.Appl, Mech., Vol,45, June 1978, 245-250. P.III, J.Appl.Mech., Vol.45, Dec.1978, 733-739.

3. Bogdanoff J.L., Kozin F., A new cumulative damage model, P.IV, J.Appl.Mech., Vol.47, March 1980, 40-44.

4. Bogdanoff J.L., Kozin F., Probabilistic models of cumulative damage, John Wiley and Sons, 1985

5. Bolotin W.W., Some mathematical and experimental models of damage (in Russian), Problems of Strength, No 2, 1971.

6. Bolotin W.W., Prediction of life-time of materials and structures (in Russian), Mashinostrojenije, Moskwa 1984.

7. Ditlevsen O, Sobczyk K., Random fatigue crack growth with retardation, Eng.Fract.Mech., Vol.24, No 6, 1986.

8. Kocańda S., Fatigue failure of metals, Sijthoff--Noorthoff, Intern.Publ. 1978.

9. Kozin F., Bogdanoff J.L., A critical analysis of some probabilistic models of fatigue crack growth, Eng.Fracture Mech, Vol.14, No 1, 1981.

10. Lin Y.K., Yang J.N., On statistical moments of fatigue crack propagation, Eng.Fracture Mech., Vol.18, No 2, pp. 243-256, 1983.

11. Madson H.O., Deterministic and probabilistic models for damage accumulation due to time varying loading, Dialog, 5-82, Denmark Ingenioakademi, Lyngby.

12. Oh K.P., A diffusion model for fatigue crack growth, Proc. Roy.Society London A367 , pp.47-58, 1979.

13. Schijve J., "Observations on the prediction of fatigue crack growth under variable amplitude loading", in: Fatigue Crack Growth under Spectrum Loads, ASTM STP 595, Amer.Soc. Testing Materials, pp.3-27, 1976.

14. Sobczyk K., On the Markovian models for fatigue accumulation, J.Mecanique Theor. et Appl., No special, 1982, 147-160.

15. Sobczyk K., On the reliability models for random fatigue damage, Proc. of the Third Swedish-Polish Symp. on "New Problems in Continuum Mech." (Eds Brulin O., Hsieh R.) Waterloo Univ.Press, 1983.

16. Sobczyk K., Modelling of random fatigue crack growth, Eng.Fract.Mech., Vol.24, No 4, 1986.

17. Sobczyk K., Stochastic models for fatigue damage of materials, Adv.Appl.Probability, 19, 652-673, 1987.

A TUTORIAL INTRODUCTION TO DIFFERENTIAL MANIFOLDS, CALCULUS OF MANIFOLDS AND LIE ALGEBRAS

M. Hazewinkel
CWI, Amsterdam, The Netherlands

ABSTRACT

The tutorial introduction is treated in two papers as follows.

1. Differential Manifolds and Calculus of Manifolds

2. Lie Algebras

A TUTORIAL INTRODUCTION TO DIFFERENTIABLE MANIFOLDS AND CALCULUS ON MANIFOLDS

M. Hazewinkel
CWI, Amsterdam, The Netherlands

In this tutorial I try by means of several examples to illustrate the basic definitions and concepts of differentiable manifolds. There are few proofs (not that there are ever many at this level of the theory). This material should be sufficient to understand the use made of these concepts in the other contributions in this volume, notably the lectures by Kliemann, and my own lectures on filtering; or at least, it should help in explaining the terminology employed. Quite generally in fact, it can be said that the global point of view, i.e. analysis on manifolds rather than on open pieces of $\mathbf{R}^n$, can have many advantages, also in areas like engineering where this approach is less traditional. This tutorial is a revised and greatly expanded version of an earlier one entitled 'A tutorial introduction to differentiable manifolds and vector fields' which appeared in M. Hazewinkel, J.C. Willems (eds), Stochastic Systems: the mathematics of filtering and identification, Reidel, 1981, 77-93.

1. Introduction and a few remarks

Roughly an n-dimensional differentiable manifold is a gadget which locally looks like $\mathbb{R}^n$, the space of all real vectors of length n, but globally perhaps not; A precise definition is given below in section 2. Examples are the sphere and the torus, which are both locally like $\mathbb{R}^2$ but differ globally from $\mathbb{R}^2$ and from each other.

Such objects often arise naturally when discussing problems in analysis (e.g. differential equations) and elsewhere in mathematics and its applications. A few advantages which may come about by doing analysis on manifolds rather than just on $\mathbb{R}^n$ are briefly discussed below.

1.1 Coordinate freeness ("Diffeomorphisms").

A differentiable manifold can be viewed as consisting of pieces of $\mathbb{R}^n$ which are glued together in a smooth (= differentiable) manner. And it is on the basis of such a picture that the analysis (e.g. the study of differential equations) often proceeds. This brings more than a mere extension of analysis on $\mathbb{R}^n$ to analysis on spheres, tori, projective spaces and the like; it stresses the "coordinate free approach", i.e. the formulation of problems and concepts in terms which are invariant under (non-linear) smooth coordinate transformations and thus also helped to bring about a better understanding even of analysis on $\mathbb{R}^n$. The more important results, concepts and definitions tend to be "coordinate free".

1.2 Analytic continuation.

A convergent power series in one complex variable is a rather simple object. It is considerably more difficult to obtain an understanding of the collection of all analytic continuations of a given power series, especially because analytic continuation along a full circuit (contour) may yield a different function value than the initial one. The fact that the various continuations fit together to form a Riemann surface (a certain kind of 2-dimensional manifold usually different from $\mathbb{R}^2$) was a major and most enlightening discovery which contributes a great deal to our understanding.

1.3 Submanifolds.
Consider an equation $\dot{x}=f(x)$ in $\mathbf{R}^n$. Then it often happens, especially in problems coming from mechanics, that the equation has the property that it evolves in such a way that certain quantities (e.g. energy, angular momentum) are conserved. Thus the equation really evolves on a subset $\{x\in\mathbf{R}^n:E(x)=c\}$ which is often a differentiable submanifold. Thus it easily could happen, for instance, that $\dot{x}=f(x)$, f smooth, is constrained to move on a (disorted) 2-sphere which then immediately tells us that there is an equilibrium point, i.e. a point where $f(x)=0$. This is the so-called hairy ball theorem which says that a vectorfield on a 2-sphere must have a zero; for vectorfields and such, cf below.

Also one might meet 2 seemingly different equations, say, one in $\mathbf{R}^4$ and one in $\mathbf{R}^3$ (perhaps both intended as a description of the same process) of which the first has two conserved quantities and the second one. It will then be important to decide whether the surfaces on which the equations evolve are diffeomorphic, i.e. the same after a suitable invertible transformation and whether the equations on these submanifolds correspond under these transformations.

1.4 Behaviour at infinity.
Consider a differential equation in the plane $\dot{x}=P(x,y)$, $\dot{y}=Q(x,y)$. To study the behaviour of the paths far out in the plane and such things as solutions escaping to infinity and coming back, Poincaré already completed the plane to real projective 2-space (an example of a differential manifold). Also the projective plane is by no means the only smooth manifold compactifying $\mathbf{R}^2$ and it will be of some importance for the behaviour of the equation near infinity whether the "right" compactification to which the equation can be extended will be a projective 2-space, a sphere, or a torus, or ..., or, whether no such compactification exists at all. A good example of a set of equations which are practically impossible to analyse completely without bringing in manifolds are the matrix Riccati equations which naturally live on Grassmann manifolds. The matrix Riccati equation is of great importance in linear Kalman-Bucy filtering. It also causes major numerical difficulties. It will therefore return below by way of example.

1.5 Avoiding confusion between different kinds of objects.
Consider an ordinary differential equation $\dot{x}=f(x)$ on $\mathbf{R}^n$, where $f(x)$ is a function $\mathbf{R}^n\to\mathbf{R}^n$. When one now tries to generalize this idea of a differential equation on a manifold one discovers that $\dot{x}$ and hence $f(x)$ is a different kind of object; it is not a function, but, as we shall see, it is a vectorfield; in other words under a nonlinear change of coordinates the right hand side of such a differential equation $\dot{x}=f(x)$ transforms not as a function, but in a different way (involving Jacobian matrices, as everyone knows).

2. Differentiable manifolds

Let U be an open subset of $\mathbf{R}^n$, e.g. an open ball. A function $f:U\to\mathbf{R}$ is said to be C^∞ or *smooth* if all partial derivatives (any order) exist at all $x\in U$. A mapping $\mathbf{R}^n\supset U\to\mathbf{R}^m$ is smooth if all components are smooth; $\phi:U\to V$, $U\subset\mathbf{R}^n$, $V\subset\mathbf{R}^n$ is called a diffeomorphism if ϕ is 1-1, onto, and both ϕ and ϕ^{-1} are smooth.

As indicated above a smooth n-dimensional manifold is a gadget consisting of open pieces of $\mathbf{R}^n$ smoothly glued together. This gives the following pictorial definition of a smooth n-dimensional manifold M (fig. 1).

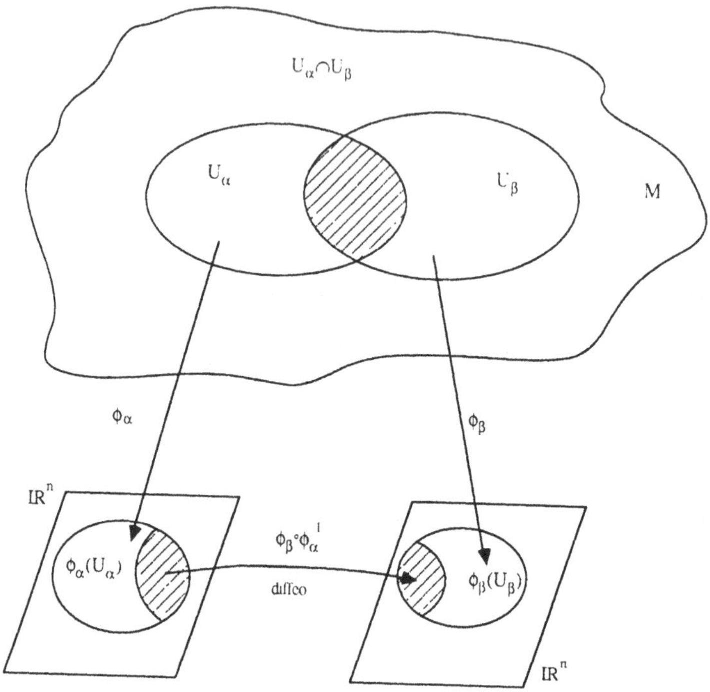

FIGURE 1. Pictorial definition of a differentiable manifold.

2.1 Example.
The circle $S^1 = \{(x_1, x_2) : x_1^2 + x_2^2 = 1\} \subset \mathbb{R}^2$

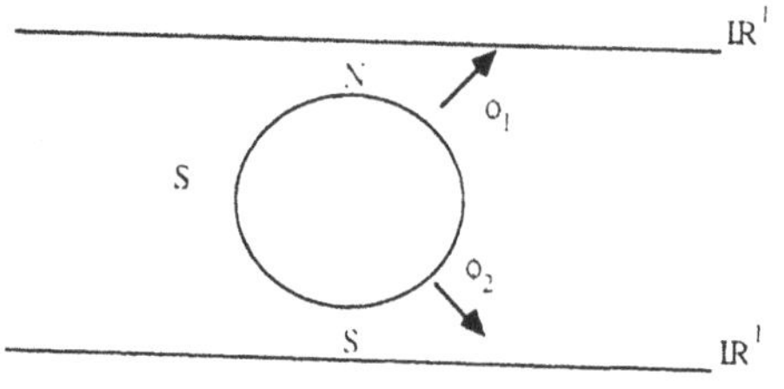

FIGURE 2. Example: the circle

$U_1 = S^1 \setminus \{S\}$, $U_2 = S^1 \setminus \{N\}$ so $U_1 \cup U_2 = S^1$. The "coordinate charts" ϕ_1 and ϕ_2 are given by

$$\phi_1(x_1,x_2) = \frac{x_1}{1+x_2} \quad , \quad \phi_2(x_1,x_2) = \frac{x_1}{1-x_2}$$

Thus $\phi_1(U_1 \cap U_2) = \mathbb{R} \setminus \{0\}$, $\phi_2(U_1 \cap U_2) = \mathbb{R} \setminus \{0\}$ and the map $\phi_2 \circ \phi_1^{-1} : \mathbb{R} \setminus \{0\} \to \mathbb{R} \setminus \{0\}$ is given by $x \mapsto x^{-1}$ which is a diffeomorphism.

2.2 Formal definition of a differentiable manifold.
The data are
- M, a Hausdorff topological space
- A covering $\{U_\alpha\}_{\alpha \in I}$ by open subsets of M
- Coordinate maps $\phi_\alpha : U_\alpha \to \phi_\alpha(U_\alpha) \subset \mathbb{R}^n$, $\phi_\alpha(U_\alpha)$ open in $\mathbb{R}^n$.

These data are subject to the following condition
- $\phi_\alpha \circ \phi_\beta^{-1} : \phi_\beta(U_\alpha \cap U_\beta) \to \phi_\alpha(U_\alpha \cap U_\beta)$ is a diffeomorphism.

Often one also adds the requirement that M be paracompact. We shall however disregard these finer points; nor shall we need them in this volume.

2.3 Constructing differentiable manifolds 1: embedded manifolds.
Let M be a subset of $\mathbb{R}^N$. Suppose for every $x \in M$ there exists an open neighbourhood $U \subset \mathbb{R}^n$ and a smooth function $\psi : U \to \mathbb{R}^N$ mapping U homeomorphically onto an open neighbourhood V of x in M. Suppose moreover that the Jacobian matrix of ψ has rank n at all $u \in U$. Then M is a smooth manifold of dimension n. (Exercise; the coordinate neighbourhoods are the V's and the coordinate maps are the ψ^{-1}; use the implicit function theorem). Virtually the same arguments show that if $\phi : U \to \mathbb{R}^k$, $U \subset \mathbb{R}^{n+k}$, is a smooth map and the rank of the Jacobian matrix $J(f)(x)$ is k for all $x \in \phi^{-1}(0)$, then $\phi^{-1}(0)$ is a smooth n-dimensional manifold. We shall not pursue this approach but concentrate instead on:

2.4 Constructing differentiable manifolds 2: gluing.
Here the data are as follows
- an index set I
- for every $\alpha \in I$ an open subset $U_\alpha \subset \mathbb{R}^n$
- for every ordered pair (α, β) an open subset $U_{\alpha\beta} \subset U_\alpha$
- diffeomorphisms $\phi_{\alpha\beta} : U_{\alpha\beta} \to U_{\beta\alpha}$ for all $\alpha, \beta \in I$

These data are supposed to satisfy the following compatibility conditions
- $U_{\alpha\alpha} = U_\alpha$, $\phi_{\alpha\alpha} = id$
- $\phi_{\beta\gamma} \circ \phi_{\alpha\beta} = \phi_{\alpha\gamma}$ (where appropriate)

(where the last identity is supposed to imply also that $\phi_{\alpha\beta}(U_{\alpha\beta} \cap U_{\beta\gamma}) \subset U_{\beta\gamma}$, so that $\phi_{\alpha\beta}(U_{\alpha\beta} \cap U_{\alpha\gamma}) = U_{\beta\gamma} \cap U_{\beta\alpha}$).

These are not yet all conditions, cf below, but the present lecturer, e.g., has often found it advantageous to stop right here so to speak, and to view a manifold simply as a collection of open subsets of $\mathbb{R}^n$ together with gluing data (coordinate transformation rules).

From the data given above one now defines an abstract topological space M by taking the disjoint union of the U_α and then identifying $x \in U_\alpha$ and $y \in U_\beta$ iff $x \in U_{\alpha\beta}$, $y \in U_{\beta\alpha}$, $\phi_{\alpha\beta}(x) = y$. This gives a natural injection $U_\alpha \to M$ with image U'_α say. Let $\phi_\alpha : U'_\alpha \to U_\alpha$ be the inverse map. The $\phi_\alpha : U'_\alpha \to U_\alpha \subset \mathbb{R}^n$ define local coordinates on M. Then this gives us a differentiable manifold M in the sense of definition 2.2 provided that M is Hausdorff and paracompact, and these are precisely the conditions which must be added to the gluing compatibility conditions above.

2.5 Functions on a "glued manifold".
Let M be a differentiable manifold obtained by the gluing process described in 2.4 above. Then a differentiable function $f:M\rightarrow\mathbf{R}$ consist simply of a collection of functions $f_\alpha:U_\alpha\rightarrow\mathbf{R}$ such that $f_\beta\circ\phi_{\alpha\beta}=f_\alpha$ on $U_{\alpha\beta}$, as illustrated in fig. 3.

Thus for example a function on the circle S^1, cf figure 2, can be described either as a function of two variables restricted to $S^1\subset\mathbf{R}^2$ or as two functions f_1,f_2 of one variable on U_1 and U_2 such that $f_1(x)=f_2(x^{-1})$. Obviously the latter approach can have considerable advantages.

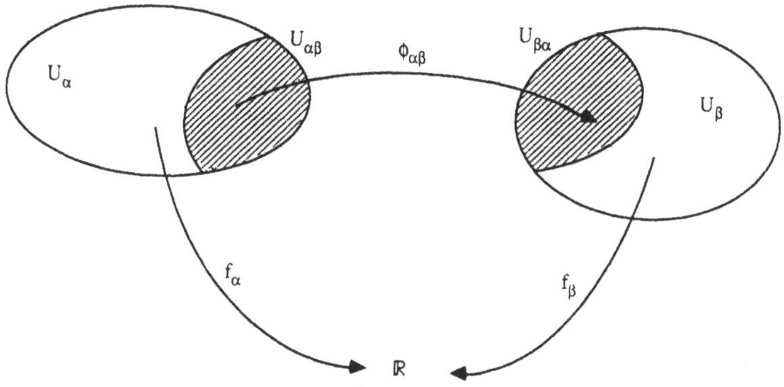

FIGURE 3. Functions on a glued manifold

2.6 Example of a 2 dimensional manifold: the Möbius band.
The (open) Möbius band is obtained by taking a strip in $\mathbf{R}^2$ as indicated below in fig. 4 without its upper and lower edges and identifying the left hand and right hand edges as indicated.

FIGURE 4. Construction of the Möbius band

The resulting manifold (as a submanifold of $\mathbf{R}^3$) looks something like the following figure 5.

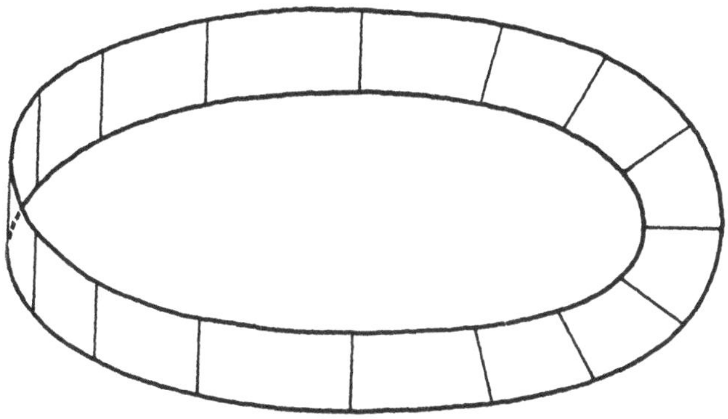

FIGURE 5. The Möbius band

It is left as an exercise to the reader to cast this description in the form required by the gluing description of 2.4 above. The following pictorial description (fig. 6) will suffice.

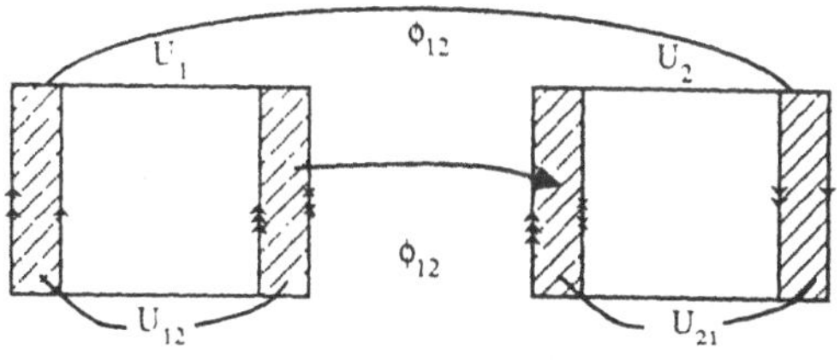

FIGURE 6. Gluing description of the Möbius band

2.7 Example: the 2-dimensional sphere.
The picture in fig. 7 below shows how the 2-sphere $S^2 = \{x_1, x_2, x_3 : x_1^2 + x_2^2 + x_3^2 = 1\}$ can be obtained by gluing two disks together. If the surface of the earth is viewed as a model for S^2 (or vice versa, which is the more customary use of the world 'model'), the first disk covers everything north of Capricorn and the second everything south of Cancer.

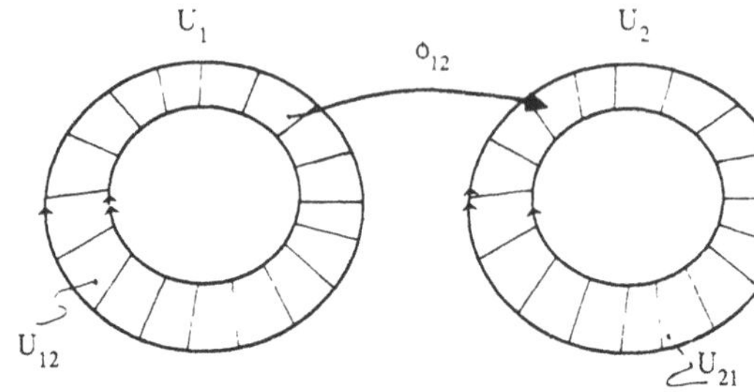

FIGURE 7. Gluing description of the 2-sphere S^2

2.8 Example. The Grassmann manifolds $Gr_k(\mathbf{R}^n)$.
As a set $Gr_k(\mathbf{R}^n)$ consists of all k-dimensional subspaces of $\mathbf{R}^n$. Thus $Gr_1(\mathbf{R}^n)$ is real projective space of dimension $n-1$ and in particular $Gr_1(\mathbf{R}^2)$ is the real projective line, i.e. the circle. We shall now also present a gluing data description of $Gr_k(\mathbf{R}^n)$. To this end it is useful to introduce the following notation. Let A be an $k \times n$ matrix, $k<n$ and let α be a subset of $\{1, \ldots, n\}$ of size k. Then A_α denotes the $k \times k$ matrix obtained from A by removing all columns whose index is not in α.

Now let U_α be the set of all $k \times n$ matrices A such that $A_\alpha = I_k$, the $k \times k$ identity matrix

$$U_\alpha = \{A \in \mathbf{R}^{k \times n} : A_\alpha = I_k\}$$

Because the entries a_{ij} with $j \in \{1, \ldots, n\} \setminus \alpha$ of these matrices are arbitrary this is clearly just a slightly crazy way of writing down all real $k \times (n-k)$ matrices or, in other words, all real $k(n-k)$ vectors, i.e. $U_\alpha \simeq \mathbf{R}^{k(n-k)}$.
The gluing data for $Gr_k(\mathbf{R}^n)$ are now as follows
- the index set I consists of all subsets α of size k of $\{1, \ldots, n\}$
- for each α, $U_\alpha = \mathbf{R}^{k(n-k)}$ realized as indicated above
- for each ordered pair of indices α, β

$$U_{\alpha\beta} = \{A \in U_\alpha : A_\beta \text{ is invertible}\}$$

- the diffeomorphsims

$$\phi_{\alpha\beta} : U_{\alpha\beta} \to U_{\beta\alpha}$$

are given by

$$A \mapsto (A_\beta)^{-1} A$$

We shall see below (in 2.12) that $Gr_k(\mathbf{R}^n)$ is indeed the space of all k-dimensional subspaces of $\mathbf{R}^n$.

2.9 Exercise.
Check that the compatibility conditions $\phi_{\alpha\alpha} = id$ and $\phi_{\beta\gamma} \circ \phi_{\alpha\beta} = \phi_{\alpha\gamma}$ of 2.4 above hold. Prove also that the manifold obtained from these gluing data is Hausdorff.

2.10 Morphism of differentiable manifolds.
Let M and N be differentiable manifolds obtained by the gluing process of section 2.4 above. Say M is obtained by gluing together open subsets U_α of $\mathbb{R}^n$ and N by gluing together open subsets V_β of $\mathbb{R}^m$. Then a smooth map $f:M\rightarrow N$ (a morphism) is given by specifying for all α,β an open subset $U_{\alpha\beta}\subset U_\alpha$ and a smooth map $f_{\alpha\beta}:U_{\alpha\beta}\rightarrow V_\beta$ such that $\bigcup_\beta U_{\alpha\beta}=U_\alpha$ and the $f_{\alpha\beta}$ are compatible under the identifications $\phi_{\alpha\alpha'}:U_{\alpha\alpha'}\rightarrow U_{\alpha'\alpha}, \phi_{\beta\beta'}:V_{\beta\beta'}\rightarrow V_{\beta'\beta}$, i.e. $f_{\alpha'\beta'}\circ\phi_{\alpha\alpha'}=\phi_{\beta\beta'}\circ f_{\alpha\beta}$ whenever appropriate. (Here the ϕ'_s are the gluing diffeomorphisms for M and the ψ'_s are the gluing diffeomorphisms for N).

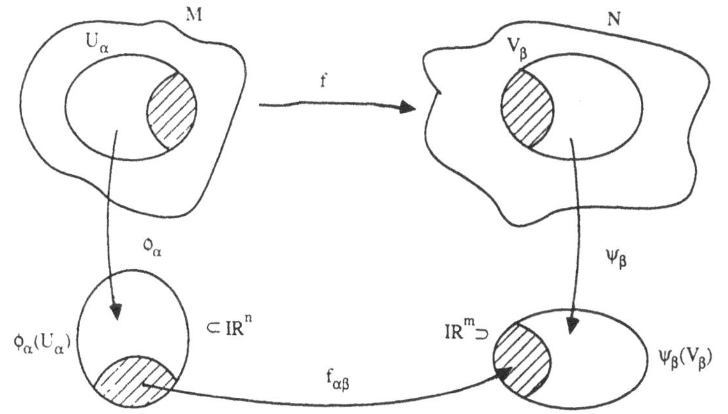

FIGURE 8. Morphisms

2.11 Exercise:
Show that the description of the circle S^1 as in 2.1 above gives an injective morphism $S^1\rightarrow\mathbf{R}^2$.

2.12 Example: Grassmann manifolds continued.
Let $\mathbf{R}^{k\times n}_{reg}$ be the open subset of $\mathbf{R}^{k\times n}=\mathbf{R}^{kn}$ consisting of all $k\times n$ matrices of maximal rank k. (Recall that $k<n$.) We are going to define a differentiable morphism

$$\pi:\mathbf{R}^{k\times n}_{reg}\rightarrow Gr_k(\mathbf{R}^n)$$

by the method of section 2.10 above. In this case $\mathbf{R}^{k\times n}_{reg}=U\subset\mathbf{R}^{kn}$ is defined by a single open subset. Thus we need for each α an open subset V_α of U and a smooth map $\pi_\alpha:V_\alpha\rightarrow U_\alpha$ where U_α is as above in 2.8. These data are defined as follows

$$V_\alpha=\{M\in\mathbf{R}^{k\times n}_{reg}: M_\alpha \text{ is invertible}\}$$

$$\pi_\alpha:V_\alpha\rightarrow U_\alpha,\ M\mapsto(M_\alpha)^{-1}M\in U_\alpha$$

It is an easy exercise (practically identical with the first part of exercise 2.9) to check that the required compatibility conditions are met.

It is now simple to see that $Gr_k(\mathbf{R}^n)$ as defined in 2.8 is indeed the space of all k-dimensional subspaces of $\mathbf{R}^n$. Indeed let W be such a subspace. Choose a basis for $W\subset\mathbf{R}^n$. These k n-vectors written as row vectors define $k\times n$ matrix $A(W)$ in $\mathbf{R}^{k\times n}_{reg}$. Taking a different basis for W amounts to replacing $A(W)$ with $SA(W)$ where S is an invertible $k \times k$ matrix. Now

$$(SA(W))_\alpha = S(A(W))_\alpha$$

and it follows that if $A(W) \in V_\alpha$ then also $SA(W) \in V_\alpha$ and that moreover

$$\pi(SA(W)) = \pi A(W)$$

Thus every k-dimensional vectorspace in $\mathbb{R}^n$ defines a unique point of $Gr_k(\mathbb{R}^n)$ and vice versa. ($A \in U_\alpha$ is of maximal rank and hence defines a k dimensional vectorspace.)

3. Differentiable vectorbundles

Intuitively a vectorbundle over a space S is a family of vectorspaces parametrized by S. Thus for example the Möbius band of example 2.6 can be viewed as a family of open intervals in $\mathbb{R}$ parametrized by the circle, cf fig. 9 below, and if we are willing to identify the open intervals with $\mathbb{R}$ this gives us a family of one dimensional vectorspaces parametrized by S^1 which locally (i.e. over small neighbourhoods in the base space S^1) looks like a product but globally is not equal to a product.

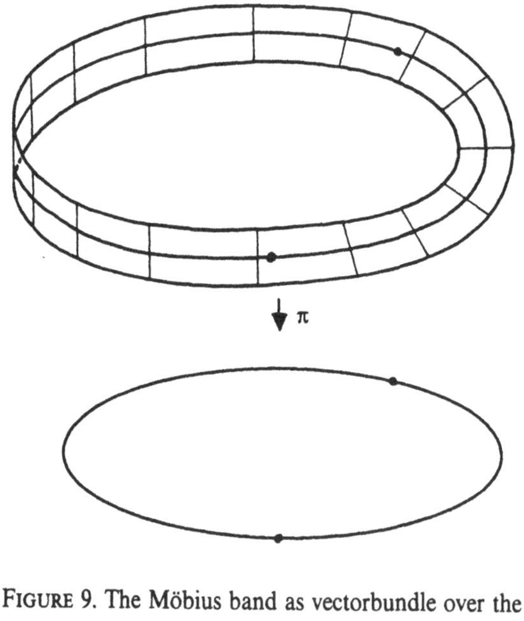

FIGURE 9. The Möbius band as vectorbundle over the circle

3.1 Formal definition of differentiable vectorbundle.

A differentiable vectorbundle of dimension m over a differentiable manifold M consists of a surjective morphism $\pi : E \to M$ of differentiable manifolds and a structure of an m-dimensional real vectorspace on $\pi^{-1}(x)$ for all $x \in M$ such that moreover there is for all $x \in M$ an open neighbourhood $U \subset M$ containing x and a diffeomorphism $\phi_U : U \times R^m \to \pi^{-1}(U)$ such that the following diagram commutes

$$\begin{array}{ccc} U\times\mathbf{R}^m & \xrightarrow{\phi_U} & \pi^{-1}(U) \\ & \searrow \quad \swarrow_{\pi} & \\ & U & \end{array}$$

where the lefthand arrow is the projection on the first factor, and such that ϕ_U induces for every $y\in U$ an isomorphism $\{y\}\times\mathbf{R}^m\rightarrow\pi^{-1}(y)$ of real vectorspaces.

3.2 Constructing vectorbundles.
The definition given above is not always particularly easy to assimilate. It simply means that a vectorbundle over M is obtained by taking an open covering $\{U_i\}$ of M and gluing together products $U_i\times\mathbf{R}^m$ by means of diffeomorphisms which are linear (i.e. vectorspace structure preserving) in the second coordinate. Thus an m-dimensional vectorbundle over M is given by the following data

- an open covering $\{U_\alpha\}_{\alpha\in I}$ of M.
- for every α,β a smooth map $\phi_{\alpha\beta}:U_\alpha\cap U_\beta\rightarrow GL_m(\mathbf{R})$ where $GL_m(\mathbf{R})$ is the space of all invertible real $m\times m$ matrices considered as an open subset of $\mathbf{R}^m$. These data are subject to the following compatibility conditions
- $\phi_{\alpha\alpha}(x)=I_m$, the identity matrix, for all $x\in U_\alpha$
- $\phi_{\beta\gamma}(x)\phi_{\alpha\beta}(x)=\phi_{\alpha\gamma}(x)$ for all $x\in U_\alpha\cap U_\beta\cap U_\gamma$

From these data E is constructed by taking the disjoint union of the $U_\alpha\times\mathbf{R}^m$, $\alpha\in I$ and identifying $(x,v)\in U_\alpha\times\mathbf{R}^m$ with $(y,w)\in U_\beta\times\mathbf{R}^m$ if and only if $x=y$ and $\phi_{\alpha\beta}(x)v=w$. The morphism π is induced by the first coordinate projections $U_\alpha\times\mathbf{R}^m\rightarrow U_\alpha$.

3.3 Constructing vectorbundles 2.
If the base manifold M is itself viewed as a smoothly glued together collection of open sets in $\mathbf{R}^n$ we can describe the gluing for M and for the vectorbundle all at once. The combined data are then as follows

- open sets $U_\alpha\times\mathbf{R}^m$, $U_\alpha\subset\mathbf{R}^n$ for all $\alpha\in I$
- open subsets $U_{\alpha\beta}\subset U_\alpha$ for all $\alpha,\beta\in I$
- diffeomorphsims $\phi_{\alpha\beta}:U_{\alpha\beta}\rightarrow U_{\beta\alpha}$
- diffeomorphisms $\tilde{\phi}_{\alpha\beta}:U_{\alpha\beta}\times\mathbf{R}^m\rightarrow U_{\beta\alpha}\times\mathbf{R}^m$ of the form $(x,v)\mapsto(\phi_{\alpha\beta}(x),A_{\alpha\beta}(x)v)$ where $A_{\alpha\beta}(x)$ is an $m\times m$ invertible real matrix depending smoothly on x.

These data are then subject to the same compatibility conditions for the $\tilde{\phi}_{\alpha\beta}$'s (and hence the $\phi_{\alpha\beta}$) as described in 2.4 above.

Again, as in the case of differentiable manifolds, it is sometimes a good idea to view a vectorbundle $\pi:E\rightarrow M$ simply as a collection of local pieces $\pi_\alpha:U_\alpha\times\mathbf{R}^m\rightarrow U_\alpha$ together with gluing data (transformation rules).

3.4 Example: the tangent vectorbundle of a smooth manifold.
Let the smooth manifold M be given by the data U_α, $U_{\alpha\beta}$, $\phi_{\alpha\beta}$ as in 2.4. Then the tangent bundle TM is given by the data

- $U_\alpha\times\mathbf{R}^n$, $U_{\alpha\beta}\times\mathbf{R}^n\subset U_\alpha\times\mathbf{R}^n$
- $\tilde{\phi}_{\alpha\beta}:U_{\alpha\beta}\times\mathbf{R}^n\rightarrow U_{\beta\alpha}\times\mathbf{R}^n$, $\tilde{\phi}_{\alpha\beta}(x,v)=(\phi_{\alpha\beta}(x),J(\phi_{\alpha\beta})(x)v)$

where $J(\phi_{\alpha\beta})(x)$ is the Jacobian matrix of $\phi_{\alpha\beta}$ at $x\in U_{\alpha\beta}$.

Exercise: check that these gluing morphisms do indeed define a vectorbundle; i.e. the compatibility. (This is the chain rule!)

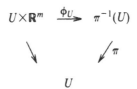

3.5 Example. The canonical bundle over a Grassmann manifold.
As said above, intuitively a vectorbundle over M is a family of vectorspaces smoothly parametrized by M. I.e. for each $x \in M$ there is given a vectorspace V_x, the fibre over x and the V_x vary smoothly with x. In this intuitive fashion the canonical bundle over $Gr_k(\mathbf{R}^n)$, the space of k dimensional subspaces of $\mathbf{R}^n$, is the bundle whose fibre over $x \in Gr_k(\mathbf{R}^n)$ "is" the vectorspace x.

In terms of gluing data, and more precisely, this vectorbundle is described as follows. Recall that $Gr_k(\mathbf{R}^n)$ was obtained from local pieces $U_\alpha \simeq \mathbf{R}^{k(n-k)}$

$$U_\alpha = \{A \in \mathbf{R}^{k \times n} : A_\alpha = I_k\}$$

Now define

$$\tilde{\phi}_{\alpha\beta}: U_{\alpha\beta} \times \mathbf{R}^k \to U_{\beta\alpha} \times \mathbf{R}^k$$
$$(A,v) \to (A_\beta^{-1}A, (A_\beta)^T v)$$

It is again the same observation that $(SA)_\alpha = SA_\alpha$ which proves the compatibility relation $\tilde{\phi}_{\beta\gamma} \circ \tilde{\phi}_{\alpha\beta} = \tilde{\phi}_{\alpha\gamma}$.

This bundle is the universal k-dimensional vectorbundle over $Gr_k(\mathbf{R}^n)$ as usually defined by topologists. The algebraic geometers often prefer to work with the dual object: the bundle over $Gr_k(\mathbf{R}^n)$ whose fibre over x is the quotient space $\mathbf{R}^n/x$. Exercise: give a gluing data description of this last bundle.

3.5 Morphisms of vectorbundles.
A morphism of vectorbundles from the vectorbundle $\pi: E \to M$ to the vectorbundle $\pi': E' \to M'$ is a pair of smooth maps $\tilde{f}: E \to E', f: M \to M'$ such that $\pi' \circ \tilde{f} = f \circ \pi$ and such that the induced map $\tilde{f}_x: \pi^{-1}(x) \to \pi^{-1}(f(x))$ is homomorphism of vectorspaces for all $x \in M$. We leave it to the reader to translate this into a local pieces and gluing data description.

As an example consider two manifolds M,N both described in terms of local pieces and gluing data. Let $f: M \to N$ be given in these terms by the $f_{\alpha\beta}: U_{\alpha\beta} \to V_\beta$ (cf 2.10 above). Then the maps $\tilde{f}_{\alpha\beta}: U_{\alpha\beta} \times \mathbf{R}^n \to V_\beta \times \mathbf{R}^m$ defined by $\tilde{f}_{\alpha\beta}(x,v) = (f_{\alpha\beta}(x), J(f_{\alpha\beta})(x)v)$ combine to define a morphism of vectorbundles $\tilde{f} = Tf: TM \to TN$.

4. VECTORFIELDS
A vectorfield on a manifold M assigns in a differentiable manner to every $x \in M$ a tangent vector at x, i.e. an element of the fibre $T_xM = \pi^{-1}(x)$ of the tangent bundle TM. Slightly more precisely this gives the

4.1 Definitions.
Let $\pi: E \to M$ be a vectorbundle. Then a section of E is a smooth map $s: M \to E$ such that $\pi \circ s = id$. A section of the tangent vectorbundle $TM \to M$ is called a vectorfield.

Suppose that M is given by a local pieces and gluing data description as in 2.4 above. Then a vectorfield s is given by "local sections" $s'_\alpha: U_\alpha \to U_\alpha \times \mathbf{R}^n$ of the form $s'_\alpha(x) = (x, s_\alpha(x))$, i.e. by a collection of functions $s_\alpha: U_\alpha \to \mathbf{R}^n$ such that $J(\phi_{\alpha\beta})(x)(s_\alpha(x)) = s_\beta(\phi_{\alpha\beta}(x))$ for all $x \in U_{\alpha\beta}$.

4.2 Derivations.
Let A be an algebra over $\mathbf{R}$. Then a derivation is an $\mathbf{R}$-linear map $D: A \to A$ such that $D(fg) = (Df)g + f(Dg)$ for all $f, g \in A$.

4.3 Derivations and vectorfields.

Now let M be a differentiable manifold and let $S(M)$ be the $\mathbf{R}$-algebra of smooth functions $M \to \mathbf{R}$. Then every vectorfield s on M defines a derivation of $S(M)$, (which assigns to a function f its derivative along s), which can be described as follows. Let M be given in terms of local pieces U_α and gluing data $U_{\alpha\beta}$, $\phi_{\alpha\beta}$. Let $f: M \to \mathbf{R}$ and the section $s: M \to TM$ given by the local functions $f_\alpha: U_\alpha \to \mathbf{R}$, $s_\alpha: U_\alpha \to \mathbf{R}^n$. Now define $g_\alpha: U_\alpha \to \mathbf{R}$ by the formula

$$g_\alpha(x) = \sum_k s_\alpha(x)_k \frac{\partial f_\alpha}{\partial x_k}(x) \tag{4.4}$$

where $s_\alpha(x)_k$ is the k-th component of the n-vector $s_\alpha(x)$. It is now an easy exercise to check that $g_\beta(\phi_{\alpha\beta}(x)) = g_\alpha(x)$ for all $x \in U_{\alpha\beta}$ (because $J(\phi_{\alpha\beta})(x)s_\alpha(x) = s_\beta(\phi_{\alpha\beta}(x))$ for these x) so that the $g_i(x)$ combine to define a function $g = D_s(f): M \to \mathbf{R}$. This defines a map $D: S(M) \to S(M)$ which is seen to be a derivation. Inversely every derivation of $S(M)$ arises in this way.

4.5 The Lie bracket of derivations and vectorfields.

Let D_1, D_2 be derivations of an $\mathbf{R}$-algebra A. Then, as is easily checked, so is

$$[D_1, D_2] = D_1 D_2 - D_2 D_1$$

So if s_1, s_2 are vectorfields on M, then there is a vectorfield $[s_1, s_2]$ on M corresponding to the derivation $[D_{s_1}, D_{s_2}]$. This vectorfield is called the Lie bracket of s_1 and $(s_1, s_2) \mapsto [s_1, s_2]$ defines a Lie algebra structure on the vectorspace $V(M)$ of all vectorfields on M.

If M is given in terms of local pieces U_α and gluing data $U_{\alpha\beta}$, $\phi_{\alpha\beta}$ then the Lie bracket operation can be described as follows. Let the vectorfields s and t be given by the local functions $s_\alpha, t_\alpha: U_\alpha \to \mathbf{R}^n$ with components s_α^i, t_α^i, $i = 1, \ldots, n$. Then $[s,t]$ is given by the local functions

$$[s,t]_\alpha^i = \sum_j s_\alpha^j \frac{\partial t_\alpha^i}{\partial x_j} - \sum_j t_\alpha^j \frac{\partial s_\alpha^i}{\partial x_j}$$

4.6 The $\frac{\partial}{\partial x}$ notation.

Let the vectorfield $s: M \to TM$ be given by the functions $s_\alpha: U_\alpha \to \mathbf{R}^n$. Then, using the symbols $\frac{\partial}{\partial x_k}$ in first instance simply as labels for the coordinates in $\mathbf{R}^n$, we can write

$$s_i = \sum s_i(x)^k \frac{\partial}{\partial x_k} \tag{4.7}$$

This is a most convenient notation because as can be seen from (4.4) this gives precisely the local description of the differential operator (derivation) D_s associated to s.

Further taking the commutator difference of the two (local) differential operators

$$D_s = \sum_k s^k \frac{\partial}{\partial x_k}, \quad D_t = \sum_l t^l \frac{\partial}{\partial x_l}$$

gives

$$\Big(\sum_k s^k \frac{\partial}{\partial x_k}\Big)\Big(\sum_l t^l \frac{\partial}{\partial x_l}\Big)(f) = \sum_{k,l} s^k t^l \frac{\partial^2 f}{\partial x_k \partial x_l} + \sum_{k,l} s^k \frac{\partial t^l}{\partial x_k} \frac{\partial f}{\partial x_l}$$

$$\Big(\sum_l t^l \frac{\partial}{\partial x_l}\Big)\Big(\sum_k s^k \frac{\partial}{\partial x_k}\Big)(f) = \sum_{k,l} t^l s^k \frac{\partial^2 f}{\partial x_l \partial x_k} + \sum_{k,l} t^l \frac{\partial s^k}{\partial x_l} \frac{\partial f}{\partial x_k}$$

so that

$$[D_s, D_t]f = (D_s D_t - D_t D_s)f = \Big(\sum_{k,l}\Big(s^k \frac{\partial t^l}{\partial x_k} - t^k \frac{\partial s^l}{\partial x_k}\Big)\frac{\partial}{\partial x_l}\Big)(f)$$

which fits perfectly with the last formula of 4.5 above.

Finally a substitution $y=\phi(x)$ in a differential operator (4.7) transforms it precisely according to the same rule as applies to the corresponding vectorfield s, cf the last formula of 4.1 above.

4.8 Differential equations on a manifold.
A differential equation on a manifold M is given by an equation

$$\dot{x} = s(x) \tag{4.9}$$

where $s:M \to TM$ is a vectorfield, i.e. a section of the tangentbundle. At every moment t, equation (4.8) tells us in which direction and how fast $x(t)$ will evolve by specifying a tangent vector $s(x(t))$ at $x(t)$.

Again it is often useful to take a local pieces and gluing data point of view. Then the differential equation (4.8) is given by a collection of differential equations $\dot{x}=s_i(x)$ in the usual sense of the word on U_i where the functions $s_\alpha(x)$ satisfy $J(\phi_{\alpha\beta}(x)s_\alpha(x)=s_\beta(\phi_{\alpha\beta}(x))$ for all $x \in U_{\alpha\beta}$.

In these terms a solution of the differential equation is simply a collection of solutions of the local equations, i.e. a collection of maps $f_\alpha:V_\alpha \to U_\alpha, V_\alpha \subset \mathbf{R}(\geq 0)$ such that $\cup V_\alpha=\mathbf{R}(\geq 0)$, $\frac{d}{dt}f_\alpha(t)=s_\alpha(f_\alpha(t))$ which fit together to define a morphism $\mathbf{R}(\geq 0) \to M$, i.e. such that $\phi_{\alpha\beta}(f_\alpha(t))=f_\beta(t)$ if $t \in V_\alpha \cap V_\beta$.

In more global terms a solution of (4.8) which passes through x_0 at time 0 is a morphism of smooth manifolds $f:\mathbf{R} \to M$ such that $Tf:T\mathbf{R} \to TM$ satisfies $Tf(t,1)=s(f(t))$ for all $t \in \mathbf{R}$ (or a suitable subset of $\mathbf{R}$), i.e. Tf takes the vectorfield $1:\mathbf{R} \to T\mathbf{R} = \mathbf{R} \times \mathbf{R}$, $t \mapsto (t,1)$ into the vectorfield (section) $s:M \to TM$.

4.10 Example. The matrix Riccati differential equation.
The simplest Riccati equation is

$$\dot{x} = 1-x^2 \tag{4.11}$$

This one has finite escape time. Indeed an initial value of $x(0)<-1$ gives a finite escape time. For this one it is still easy to figure out what happens near infinity and whether and how the trajectory goes through infinity and comes back. The general matrix Riccati equation is

$$\dot{K} = KA + KBK + C + DK \tag{4.12}$$

where K is an $m \times n$ matrix and A,B,C,D are known constant matrices of sizes $n \times n, n \times m, m \times n$ and $m \times m$ respectively. This one is very hard to understand directly. The first step of a somewhat indirect approach is as follows. Consider $n \times (n+m)$ matrices partitioned into two blocks of sizes $m \times n$ $m \times m$ respectively. Now consider the linear system of equations

$$\frac{d}{dt}(X \;\; Y) = (XY)P, \quad P = \begin{pmatrix} A & -B \\ C & -D \end{pmatrix} \tag{4.13}$$

Let $K=Y^{-1}X$ (assuming for the moment that Y^{-1} exists). Then

$$\frac{d}{dt}K = -Y^{-1}\dot{Y}Y^{-1}X+Y^{-1}\dot{X}=Y^{-1}(XB+YD)Y^{-1}X+Y^{-1}(XA+YC)$$

$$= KBK + DK + KA + C$$

In other words the matrix Riccati equation lifts to a linear equation on $\mathbf{R}^{n \times (n+m)}$. If $(X(0)\; Y(0))$ is a full rank matrix, then so is $(X(t)\; Y(t))=(X(0)\; Y(0)e^{tP}$ for all t. But even so $Y(t)$ may very well become noninvertible and that accounts for finite escape time phenomena of the Riccati equation. As already noted if $K(0) \in \mathbf{R}_{reg}^{m \times (n+m)}$ then K remains in this subspace. Now we have already seen the projection map

$$\pi:\mathbf{R}_{reg}^{m\times(n+m)} \to Gr_m(\mathbf{R}^{n+m})$$

in section 2.12 above. I claim that the differential equation (4.13) descends (i.e. induces) a differential equation on the manifold $Gr_m(\mathbf{R}^{n+m})$. To prove this one must show that if $M=(X,Y)$ and $M_1=(X_1,Y_1)$ in $\mathbf{R}_{\text{reg}}^{mx\times(n+m)}$ both map to the same point $x\in Gr_m(\mathbf{R}^{n+m})$ then $T\pi$ takes the respective tangent vectors at M and M_1 into the same tangent vector at x. This is essentially the same calculation as we already did (several times). Indeed let $x\in U_\alpha$. The map π is given locally by $M\mapsto M_\alpha^{-1}M$. Now $M(t)=Me^{tP}$ and $M_1(t)=M_1e^{tP}$. Now if M and M_1 both map to the same $x\in Gr_m(\mathbf{R}^{n+m})$ then $M_1=SM$ for some constant matrix S. But then

$$M_1(t) = M_1e^{tP}=SMe^{tP}=SM(t)$$

for all t. So that $M_1(t)$ and $M(t)$ map to the same point $x(t)\in Gr_n(\mathbf{R}^{n+m})$ for all t. This proves the claim. However $Gr_m(\mathbf{R}^{n+m})$ is a smooth *compact* manifold (a fact I did not prove), so $x(t)\in Gr_m(\mathbf{R}^{n+m})$ of all t. The finite escape time phenomena of the matrix Riccati are now analyzed and understood in terms of the embedding

$$K \mapsto m-\text{dim subspace of } \mathbf{R}^{n+m} \text{ spanned by the rows of } \quad (K \quad I_m)$$

$$\mathbf{R}^{m\times n} \hookrightarrow Gr_m(\mathbf{R}^{n+m})$$

The matrix Riccati equation is at first only defined on $\mathbf{R}^{m\times n}$. It extends to a equation on the smooth compactification $Gr_m(\mathbf{R}^{n+m})$ (but not to other compactifications such as the projective space $\mathbf{P}^{mn}$ (unless $m=1$) or the sphere S^{nm}). From time to time $x(t)$ may exit from the open dense subset $\mathbf{R}^{m\times n}$ in $Gr_m(\mathbf{R}^{n+m})$ to cross the set at infinity $Gr_m(\mathbf{R}^{n+m})\setminus\mathbf{R}^{m\times n}$.

4.14 Compatibility of vectorfields under differentiable maps.
Let $\phi:M\to N$ be a map of differentiable manifolds. Then, as we have seen, 3.5, for each $x\in M$ we have the induced map $T\phi(x):T_xM\to T_{\phi(x)}N$ of the tangent vectorspace of M at x to the tangent vectorspace of N at $\phi(x)$. All together these map define the vectorbundle map $T\phi:TM\to TN$ which is also often denoted $d\phi$.

Now let $\alpha:M\to TM$ be a vectorfield on M, $\alpha(x)\in T_xM$. Then we have the various tangent vectors $T\phi(x)(\alpha(x))\in T_{\phi(x)}N$. These may or may not define a vectorfield on N. Firstly because not every $y\in N$ need to be of the form $\phi(x)$ for some $x\in M$ and secondly because if x and x' both map to the same $y\in N$ then it may very well happen that $T\phi(x)(\alpha(x))\neq T\phi(x')(\alpha(x'))$. (As well shall see below the dual notion to that of a vectorfield, i.e. the notion of a differentiable 1-form, is much better behaved in this respect: for each 1-form ω on M and differentiable map $\phi:M\to N$ there is a canonically associated (induced) 1-form $\phi^*\omega$ on M.)

If two vectorfields α on M and β on N are such that $T\phi(x)(\alpha(x))$ for all $x\in M$ then we say that α and β are compatible under ϕ. If $\phi:M\to N$ is a diffeomorphism then $\beta(y) = T\phi(\phi^{-1}(y))\alpha(\phi^{-1}(y))$ defines a unique vectorfield on N compatible with α on M under ϕ.

The Lie bracket of vectorfields $[\alpha,\alpha']$ is 'functorial' with respect to transforms of vectorfields in the following sense.

4.15 Proposition. Let α,β and α',β' be compatible pairs of vectorfields under $\phi:M\to N$. Then $[\alpha,\alpha']$ and $[\beta,\beta']$ are also compatible under ϕ.

The easiest way to see this is first to do the following exercise. Let $\phi:M\to N$ be differentiable and let the vectorfield α on M and β on N be compatible under ϕ. Let D_α be the derivation on $\mathcal{F}(M)$ corresponding to α and D_β the one on $\mathcal{F}(N)$ corresponding to β. (Here $\mathcal{F}(M)$ is the ring of smooth functions on M.) Then for all $f\in\mathcal{F}(N)$

$$D_\alpha(f\circ\phi) = D_\beta(f)\circ\phi \tag{*}$$

Indeed in local coordinates x on M and y on N and ϕ given by $y_j=\phi_j(x)$

$$D_\alpha(f\circ\phi)(x) = \sum_i \alpha_i \frac{\partial}{\partial x_i}(f\circ\phi)(x) =$$

$$\sum_{i,j}\alpha_i\frac{\partial f}{\partial y_j}(\phi(x))(T\phi(x))_{ji} = \sum\beta_j\frac{\partial f}{\partial y_j}(\phi(x)) = (D_\beta f)(\phi(x)) = (D_\beta f\circ\phi)(x).$$

And inversely if (*) holds then α and β are compatible under ϕ. (This is the chain rule of course.) Now let α,β and α',β' be compatible pairs. Then by the exercise we just did

$$D_\alpha' D_\alpha(f\circ\phi) = D_\alpha'(D_\beta(f)\circ\phi) = D_\beta' D_\beta(f)\circ\phi$$

Thus $[D_\alpha',D_\alpha](f\circ\phi) = [D_\beta',D_\beta](f)\circ\phi$ so that the vectorfields belonging to $[D_\alpha',D_\alpha]$, i.e. $[\alpha',\alpha]$, and $[D_\beta',D_\beta]$, i.e. $[\beta',\beta]$, are also compatible.

4.16 Distributions.
A distribution Δ on a manifold specifies for each x a subspace $\Delta_x \subset T_xM$ of the tangent bundle at x. They arise naturally in several contexts. E.g. in control systems of the following kind

$$\dot{x} = \sum_{i=1}^{m} u_i g_i(x) \tag{4.17}$$

where the $u_i \in \mathbf{R}$ are controls (and the $g_i(x)$ are vectorfields). The corresponding distribution is (of course) defined by $\Delta_x=$ linear subspace of T_xM spanned by the tangent vectors $g_1(x), \ldots, g_m(x)$. In this setting Δ_x responds the totality of directions in which the state x can be made to move infinitesimally by suitable (constant in time) control vectors $(u_1, \ldots, u_m)$.

This does not mean that by taking suitable functions $u_1(t), \ldots, u_m(t)$ the vector x can not be made to move in still more directions, as we shall see immediately below. To get some feeling for this consider the special case

$$\dot{x} = \sum_{i=1}^{m} u_i A_i x\,, \quad x\in\mathbf{R}^n \tag{4.18}$$

where the A_i are constant $n\times n$ matrices. As is well known the solution of $\dot{x}=Ax$ is $x(t)=e^{At}x(0)$, where

$$e^{At} = I+\frac{At}{1!}+\frac{A^2t^2}{2!}+\cdots$$

Now let us take in (4.18)

$$u_1 = 1,\ u_2=\cdots=u_m=0,\ \text{ for } t\in[0,\epsilon)$$

$$u_2 = 1,\ u_1=u_3=\cdots=u_m=0, \text{for } t\in[\epsilon,2\epsilon)$$

$$u_1 = -1,\ u_2=\cdots u_m=0,\ \text{ for } t\in[2\epsilon,3\epsilon)$$

$$u_2 = -1,\ u_1=u_3=\cdots=u_m=0,\ \text{ for } t\in[3\epsilon,4\epsilon]$$

then at time $t=4\epsilon$, we have

$$x(t) = e^{-A_2\epsilon}e^{-A_1\epsilon}e^{A_2\epsilon}e^{A_1\epsilon}x(0)$$

which is equal to

$$x(t) = x(0)+\epsilon^2(A_2A_1-A_1A_2)x(0)+O(\epsilon^3)$$

Thus from $x(0)$, $x(t)$ can also be made to move in the direction

$$[A_2,A_1]x(0) = (A_2A_1-A_1A_2)x(0)$$

Now, as follows from the formula at the end of (4.5) the vectorfield

$$(A_2A_1-A_1A_2)x$$

is precisely the Lie bracket of the two vectorfields A_1x, A_2x

$$[A_1x,A_2x] = (A_2A_1-A_1A_2)x.$$

(Note the reversal of order with respect to the usual commutator difference of matrices).

The same holds for arbitrary vectorfields, i.e. for equations like (4.17): in addition to the immediately given directions $g_i(x), \ldots, g_m(x)$ the control system can be made to evolve in the directions $[g_i,g_j](i,j=1,\ldots,m)$, and $[g_i,[g_j,g_k]](i,j,k=1,\ldots,m)$, and $[[g_i,g_j],[g_k,g_l]](i,j,k,l=1,\ldots,m)$, etc., etc.; that is in all directions which can be constructed by taking repeated Lie brackets of the given vectorfields $g_1, \ldots, g_m$.

This leads to the notion of an involutive distribution. A vectorfield $s:N \to TM$ is said to belong to (or be in) the distribution Δ if $s(x) \in \Delta_x$ for all x. A distribution Δ is said to be *involutive* if for all vectorfields s and t in Δ the vectorfield $[s,t]$ is also in Δ.

Natural examples of involutive distributions arise as follows. A *foliation* of codimension r of M is a decomposition of M into subsets (called *leaves*) such that locally the decomposition looks like the decomposition $\mathbf{R}^n = \bigcup_{a \in \mathbf{R}^r} a+\mathbf{R}^{n-r}$ where $\mathbf{R}^r$ is viewed as the subvectorspace of $\mathbf{R}^n$ of vectors whose last $n-r$ coordinates are zero, and $\mathbf{R}^{n-r} \subset \mathbf{R}^r$ is the subspace of vectors whose first r coordinates are zero. Here the phrase 'locally looks like' means that for each $x \in M$, there is an open neigbourhood U of x and a diffeomorphism of U to an open subset V of $\mathbf{R}^n$ such that ϕ applied to the decomposition of U gives a decomposition of V as given above.

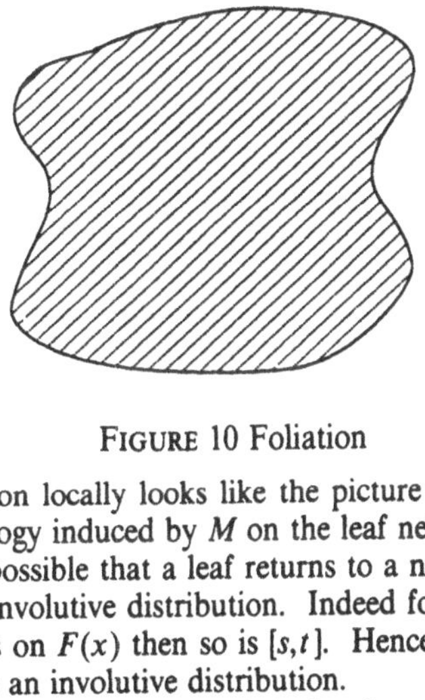

FIGURE 10 Foliation

Thus for codimension 1 a foliation locally looks like the picture of fig. 10. Note that each leaf is a submanifold. However the topology induced by M on the leaf need not be that of the leaf as a manifold in its own right. It is very possible that a leaf returns to a neigbourhood infinitely often. A foliation of M naturally defines an involutive distribution. Indeed for $x \in M$ let $F(x)$ be the leaf through x. If s and t are two vectorfields on $F(x)$ then so is $[s,t]$. Hence if Δ_x is the subspace of T_xM of all vectors tangent to $F(x)$ then Δ is an involutive distribution.

The converse is also true. If Δ is involutive and Δ_x is of constant rank, i.e. $\dim\Delta_x = k$ for all x, then there exists a foliation $\mathcal{F}$ of M such that Δ_x is the tangent space to $F(x)$ for all x. This is *Frobenius' theorem* and it is a sort of multi-time or multi-control variant of the existence of solutions theorem for ordinary differential equations.

5. Riemannian manifolds

A differentiable manifold as defined above is still a rather floppy (topological) structure. To have real fun and do real analysis, including stochastic analysis, some more structure is needed. One of the more popular is a Riemannian structure. Intuitively this means that each tangent space T_xM is provided with an inner product and these inner products are supposed to vary smoothly with x. As usual this is made precise by providing a local pieces and patching data description. Locally the manifold and its tangent bundle look like $U_\alpha \times \mathbf{R}^n, U_\alpha \subset \mathbf{R}^n$. Let P be the space of positive definite inner products, i.e. positive definite symmetric matrices. Then a Riemannian structure on M is given in terms of local data by a collection of smooth maps $g(\alpha): U_\alpha \to P$ with the following transformation properties: if $x \in U_{\alpha\beta}$ then

$$g(\beta)(\phi_{\alpha\beta}(x)) = (J(\phi_{\alpha\beta}(x))^T)^{-1} g(\alpha)(x) J(\phi_{\alpha\beta})(x)^{-1}$$

Because P is convex it is not difficult to see that these always exist Riemannian matrices. On a Riemannian manifold one can define the length and energy of a curve and one can relate the tangent vectors at one point of M to those at another point thus making all kinds of analysis and estimates possible.

6. Calculus

So far we have mainly dealt with the topology of manifolds, i.e. those gadgets which locally look like $\mathbf{R}^n$ and we have done a bit of differential calculus. E.g. if $f: M \to N$ is a differentiable mapping of M into N we know what the 'derivative' of f is, viz. the mapping $Tf: TM \to TN$ of the tangent bundle TM of M into the tangent bundle TN. And indeed if $x \in M$, then $Tf(x): T_xM \to T_xN$ is the linear part (approximation) of $f: x \mapsto f(x)$ at x.

Naturally we would also like to do the integral bit, that is to give the right kind of meaning to such things as the integral of a function on the sphere, say, over that sphere. This requires some more preparations having mainly to do with 'what (variables) to integrate against' or, more generally, 'what can be integrated over what'. Also, as we shall see, to integrate functions one needs more structure than just a manifold; e.g. a Riemannian metric will do.

6.1. Chains and cubes.

What we want to do is to define integrals over (broken) curves, surfaces etc. in arbitrary manifolds. Curves and surfaces etc. can be thought of as made up from pieces which are images of intervals, filled squares, filled cubes, etc. It turns out to be convenient to define integrals initially as 'something over a map of an interval, square, ... into M' rather than as something over the image of that map.

The standard n-cube Δ_n is $[0,1]^n \subset \mathbf{R}^n$ e.g. the square or the familiar 3-cube depicted below. The boundary of Δ_n is made up of various pieces isomorphic (but not identical) with Δ_{n-1}. More precisely for each $i = 1,...,n$ we define two maps $\alpha_0^i: \Delta_{n-1} \to \Delta_n$, $\alpha_1^i: \Delta_{n-1} \to \Delta_n$ as follows

$$\alpha_0^i(x_1,...,x_{n-1}) = (x_1,...,x_{i-1},0,x_i,...,x_{n-1})$$

$$\alpha_1^i(x_1,...,x_{n-1}) = (x_1,...,x_{i-1},1,x_i,...,x_{n-1})$$

The images of these maps make up the boundary of Δ_n.

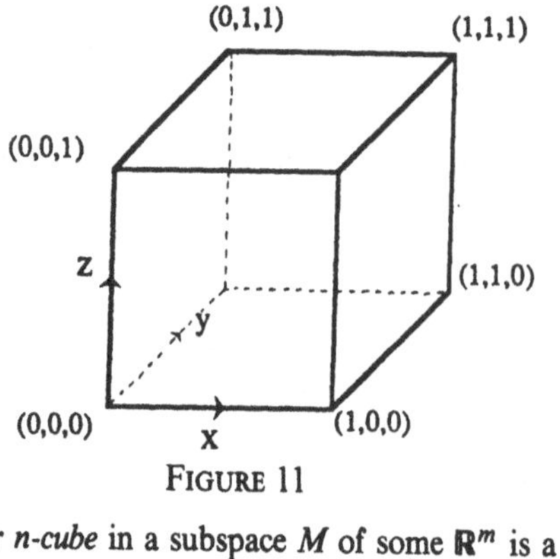

Figure 11

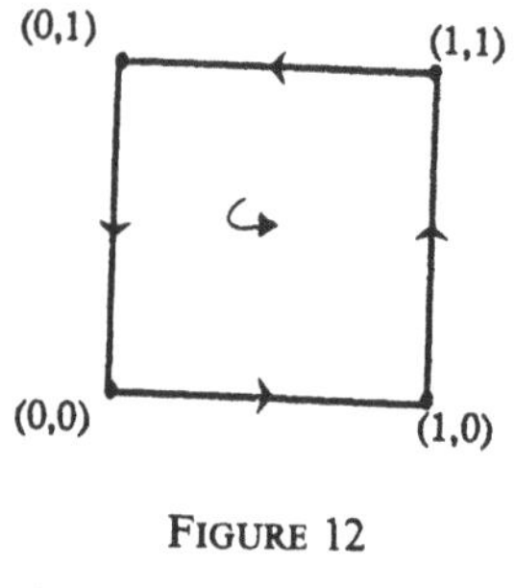

Figure 12

A *singular n-cube* in a subspace M of some $\mathbf{R}^m$ is a mapping $s: \Delta_n \to M$. (Here it is good to think of m as larger of equal to n.) A *singular n-chain* is a finite formal sum $\sum n_i s_i$ where the n_i are integers and the s_i are singular n-cubes. The boundary ∂s of a singular n-cube $s: \Delta_n \to A$ is by definition the $(n-1)$-chain $\partial s = \sum_{i=1}^{n} (-1)^i (s \circ \alpha_0^i - s \circ \alpha_1^i)$. Thus the boundary of the 3-cube id$: \Delta_3 \to \Delta_3 \subset \mathbf{R}^n$ is (in terms of images of Δ_2) equal to + (front square) - (back square) - (left square) + (right square) - (bottom) + (top), and the boundary of $\mathrm{id}:\Delta_2 \to \Delta_2 \subset \mathbf{R}^2$ is the sum of intervals: $[(0,0),(1,0)]+[(1,0),(1,1)]-[(0,1),(1,1)]-[(0,0),(0,1)]$ which fits our intuitive idea of the (oriented) boundary of the square. The boundary of a singular n-chain $c \cong \sum n_i s_i$ is by definition equal to $\partial c = \sum n_i \partial s_i$.

These are the formal definitions. In practices one tends to think of a singular n-chain in terms of the images (with multiplicities) of the singular n-cubes making up the chain as illustrated below. Intuitively the boundary of the piece of surface (corresponding to the chain s_1+s_2) depicted in fig. 13 ought to be the outer circle. And if s_1 and s_2 are chosen such that maps $\Delta_1 \to M$ induced by s_1 and s_2 for the piece of boundary in the middle are the same then this will indeed be the case (thanks to the orientations chosen). Moreover s_1 and s_2 can always be chosen in such a way. However if s_1 and s_2 are just any differentiable maps whose images happen to fit together as indicated, then the boundary of s_1+s_2 will be more complicated. It turns out that for integration purposes (and the multidimensional generalization of the fundamental theorem of calculus: Stokes theorem) this matters little.

For clarities sake let us remark that $\Delta_0 = [0,1]^0$ is a single point and that, thus, a singular 0-chain in M is just a finite set of points in M with multiplicities.

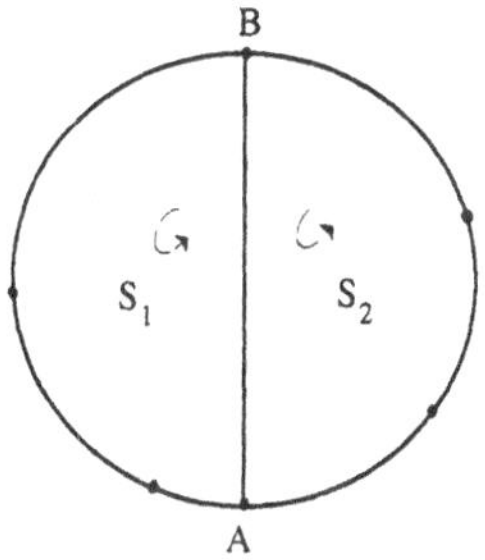

FIGURE 13

6.3. Forms

The next thing to decide is what kinds of animals can be integrated. As everyone knows functions cannot be quite the right answer. Simply because under a change of variables the things under an integral sign do not transform as functions. Indeed if $\phi:\mathbf{R}^n \to \mathbf{R}^n$ is a diffeomorphism (change of variables $y=\phi(x)$). Then $f:\mathbf{R}^n \to \mathbf{R}$, $x \mapsto f(x)$ transforms as $y \mapsto f(\phi^{-1}(y))$, i.e. $f \mapsto f \circ \phi^{-1}$. But for an integral we have

$$\int_A f = \int_{\phi(A)} f(\phi^{-1}(y)) \,|\det J(\phi)(\phi^{-1}(y)|^{-1}$$

which is, of course, the reason one writes fdx or something like that under an integral sign.

The kinds of things which belong under an integral sign turn out to be *differential forms*. These things we now proceed to define.

If V is a vectorspace, a k-form on V is a k-multilinear mapping

$$\omega: V \times \cdots \times V \to \mathbf{R}$$

such that moreover for each $i \neq j$,

$$\omega(\cdots, v_i, \cdots, v_j, \cdots) = -\omega(\cdots, v_j, \cdots, v_i, \cdots)$$

i.e. interchanging two arguments just causes a sign change. One particular n-form on $\mathbf{R}^n$ is very well-known, the determinant, where $\det(v_1,\ldots,v_n)$, $v_i \in \mathbf{R}^n$ is the determinant of the $n \times n$ matrix obtained by writing out the n-vectors $v_1,...,v_n$ as column vectors in the standard basis. It is moreover just about unique as most everyone knows: if ω is a n-form on $\mathbf{R}^n$ then $\omega =$ a det for some constant $a \in \mathbf{R}$.

More generally the space $\Omega^k(V)$ of k-forms on a n-dimensional vectorspace V has dimension $\binom{n}{k}$. It will be useful to have a basis for $\Omega^k(V)$. Let $e_1,...,e_n$ be a basis for V, and let $\phi_1,...,\phi_n$ be the dual basis, i.e. $\phi_i(e_j)=\delta_{ij}$. Then a basis for $\Omega^k(V)$ is given by the functions $\phi_{i_1} \wedge \cdots \wedge \phi_{i_k}$, $i_1 < \cdots < i_k$ defined by the formula

$$(\phi_{i_1} \wedge \cdots \wedge \phi_{i_k})(e_{j_1}, \cdots, e_{j_k}) = \begin{cases} 0 \text{ if } \{i_1,...,i_k\} \neq \{j_1, \cdots, j_k\} \\ \text{sign } \tau \text{ if there is a permutation } \tau \text{ of } i_1,...,i_k \\ \text{such that } \tau(i_r)=j_r, r=1,...,k \end{cases}$$

Thus for example $\phi_1 \wedge \phi_2 \in \Omega^2(\mathbf{R}^3)$ takes the values

	(e_1,e_1)	(e_1,e_2)	(e_1,e_3)	(e_2,e_1)	(e_2,e_2)	(e_2,e_3)	(e_3,e_2)	(e_3,e_3)
$\phi_1 \wedge \phi_2$	0	1	0	-1	0	0	0	0

It is useful to declare by definition that for arbitrary $i_1,...,i_k \in \{1,...,n\}$

$$\phi_{j_1} \wedge \cdots \wedge \phi_{j_k} = \text{sign}\, \sigma\, \phi_{i_1} \wedge \cdots \wedge \phi_{i_k}$$

if the $j_1,...,j_k$ are all different where $(i_1,...,i_k)$ is the unique permutation of $(j_1,...,j_k)$ such that $i_1 < i_2 < \cdots < i_k$ and $\sigma(j_k) = i_k$, and to set $\phi_{j_1} \wedge \cdots \wedge \phi_{j_k} = 0$ if two or more of the ϕ_{j_r} are equal.

Now let M be a manifold. Then a differentiable k-form ω on M consists of giving an k-form $\omega(x)$ on T_xM for all x such that $\omega(x)$ varies smoothly with x. As usual this can be given a local and gluing data description. Let M be obtained by patching together pieces $U_\alpha \subset \mathbf{R}^n$ with the help of gluing functions $\phi_{\alpha\beta}$. On U_α the k-form ω is specified by giving functions

$$\omega_\alpha^{i_1 \cdots i_k}, \; i_1,...,i_k \in \{1,...,n\}$$

The corresponding form is then defined by

$$\omega_\alpha(v_1,...,v_k) = \sum_{i_1,...,i_k} \omega_\alpha^{i_1 \cdots i_k} v_{1i_1} \cdots v_{ki_k}$$

where $v_j \in \mathbf{R}^n$ and v_{ji_j} is the i_j-th component of the vector v_j. For the $\omega_\alpha^{i_1 \cdots i_k}$ to define a k-form i.e. an *alternating* k-multilinear function, it is necessary and sufficient that $\omega^{\cdots i \cdots j \cdots} = -\omega^{\cdots j \cdots i \cdots}$. Thus it suffices to specify the $\omega_\alpha^{i_1 \cdots i_k}$ for $i_1 < \cdots < i_k$. A collection of 'local' k-forms ω_α on U_α defines a k-form on all of M provided the ω_α are compatible in the sense that one must have

$$\omega_\beta(\phi(x))(J(\phi_{\alpha\beta}(x)v_1,...,J(\phi_{\alpha\beta}(x)v_k) = \omega_\alpha(x)(v_1,...,v_k)$$

for all $v_1,...,v_k \in T_xM$. This means that if $(s_{ij}) = J(\phi)(x)$

$$\omega_\alpha(x)^{i_1...i_k} = \sum_{j_1,...,j_k} \omega_\beta(\phi(x))^{j_1...j_k} s_{j_1i_1} \ldots s_{j_ki_k} \tag{6.4}$$

Note the similarity with the compatibility requirements for the local pieces and gluing data description of a Riemannian metric. Indeed both are examples of contravariant tensors, g is a symmetric 2-tensor and ω is an alternating (= antisymmetric) k-tensor. For a local piece $U \subset \mathbf{R}^n$ of the manifold M let us use again the notation $\partial/\partial x_1,...,\partial/\partial x_n$ for the canonical basis of T_xU, $x \in U$. Let us use the symbols $dx_1,...,dx_n$ to denote the dual basis; i.e. $dx_i(\partial/\partial x_j)=\delta_{i,j}$. Then a differential k-form ω with components $\omega_\alpha^{i_1,...,i_k}$ can be written as

$$\omega_\alpha = \sum_{i_1 < \cdots < i_k} \omega_\alpha^{i_1,...,i_k} dx_{i_1} \wedge \cdots \wedge dx_{i_k}$$

This is a rather good notation because (exercise!) it fits perfectly with the transformation rules (6.4).

Let $f: M \to N$ be a differentiable map of manifolds. Than there is an induced map $(Tf)(x): T_xM \to T_xN$ for all $x \in M$. Now let ω be a k-form on N i.e. for every $y \in N$ there is an antisymmetric k-linear mapping

$$\omega(y): T_yN \times \cdots \times T_yN \to \mathbf{R}$$

Then there is a natural k-form $f^*\omega$ on M defined by

$$f^*(\omega)(x): T_xM \times \cdots \times T_xM \to \mathbf{R}$$
$$(v_1, \ldots, v_k) \mapsto \omega(f(x))((Tf)(x)v_1, \ldots, (Tf)(x)v_k)$$

6.4. Integrals 1.

The geometric preparations above are enough to enable us to make a first attempt at defining integrals. It turns out that what one can integrate is k-forms over k-chains. The first step is as follows.

Let ω be a k-form on the standard cube Δ_k. Then ω is given by a function f on Δ_k

$$\omega = f dx_1 \wedge \cdots \wedge dx_k$$

One now defines

$$\int_{\Delta_k} \omega = \int_{\Delta_k} f \tag{6.5}$$

where the right hand side of (6.5) is the usual Lebesgue integral. The next step is to define integrals over a singular cube. Thus let $s: \Delta_n \to M \subset \mathbf{R}^m$ be a smooth singular cube and let ω be a k-form on M. Then one defines

$$\int_s \omega = \int_{\Delta_k} s^*\omega \tag{6.6}$$

and for a singular k-chain $c = \sum n_i s_i$ one takes of course

$$\int_c \omega = \sum_i n_i \int_{s_i} \omega \tag{6.7}$$

Formula (6.6) defines an integral over each singular cube $s: \Delta_k \to M$. This is definitely not yet something like an integral over the subset $s(\Delta_k)$ of M. Nor can it be. For one thing if $k=1$ and the curve $s(\Delta_1)$ runs from A to B say, we definitely want the integral from A to B along the curve to be equal to minus the integral from B to A along the curve. This brings in the point of orientations, cf 6.8 below. For another if say $s': \Delta_1 \to M$ is defined by $s'(t) = s(2t)$ for $0 \leq t \leq \frac{1}{2}$ and $s'(t) = B$ for $\frac{1}{2} \leq t \leq 1$, then, as is very easy to see, as a rule the integral over s' of ω will be different from the integral over s of ω. However as we shall see below for nice enough singular chains $c = \sum n_i s_i$, the integral of ω over c will only depend on the image of c understood in the sense of a family $s_i(\Delta_n)$ of twisted (smooth) cubes with multiplicities n_i and then one can truly speak of an integral of ω over the "subset" $c(\Delta_n)$ of M. Here nice enough will turn out to mean that each s_i must be orientation preserving and define a smooth imbedding $s_i: \Delta_n \to M$.

6.8. Orientations.

Consider all bases $(a_1, \ldots, a_n)$ of a vectorspace V of dimension n. We say that two bases $(a_1, \ldots, a_n)$ $(b_1, \ldots, b_n$ are in the same orientation class if the matrix (s_{ij}) defined by $b_j = \sum s_{ij} a_i$ has positive determinant. Thus there are two orientation classes often denoted + and -. Giving an orientation on V means specifying one of these classes which is often done by specifying one particular basis in that class. The usual ('counterclockwise in case $n=2$') orientation on $\mathbf{R}^n$ is given by the standard basis $(e_1, \ldots, e_n)$. An isomorphism $\phi: V \to W$ of oriented vectorspaces is orientation preserving if ϕ takes the orientation class of V into that of W, i.e. if $(a_1,...,a_n)$ is a basis from the orientation class of V then $(f(a_1),...,f(a_n))$ must be a basis of the orientation class of W.

Let M be a manifold, thought of, as usual, as obtained by gluing together local pieces $U_\alpha \subset \mathbf{R}^n$. An orientation on M is now specified by choosing an orientation on each $x \times \mathbf{R}^n$, $x \in U_\alpha$, all α, such that

(i) if $x,y \in U_\alpha$, $x \times \mathbf{R}^n$ and $y \times \mathbf{R}^n$ have the same orientation (i.e. $(x,v) \mapsto (y,v)$ is orientation preserving.

(ii) if $x \in U_\alpha, y \in U_\beta, \phi_{\alpha\beta}(x)=y$ then $\det(J(\phi_{\alpha\beta}(x)))>0$

This can not always be done. A classic example of a non-orientable manifold is the Möbius strip defined above. Exercise: prove this. Another example is the projective plane $\mathbf{P}^2_{\mathbf{R}} = Gr_1(\mathbf{R}^3)$.

A manifold together with an orientation is an *oriented manifold.* If $f: M \to N$ is a differentiable immersion of an oriented manifold M into an oriented manifold N of the same dimension then f is called *orientation preserving* if $Jf(x): T_xM \to T_{f(x)}N$ is orientation preserving for all $x \in M$. A smooth singular n-cube $s: \Delta_n \to M$, $\dim M = n$, is orientation preserving if there exists an extension of s to some open neighbourhood U of Δ_n in $\mathbf{R}^n$ such that this extension is orientation preserving.

6.9. Integrals 2.

Now consider an oriented submanifold N of dimension k of a manifold M. Let $c = \sum n_i s_i$, $c' = \sum n_i s_i'$ (same n_i) be two singular k-chains in N such that $s_i(\Delta_k) = s_i'(\Delta_k)$ for all i and such that both s_i and s_i' are orientation preserving for all i. Let ω be a k-form on M and (hence) on N. Then

$$\int_c \omega = \int_{c'} \omega$$

In particular if all the n_i are $+1$ and the images of the s_i fit together to define a piecewise differentiable submanifold with boundary N' of N as indicated in fig. 14 then we can truly speak of $\int_{N'} \omega$, the integral of ω over N'. To define this integral of course we first reduce to the case of one singular k-cube, cf above.

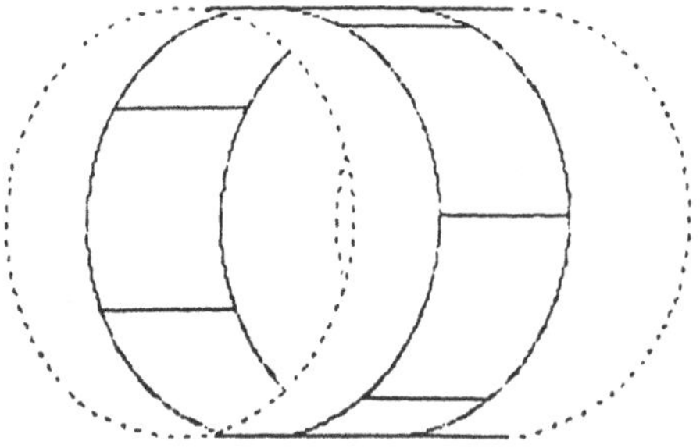

FIGURE 14

In that case we needed to assume $N \subset \mathbf{R}^m$ for some m, cf 6.4. Since every manifold can be embedded in an $\mathbf{R}^m$ for large enough m ($m \geqslant 2$ dim (manifold) $+1$ suffices) this is no real restriction. However, this does not fit well with our overriding attitude of viewing a manifold simply as a collection of local pieces U_α to be fitted together.

Let M be the manifold obtained by gluing the U_α; let $U_\alpha' \subset M$ be the piece corresponding to U_α. By cutting up Δ_k into smaller cubes if necessary we can see to it that the image of the chain c is such that it is made up of singular cubus which each lie completely into some coordinate neighbourhood U_α'. Then c is specified by a corresponding map $s': \Delta_k \to U_\alpha$ (such that the diagram of fig. 15 commutes) and the integral is defined entirely in terms of the local descriptions $s_\alpha: \Delta_k \to U_\alpha$.

$$\begin{array}{ccc} \Delta_k & \xrightarrow{s_\alpha} & U_\alpha \subset \mathbf{R}^n \\ & {\scriptstyle s}\searrow \quad \swarrow & \\ & U'_\alpha \subset M & \end{array}$$

FIGURE 15

A zero-chain $c = \sum n_i P_i$ is a collection of points with multiplicities. A zero-form is a function $F: M \to \mathbf{R}$. The integral of a zero-form F over a zero-chain c is defined as $\sum n_i F(P_i)$.

6.10. The fundamental theorem of calculus.

The fundamental theorem of calculus (one variable) says that if F is a function with derivative $f = F'$ then

$$\int_a^b f dx = F(b) - F(a)$$

In our setting fdx is a one-form, F is a function, i.e. a zero-form. The 'chain' over which we integrate is an interval $[a,b]$ with boundary the 0-chain 'b'-'a' meaning the formal sum of 1 times the point b minus 1 times the point a. Thus the integral of $dF = fdx$ over $[a,b]$ is the integral of F over the boundary 'b'-'a'. This generalizes. To that end we need to define $d\omega$ of an arbitrary k-form ω. As usual let the manifold M be obtained by gluing together local pieces U_α and let the k-form ω be given locally by the ω_α

$$\omega_\alpha = \sum_{i_1 < \cdots < i_k} \omega_\alpha(x)^{i_1 \cdots i_k} dx_{i_1} \wedge \cdots \wedge dx_{i_k}$$

In case $k = 0$, ω is a function f and one defines the 1-form

$$df_\alpha = \sum_i \frac{\partial f_\alpha}{\partial x_i} dx_i$$

For $k > 0$, this generalizes to

$$d\omega_\alpha = \sum_i \sum_{i_1 < \cdots < i_k} \frac{\partial \omega_\alpha^{i_1 \cdots i_k}}{\partial x_i} dx_i \wedge dx_{i_1} \wedge \cdots \wedge dx_{i_k}$$

where the right hand side is brought into the right form by the calculation rules $dx_i \wedge dx_{i_1} \wedge \cdots \wedge dx_{i_k} = 0$ if $i \in \{i_1, \ldots, i_k\}$ and $dx_i \wedge dx_{i_1} \wedge \cdots \wedge dx_{i_k} = (-1)^l dx_{i_1} \wedge \cdots \wedge dx_{i_l} \wedge dx_i \wedge dx_{i_{l+1}} \wedge \cdots \wedge dx_{i_k}$ if $i_1 < \cdots < i_l < i < i_{l+1} < \cdots < i_k$.

It is a not too difficult exercise to check that the local $(k+1)$-forms $d\omega_\alpha$ fit together to define a $(k+1)$-form on all of M. Two other exercises are: $d(d\omega) = 0$ and $d(f^* \omega) = f^*(d\omega)$ if $f: N \to M$ is a differentiable map.

The fundamental theorem of calculus now generalizes in the case of integrals over k-chains to the Stokes theorem

$$\int_c d\omega = \int_{\partial c} \omega$$

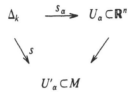

where c is a singular k-chain and ω is a $(k-1)$-form on M.

6.11. Manifolds with boundary

We defined a smooth manifold (without boundary) as a collection of subsets $U_\alpha \subset \mathbf{R}^n$ together with gluing data. This yields such things as the sphere surface. But not such things as the solid unit ball and the solid torus. These are manifolds with boundary which we now proceed to define. Let $H = \{x \in \mathbf{R}^n : x_n \geqslant 0\}$.

Now a manifold with boundary M is defined as a collection of open subsets U_α open in $\mathbf{R}^n$ or open in H with gluing data $\phi_{\alpha\beta}$: $U_{\alpha\beta} \to U_{\beta\alpha}$ as before which the additional requirement that

$$\phi_{\alpha\beta}(U_{\alpha\beta} \cap \partial H) = (U_{\beta\alpha} \cap \partial H)$$

where $\partial H = \{x \in \mathbf{R}^n : x_n = 0\}$, the boundary of H. (If $U_{\alpha\beta} \cap \partial H \neq \varnothing$ then differentiability of $\phi_{\alpha\beta}$ means (as always) that $\phi_{\alpha\beta}$ extends to a differentiable mapping on some subset open in $\mathbf{R}^n$ which contains $U_{\alpha\beta}$).

The $U_\alpha \cap \partial H$ and $U_{\alpha\beta} \cap \partial H$ are open subsets in $\mathbf{R}^{n-1}$ and the $\phi_{\alpha\beta}$ restricted to these subsets then define an $(n-1)$-manifold (without boundary) ∂M, the boundary of M. The tangent spaces (bundle) to M are again defined by means of the local pieces $U_\alpha \times \mathbf{R}^n$ (also for the points in $U_\alpha \cap \partial H$), and a Riemannian metric on M means again an inner product on all of the T_xM.

Let M be a Riemannian manifold with boundary ∂M. For $x \in \partial M$, $T_x\partial M$ is an $(n-1)$-dimensional subspace. Thus there are two vectors of unit length in T_xM perpendicular to $T_x\partial M$. Precisely one of these points outwards (seen, as always, locally by going back to a $U_\alpha \subset H$). This defines the *outward normal* to ∂M at $x \in \partial M$.

An outward normal can also be defined in a slightly different setting. Let N be a oriented $(n-1)$-dimensional submanifold of an oriented Riemannian n-manifold M. For each $x \in M$ let $(v_1,...,v_{n-1})$ be an orthonormal basis of T_xN with the given orientation on N. Then there is precisely one unit length vector $v_n \in T_xM$ such that $(v_1,...,v_{n-1},v_n)$ is an orthonormal basis of T_xM with the given orientation on M. In this setting v_n is also called the outward normal to N at $x \in N$.

6.12. The volume form.

We now know how to integrate k-forms over k-chains and in particular n-forms over n-manifolds. This still does not give meaning to, say, the integral over a sphere of a function on that sphere. For that we must find a good way of assigning n-forms to functions much like in ordinary one dimensional calculus one assigns the 1-form fdx to the function f.

The multidimensional analogue of this for manifolds is the volume form. Let V be a n-dimensional vectorspace with inner product and an orientation. An n-form on V is of the form $\omega = a\det$. For each orthonormal basis $(v_1,...,v_1)$ we have $\omega(v_1,...,v_n) = \pm a$. Thus there is precisely one n-form on V with the additional property that it takes the value 1 on each orthonormal basis with the given orientation. This one is called the volume element of V (determined by the inner product and the orientation).

Now let M be an oriented Riemannian manifold (with or without boundary). Then the volume form ω_M on M is defined by setting $\omega(x): T_xM \times \cdots \times T_xM \to \mathbf{R}$ equal to the unique volume element of T_xM for each $x \in M$ (determined by the given inner product on T_xM defined by the Riemannian metric and the given orientation).

More explicitly in terms of local coordinate patches U_α, the volume form can be described as follows. Let $\epsilon_\alpha = 1$ or -1 depending on whether the given orientations on $\{x\} \times \mathbf{R}^n$ agree with the standard orientation or not. For each x apply Gram-Schmidt orthonormalization with respect to the given inner product on $\{x\} \times \mathbf{R}^n$ to the standard basis $(e_1,...,e_n)$, to obtain a differentiable family of orthonormal bases $\{v_1(x),...,v_n(x)\}$. Now set

$$\omega_\alpha = \epsilon_\alpha \det(v_1(x),...,v_n(x))^{-1} dx_1 \wedge \cdots \wedge dx_n$$

These ω_α are then the local pieces and gluing data description of the volume form ω_M often written as dV (even enough there may not be an $(n-1)$-form V such that $\omega_M = dV$).

Now of course if $(v_1,...,v_n)$ is the Gram-Schmidt orthonormalization of $(e_1,...,e_n)$, $v_i = Se_i$, then $S^T gS = I_n$ and hence $\det(S) = \det(g)^{-\frac{1}{2}}$ so that the volume form is equal to

$$dV = \epsilon \det(g)^{\frac{1}{2}} dx_1 \wedge \cdots \wedge dx_n$$

A function f on M is now integrated as

$$\int_M f = \int_M f\omega_M = \int_M f dV$$

6.13. Classical Stokes' type theorems.

A number of classical theorems now follow more or less directly from the general Stokes theorem 6.10.

Let $M \subset \mathbf{R}^2$ be a compact 2-dimensional manifold with boundary. E.g. a disk or an annulus. Let $f,g:M \to \mathbf{R}$ be differentiable. Then (Green's theorem)

$$\int_{\partial M} fdx + gdy = \int\int_M (\frac{\partial g}{\partial x} - \frac{\partial f}{\partial y})dxdy$$

This results from the general Stokes' theorem of 6.10 (and the various remarks on defining integrals over manifolds instead of chains, cf 6.9), because

$$d(fdx + gdy) = \frac{\partial f}{\partial x}dx \wedge dx + \frac{\partial f}{\partial y}dy \wedge dx + \frac{\partial g}{\partial x}dx \wedge dy + \frac{\partial g}{\partial y}dy \wedge dy =$$

$$\frac{\partial g}{\partial x}dx \wedge dy - \frac{\partial f}{\partial y}dx \wedge dy.$$

For a vectorfield ψ on $\mathbf{R}^n$, $\psi = \sum \psi^i \frac{\partial}{\partial x_i}$ one defines the divergence by $\mathrm{div}(\psi) = \sum \frac{\partial \psi^i}{\partial x_i}$. The divergence theorem now says that for an oriented manifold with boundary $M \subset \mathbf{R}^3$ one has

$$\int_M \mathrm{div}(\psi)dV = \int_{\partial M} <\psi,n> dA$$

where dV is the volume form of the three dimensional manifold M, dA the volume form (area form) of the two-dimensional manifold ∂M. Here n is the outward normal to ∂M, and the inner products (i.e. the Riemannian structure) are induced from the standard ones on $\mathbf{R}^3$.

The curl of a vectorfield $\psi = \Sigma \psi^i \frac{\partial}{\partial x_i}$ on $\mathbf{R}^3$ is defined by

$$\mathrm{curl}(\psi) = (\frac{\partial \psi^3}{\partial x_2} - \frac{\partial \psi^2}{\partial x_3})\frac{\partial}{\partial x^1} + (\frac{\partial \psi^1}{\partial x_3} - \frac{\partial \psi^3}{\partial x_1})\frac{\partial}{\partial x_2} + (\frac{\partial \psi^1}{\partial x_2} - \frac{\partial \psi^2}{\partial x_1})\frac{\partial}{\partial x_3}$$

Let $M \subset \mathbf{R}^3$ be a compact, oriented, 2-dimensional manifold with boundary. Give ∂M an orientation such that together with the outward normal n its oriented bases give back the orientation of M. Let s parametrize ∂M and let ϕ be a vectorfield in ∂M such that $ds(\phi) = 1$ (everywhere). Then the classical Stokes' formula says that

$$\int_M <\mathrm{curl}(\psi),n> dA = \int_{\partial M} <\psi,\phi> ds$$

All these theorems hold in greater generality. E.g. M could be a cube in the divergence theorem. To obtain those one uses either approximation arguments (smooth the corner and edges of the cube) or one can do the whole theory again with manifolds with corners and worse (which is possible).

All these theorems also generalize both to more general situations and to higher dimensions. To describe and discuss those, however, would bring in still more machinery such as the $*$-operator and contractions, though of course locally on the U_α it can all be done in terms of explicit formulas. For example the divergence of a vectorfield $\psi = \sum \psi^i \frac{\partial}{\partial x_i}$ (locally) on a Riemannian manifold is defined as the function

$$\operatorname{div}(\psi) = \sum_i \det(g)^{-\frac{1}{2}} \frac{\partial}{\partial x_i} (\det(g)^{\frac{1}{2}} \psi^i)$$

which of course fits the standard definition in the case of the Riemannian manifold $\mathbb{R}^n$ with $g = I_n$. One has $d(*\psi) = (dV)\operatorname{div}(\psi)$ and there results a higher dimensional divergence theorem.

7. Conclusion

The above is sort of a 'bare-bones-with-decorations' outline of manifolds and calculus on manifolds with a number of important omissions, notably contractions, the Poincaré lemma, the $*$-duality operator, connections and covariant differentiation, and curvature. It is at this point that things start getting interesting and it is at this point that this tutorial stops. Several lecture series in this volume will testify to the usefulness and power of all this machinery.

A SHORT TUTORIAL ON LIE ALGEBRAS

M. Hazewinkel
CWI, Amsterdam, The Netherlands

Lie algebras play an increasingly important role in several areas of modern control theory and filtering and estimation theory, and also, in fact, in many (other) areas of (stochastic) mechanics. This is especially true for the approach to (nonlinear) filtering theory via the Duncan-Mortensen-Zakai equation (also called the reference probability approach). The material in this tutorial corresponds to the second lecture of my series of lectures on filtering. It forms a separate coherent unit and contains all the definitions concepts and results specific to Lie algebra theory that are needed (so far) in this approach. This tutorial is a revised and expanded version of an earlier one with the same title which appeared in M. Hazewinkel, J.C. Willems (eds), Stochastic systems: the mathematics of filtering and identification, Reidel 1981, 95-108.

1. Definition of Lie Algebras. Examples

Let k be a field and V a vectorspace over k. (For the purpose of this volume it suffices to take $k=\mathbf{R}$ or (rarely) $k=\mathbf{C}$; the vectorspace V over k need not be finite dimensional.) A Lie algebra structure on V is then a bilinear map (called bracket multiplication)

$$[\ ,\ ]: V\times V\to V \tag{1.1}$$

such that the two following conditions hold

$$[u,u]=0 \text{ for all } u\in V \tag{1.2}$$

$$[u,[v,w]]+[v,[w,u]]+[w,[u,v]]=0 \text{ for all } u,v,w,\in V. \tag{1.3}$$

The last identity is called the Jacobi identity. Of course the bilinearity of (1.1) means that $[au+bv,w]=a[u,w]+b[v,w]$, $[u,bv+cw]=b[u,v]+c[u,w]$. From (1.2) it follows that

$$[u,v]=-[v,u] \tag{1.4}$$

by considering $[u+v,u+v]=0$ and using bilinearity.

1.5. Example. The Lie algebra associated to an associative algebra
Let A be an associative algebra over k. Now define a new multiplication (bracket) on A by the formula

$$[v,w]=vw-wv,\ w,v\in A \tag{1.6}$$

Then A with this new multiplication is a Lie algebra. (Exercise: check the Jacobi identity (1.3).)

1.7. Remark
In a certain precise sense all Lie algebras arise in this way. That is for every Lie algebra L there is an associative algebra A containing L such that $[u,v]=uv-vu$. I.e. every Lie algebra arises as a subspace of an associative algebra A which happens to be closed under the operation $(u,v)\mapsto uv-vu$. Though this "universal enveloping algebra" construction is quite important it will play no role in the following and the remark is just intended to make Lie algebras easier to understand for the reader.

1.8. Example
Let $M_n(k)$ be the associative algebra of all $n \times n$ matrices with coefficients in k. The associated Lie algebra is written $gl_n(k)$; i.e. $gl_n(k)$ is the n^2-dimensional vectorspace of all $n \times n$ matrices with the bracket multiplication $[A,B] = AB - BA$.

1.9. Example
Let $sl_n(k)$ denote the subspace of all $n \times n$ matrices of trace zero. Because $Tr(AB - BA) = 0$ for all $n \times n$ matrices A,B, we see that $[A,B] \in sl_n(k)$ if $A,B \in sl_n(k)$ giving us an $(n^2 - 1)$-dimensional sub-Lie-algebra of $gl_n(k)$.

1.10. Example. The Lie algebra of first order differential operators with C^∞-coefficients.
Let V_n be the space of all differential operators (on the space $F(\mathbf{R}^n)$ of C^∞-functions (i.e. arbitrarily often differentiable functions in $x_1,\ldots,x_n$)) of the form

$$X = \sum_{i=1}^{n} f_i(x_1,\ldots,x_n)\frac{\partial}{\partial x_i} \tag{1.11}$$

where the $f_i, i = 1, \ldots, n$ are C^∞-functions. Thus $X: F(\mathbf{R}^n) \to F(\mathbf{R}^n)$ is the operator $X(\phi) = \sum_{i=1}^{n} f_i \frac{\partial \phi}{\partial x_i}$.
Now define a bracket operation on V_n by the formula

$$[X,Y] = \sum_{i,j}(f_i \frac{\partial g_j}{\partial x_i}\frac{\partial}{\partial x_j} - g_j \frac{\partial f_i}{\partial x_j}\frac{\partial}{\partial x_i}) \tag{1.12}$$

if $X = \sum f_i \frac{\partial}{\partial x_i}, Y = \sum g_j \frac{\partial}{\partial x_j}$. This makes V_n a Lie algebra. Check that $[X,Y](\phi) = X(Y(\phi)) - Y(X(\phi))$ for all $\phi \in F(\mathbf{R}^n)$.

More generally, cf the tutorial in differentiable manifolds and calculus in manifolds on this volume, one has the infinite dimensional Lie algebra of vectorfields $V(M)$ on any differentiable manifold M.

1.13. Example. Derivations
Let A be any algebra (i.e. A is a vectorspace together with any bilinear map (multiplication) $A \times A \to A$; in particular A need not be associative). A *derivation* on A is a linear map $D: A \to A$ such that

$$D(uv) = (Du)v + u(Dv) \tag{1.14}$$

For example let $A = \mathbf{R}[x]$ and D the operator $\frac{d}{dx}$. Then D is a derivation. The operators (1.11) of the example above are the derivations on $F(\mathbf{R}^n)$. Also in the case of the Lie algebra of smooth vectorfields $V(M)$ on a differentiable manifolds M, one has that $V(M)$ is the Lie algebra of derivations on the algebra $F(M)$ of smooth, i.e. C^∞, functions on M.

Let $Der(A)$ be the vectorspace of all derivations. Define $[D_1,D_2] = D_1D_2 - D_2D_1$. Then $[D_1,D_2]$ is again a derivation and this bracket multiplication makes $Der(A)$ a Lie algebra over k.

1.15. Example. The Weyl algebra W_1
Let W_1 be the vectorspace of all (any order) differential operators in one variable with polynomial coefficients. I.e. W_1 is the vectorspace with basis $x^i\frac{d^j}{dx^j}$, $i,j \in \mathbf{N} \cup \{0\}$. ($x^i$ is considered as the operator $f(x) \mapsto x^i f(x)$.) Consider W_1 as a space of operators acting, say, on $k[x]$. Composition of operators makes W_1 an associative algebra and hence gives W_1 also the structure of a Lie algebra; cf. example 1.5 above. For example one has

$$[x\frac{d^2}{dx^2}, x^2\frac{d}{dx}] = 5x^2\frac{d^2}{dx^2} + 2x\frac{d}{dx}, \quad [x\frac{d}{dx}, x^i\frac{d^j}{dx^j}] = (i-j)x^i\frac{d^j}{dx^j}$$

1.16. Example. The oscillator algebra

Consider the four dimensional subspace of W_1 spanned by the four operators $\frac{1}{2}\frac{d^2}{dx^2}-\frac{1}{2}x^2, x, \frac{d}{dx}, 1$.

One easily checks that (under the bracket multiplication of W_1)

$$\begin{aligned}
&[\frac{1}{2}\frac{d^2}{dx^2}-\frac{1}{2}x^2, x]=\frac{d}{dx},\ [\frac{1}{2}\frac{d^2}{dx^2}-\frac{1}{2}x^2, \frac{d}{dx}]=x,\\
&[\frac{d}{dx}, x]=1, \qquad\qquad\qquad\qquad\qquad\qquad\qquad\qquad (1.17)\\
&[\frac{1}{2}\frac{d^2}{dx^2}-\frac{1}{2}x^2, 1]=[x,1]=[\frac{d}{dx},1]=0
\end{aligned}$$

Thus this four dimensional subspace is a sub-Lie-algebra of W_1. It is called the *oscillator Lie algebra* (being intimately associated to the harmonic oscillator).

2. Homomorphisms, isomorphisms, subalgebras and ideals

2.1. Sub-Lie-algebras

Let L be a Liealgebra over k and V a subvectorspace of L. If $[u,v]\in V$ for all $u,v\in L$, then V is a sub-Lie-algebra of L. We have already seen a number of examples of this, e.g. the oscillator algebra of example 1.16 as a sub-Lie-algebra of the Weyl algebra W_1 and the Lie-algebra $sl_n(k)$ as a sub-Lie-algebra of $gl_n(k)$. Some more examples follow.

2.2. The Lie-algebra $so_n(k)$.

Let $so_n(k)$ be the subspace of $gl_n(k)$ consisting of all matrices A such that $A+A^T=0$ (where the upper ^{T} denotes transposes). Then if $A,B\in so_n(k)$ $[A,B]+[A,B]^T = AB-BA+(AB-BA)^T = A(B+B^T)-B(A+A^T)+ (B^T+B)A^T-(A^T+A)B^T=0$, so that $[A,B]\in so_n(k)$. Thus $so_n(k)$ is a sub-Lie-algebra of $gl_n(k)$.

2.3. The Lie-algebra $t_n(k)$.

Let $t_n(k)$ be the subspace of $gl_n(k)$ consisting of all upper triangular matrices. Because product and sum of upper triangular matrices are again upper triangular $t_n(k)$ is a sub-Lie-algebra of $gl_n(k)$.

2.4. The Lie-algebra $sp_n(k)$.

Let Q be the $2n\times 2n$ matrix $Q=\begin{pmatrix}0 & I_n\\ -I_n & 0\end{pmatrix}$. Now let $sp_n(k)$ be the subspace of all $2n\times 2n$ matrices A such that $AQ+QA^T=0$. Then as above in example 2.2 one sees that $A,B\in sp_n(k)\Rightarrow [A,B]\in sp_n(k)$ so that $sp_n(k)$ is a sub -Lie-algebra of $gl_{2n}(k)$.

2.5. Ideals

Let L be a Lie-algebra over k. A subvectorspace $I\subset L$ with the property that for all $u\in I$ and all $v\in L$ we have $[u,v]\in I$ is called an ideal of L. An example is $sl_n(k)\subset gl_n(k)$, cf example 1.8 above. Another example follows.

2.6. Example. The Heisenberg Lie-algebra

Consider the 3-dimensional subspace of W_1 spanned by the operators $x, \frac{d}{dx}, 1$. The formulas (1.17) show that this subspace is an ideal in the oscillator algebra.

2.7. Example. The centre of a Lie algebra
Let L be a Lie algebra. The centre of L defined as the subset $Z(L)=\{z\in L:[u,z]=0 \text{ for all } u\in L\}$. Then $Z(L)$ is a subvector space of L and in fact an ideal of L. As an example it is easy to check that the centre of $gl_n(k)$ consists of scalar multiples of the unit matrix I_n.

2.8. Homomorphisms and isomorphisms.
Let L_1 and L_2 be a two Lie algebras over k. A morphism $\alpha:L_1\to L_2$ of vectorpaces (i.e. a k-linear map) is a *homomorphism of Lie algebras* if $\alpha[u,v]=[\alpha(u),\alpha(v)]$ for all $u,v\in L_1$. The homomorphism α is called an isomorphism if it is also an isomorphism of vectorspaces.

2.9. Example
Consider the following three first-order differential operators in two variables x,P

$$a=(1-P^2)\frac{\partial}{\partial P}-Px\frac{\partial}{\partial x},\quad b=P\frac{\partial}{\partial x},\quad c=\frac{\partial}{\partial x}$$

Then one easily calculates (cf. (1.9)) $[a,b]=c$, $[a,c]=b$, $[b,c]=0$. Now define α from the oscillator algebra of example 1.16 to this 3-dimensional Lie algebra as the linear map $\frac{1}{2}\frac{d^2}{dx^2}-\frac{1}{2}x^2\mapsto a$, $x\mapsto b$, $\frac{d}{dx}\mapsto c$, $1\mapsto 0$. Then the formulas above and (1.17) show that α is a homomorphism of Lie algebras.

2.10. Kernel of a homomorphism
Let $\alpha:L_1\to L_2$ be a homomorphism of Lie algebras. Let $\mathrm{Ker}(\alpha)=\{u\in L_1:\alpha(u)=0\}$. Then $\ker(\alpha)$ is ideal in L_1.

2.11 Quotient Lie algebras
Let L be a algebra and I an ideal in L. Consider the quotient vector space L/I and the quotient morphisms of vector spaces $L\xrightarrow{\alpha}L/I$. For all $\bar{u},\bar{v}\in L/I$ choose $u,v\in L$ such that $\alpha(u)=\bar{u}$, $\alpha(v)=\bar{v}$. Now define $[\bar{u},\bar{v}]=\alpha[u,v]$. Check that this does not depend on the choice of u,v.

This then defines a Lie-algebra structure on L/I and $\alpha:L\to L/I$ becomes a homomorphism of Lie-algebras.

2.12. Image of a homomorphism
Let $\alpha:L_1\to L_2$ be a homomorphism of Lie algebras. Let $\mathrm{Im}(\alpha)=\alpha(L_1)=\{u\in L_2:\exists v\in L_1,\alpha(v)=u\}$. Then Im (α) is a sub-Lie-algebra of L_2 and α induces an isomorphism $L_1/Ker(\alpha)\cong\mathrm{Im}(\alpha)$.

2.13. Exercise
Consider the 3-dimensional vector space of all real upper triangular 3×3 matrices with zero's on diagonal. Show that this a sub-Lie-algebra of $gl_3(\mathbf{R})$, and show that it is isomorphic to the 3-dimensional Heisenberg-Lie-algebra of example 2.6 but that is not isomorphic to the 3-dimensional Lie-algebra $sl_2(\mathbf{R})$ of example 1.8.

2.14. Exercise
Show that the four operators x^2, $\frac{d^2}{dx^2}$, $x\frac{d}{dx}$, 1 span a 4-dimensional subalgebra of W_1, and show that this 4-dimensional Lie algebra contains a three dimensional Lie algebra which is isomorphic to $sl_2(\mathbf{R})$.

2.15. Exercise

Show that the six operators $x^2, \frac{d^2}{dx^2}, x, \frac{d}{dx}, x\frac{d}{dx}$, 1 span a six dimensional sub-Lie-algebra of W_1. Shows that $x, \frac{d}{dx}$, 1 span a 3-dimensional ideal in this Lie-algebra and shows that the corresponding quotient algebra is $sl_2(\mathbf{R})$.

3. LIE ALGEBRAS OF VECTORFIELDS

Let M be a C^∞-manifold (cf. the tutorial on manifolds and calculus on manifolds in this volume). Intuitively a vectorfield on M specifies a tangent vector $t(m)$ at every point $m \in M$. Then given a C^∞-function f on M we can for each $m \in M$ take the derivation of f at m in the direction $t(m)$, giving us a new function g on M. This can be made precise in varying ways; e.g. as follows.

3.1. The Lie algebra of vectorfields on a manifold M

Let M be a C^∞-manifold, and let $F(M)$ be the $\mathbf{R}$-algebra (pointwise addition and multiplication) of all smooth $(=C^\infty)$ functions $f:M \to \mathbf{R}$. *By definition* a *C^∞-vectorfield* on M is a derivation $X:F(M) \to F(M)$. The Lie algebra of derivations of $F(M)$, cf. example 1.13, i.e. the Lie-algebra of smooth vectorfields on M, is denoted $V(M)$.

3.2. Derivations and vectorfields

Now let $M = \mathbf{R}^n$ so that $F(M)$ is simply the $\mathbf{R}$-algebra of C^∞-functions in $x_1, \ldots, x_n$. Then it is not difficult to show that every derivation $X:F(\mathbf{R}^n) \to F(\mathbf{R}^n)$ is necessarily of the form

$$X = \sum_{i=1}^{n} g_i \frac{\partial}{\partial x_i} \tag{3.3}$$

with $g_i \in F(\mathbf{R}^n)$. For a proof cf. [4, Ch.I, § 2]. The corresponding vectorfield on $\mathbf{R}^n$ now assigns to $x \in \mathbf{R}^n$ the tangent vector $(g_1(x), \ldots, g_n(x))^T$.

On an arbitrary manifold we have representations (3.3) locally around every point and these expressions turn out to be compatible in precisely the way needed to define a vectorfield as described in the tutorial on differentiable manifolds and calculus on manifolds in this volume.

3.4. Homomorphisms of Lie algebras of vectorfields

Let M and N be C^∞-manifolds and let $\alpha:L \to V(N)$ be a homomorphism of Lie algebras where L is a sub-Lie-algebra of $V(M)$. Let $\phi:M \to N$ be a smooth map. Then α and ϕ are said to be *compatible* if

$$\phi^*(\alpha(X)f) = X(\phi^*(f)) \quad \text{for all } f \in F(N) \tag{3.5}$$

where ϕ^* is the homomorphism of algebras $F(N) \to F(M), f \mapsto \phi^*(f) = f \circ \phi$.

In terms of the Jacobian of ϕ (cf. [3]), this means that

$$J(\phi)(X_m) = \alpha(X)_{\phi(m)} \tag{3.6}$$

where X_m is the tangent vector at m of the vectorfield X.

If $\phi:M \to N$ is an isomorphism of C^∞-manifolds there is always precisely one homomorphism of Lie-algebras $\alpha:V(M) \to V(N)$ compatible with ϕ (which is then an isomorphism). It is defined (via formula (3.5)) by

$$\alpha(X)(f) = (\phi^*)^{-1} X(\phi^* f), \quad f \in F(N). \tag{3.7}$$

3.8. Isotropy subalgebras
Let L be a sub-Lie-algebra of $V(M)$ and let $m \in M$. The isotropy subalgebra L_m of L at m consists of all vectorfields in L whose tangent vector vector in m is zero, or, equivalently

$$L_m = \{X \in L : Xf(m) = 0 \text{ all } f \in F(M)\} \tag{3.9}$$

Now suppose that $\alpha: L \rightarrow V(N)$ and $\phi: M \rightarrow N$ are compatible in the sense of 3.4 above. Then it follows easily from (3.5) that

$$\alpha(L_m) \subset V(N)_{\phi(m)} \tag{3.10}$$

i.e. α takes isotropy subalgebras into isotropy subalgebras. Inversely if we restrict our attention to analytic vectorfields then condition (3.10) on α at m implies that locally there exists a ϕ which is compatible with α [7].

4. Simple, Nilpotent, and solvable algebras

4.1. Nilpotent Lie algebras.
Let L be a Lie-algebra over k. The *descending central series* of L is defined inductively by

$$C^1L = L, \quad C^{i+1}L = [L, C^iL], \quad i \geq 1 \tag{4.2}$$

It is easy to check that the C^iL are ideals. The Lie algebra L is called *nilpotent* if $C^nL = \{0\}$ for n big enough.

For each $x \in L$ we have the endomorphism $adx: L \rightarrow L$ defined by $y \mapsto [x,y]$. It is now a theorem that if L is finite dimensional then L is nilpotent iff the endomorphisms adx are nilpotent for all $x \in L$. Whence the terminology.

4.3. Solvable Lie algebras.
The *derived series of Lie algebras* of a Lie algebra of a Lie algebra L is defined inductively by

$$D^1L = L, \quad D^{i+1}L = [D^iL, D^iL], \quad i \geq 1 \tag{4.4}$$

It is again easy to check that the D^iL are ideals. The Lie algebra L is called solvable if $D^nL = \{0\}$ for n large enough.

4.5. Examples
The Heisenberg Lie algebra of example 2.6 is nilpotent. The Oscillator algebra of example 1.16 is solvable but not nilpotent. The sub-Lie-algebra of W_1 with vector-space basis $x^2, \frac{d^2}{dx^2}, x, \frac{d}{dx}, 1, x\frac{d}{dx}$ is neither nilpotent, nor solvable. The Lie-algebra $t_n(k)$ of example 2.3 is solvable and in a way is typical of finite dimensional solvable Lie algebras in the sense that if k is algebraically closed (e.g. $K = \mathbf{C}$), then every finite dimensional solvable Lie algebra over k is isomorphic to a sub-Lie-algebra of some $t_n(k)$.

4.6. Exercise
Show that sub-Lie-algebras and quotient-Lie-algebras of solvable Lie algebras (resp. nilpotent Lie algebras) are solvable (resp. nilpotent).

4.7. Abelian Lie-algebras
A Lie algebra L is called *abelian* if $[L, L] = \{0\}$, i.e. if every bracket product is zero.

4.8. Simple Lie-algebras
A Lie algebra L is called *simple* if it is not abelian and if it has no other ideals than 0 and L. (Given the second condition the first one rules out the zero- and one-dimensional Lie algebras.) These simple-Lie-algebras and the abelian ones are in a very precise sense the basic building blocks of all Lie algebras.

The finite dimensional simple Lie algebras over $\mathbf{C}$ have been classified. They are the Lie algebras $sl_n(\mathbf{C}), sp_n(\mathbf{C}), so_n(\mathbf{C})$ of examples 1.8, 2.4 and 2.2 above and five additional exceptional Lie algebras. For infinite dimensional Lie algebras things are more complicated. The socalled filtered, primitive, transitive simple Lie algebras have also been classified (cf. e.g. [2]). One of these is the Lie-algebra V_n of all formal vector fields $\sum f_i(x_i, \ldots, x_n)\frac{\partial}{\partial x_i}$, where the $f_i(x)$ are (possibly non converging) formal power series in $x_1, \ldots, x_n$. This class of infinite dimensional simple Lie algebras by no means exhausts all possibilities. E.g. the quotient-Lie algebras $W_n/\mathbf{R}\cdot 1$ are simple and non-isomorphic to any of those just mentioned.

4.9. Exercise
Let $V_{alg}(\mathbf{R}^n)$ be the Lie algebra of all differential operators (vector fields) of the form $\sum f_i(x_1, \ldots, x_n)\frac{\partial}{\partial x_1}$ with $f_i(x_1, \ldots, x_n)$ polynomial. Prove that $V_{alg}(\mathbf{R}^n)$ is simple.

5. Representations
Let L be a Lie algebra over k and M a vectorspace over k. A representation of L in M is a homomorphism of Lie algebras.

$$\rho: L \rightarrow End_k(M) \tag{5.1}$$

where $End_k(M)$ is the vectorspace of all k-linear maps $M \rightarrow M$ which is of course given the Lie algebra structure $[A,B]=AB-BA$. Equivalently a representation of L in M consists of a k-bilinear map

$$\sigma: L \times M \rightarrow M \tag{5.2}$$

such that, writing xm for $\sigma(x,m)$, we have $[x,y]m = x(ym) - y(xm)$ for all $x,y \in L$, $m \in M$. The relation between the two definitions is of course $\sigma(x,m)=\rho(x)(m)$.

Instead of speaking of a representation of L in M we also speak (equivalently) of the L-module M.

5.3. Examples
The Lie algebra $gl_n(k)$ of all $n \times n$ matrices naturally acts on k^n by $(A,v) \mapsto Av \in k^n$ and this defines a representation $gl_n(k) \times k^n \rightarrow k^n$. The Lie algebra $V(M)$ of vectorfields on a manifold M acts (by its definition) on $F(M)$ and this is a representation of $V(M)$. A quite important theorem concerning the existence of representations is

5.4. Ado's theorem
Cf. e.g. [1, § 7]. If k is a field of characteristic zero, e.g. $k = \mathbf{R}$ or $\mathbf{C}$ and L is finite dimensional then there is a faithful representation $\rho: L \rightarrow End(k^n)$ for some n. (Here faithful means that ρ is injective.)

Thus every finite dimensional Lie algebra L over k (of characteristic zero) can be viewed as a subalgebra of some $gl_n(k)$, and this subalgebra can then be viewed as a more concrete matrix "representation" of the "abstract" Lie algebra L.

5.5. Realizing Lie-algebras in $V(M)$

A question of some importance for filtering theory is when a Lie algebra L can be realized as a sub-Lie-algebra of $V(M)$, i.e. when L can be represented in $F(M)$ by means of derivations. For finite dimensional Lie algebras Ado's theorem gives the answer because $(a_{ij}) \mapsto \sum a_{ij} x_i \frac{\partial}{\partial x_j}$ defines an injective homomorphism of Lie-algebras $gl_n(\mathbf{R}) \to V(\mathbf{R}^n)$ (Exercise: check this).

6. Lie algebras and Lie groups

6.1. Lie groups

A (finite dimensional) Lie group is a finite dimensional smooth manifold G together with smooth maps $G \times G \to G$, $(x,y) \mapsto xy$, $G \to G$, $x \mapsto x^{-1}$ and a distinguished element $e \in G$ which make G a group. An example is the open subset of $\mathbf{R}^{n^2}$ consisting of all invertible $n \times n$ matrices with the usual matrix multiplication.

6.2. Left invariant vectorfields and the Lie algebra of a Lie group

Let G be a Lie group. Let for all $g \in G$, $L_g: G \to G$ be the smooth map $x \to gx$. A vectorfield $X \in V(G)$ is called left invariant if $X(L_g^* f) = L_g^*(Xf)$ for all functions f on G, where, of course, the left translate $L_g^* \gamma$ of a function γ is defined by $(L_g^* \gamma)(x) = \gamma(gx)$. Or, equivalently, if $J(L_g)X_x = X_{gx}$ for all $x \in G$, cf. section 3.4 above. Especially from the last condition it is easy to see that $X \mapsto X_e$ defines an isomorphism between the vectorspace of left invariant vectorfields on G and the tangent space of G at e. Now the bracket product of two left invariant vectorfields is easily seen to be left invariant again so the tangent space of G at e (which is $\mathbf{R}^n$ if G is n-dimensional) inherits a Lie algebra structure. This is the Lie algebra Lie(G) of the Lie group G. A main reason for the importance of Lie algebras in many parts of mathematics and its applications is that this construction is reversible to a great extent making it possible to study Lie groups by means of their Lie algebras.

6.3. Exercise

Show that the Lie algebra of the Lie group $GL_n(\mathbf{R})$ of invertible real $n \times n$ matrices is the Lie algebra $gl_n(\mathbf{R})$.

7. The adjoint representation

Let L be a Lie algebra. Then there is a natural representation of L into the vectorspace L given by

$$ad: L \to End(L), \quad ad(x)(y) = [x,y] \tag{7.1}$$

The Jacobi identity is precisely what is needed to show that $ad([x,y]) = ad(x)ad(y) - ad(y)ad(x)$. This representation is called the adjoint representation. It is the infinitesimal part of a representation denoted Ad of the group G of L in L.

In the case G is a connected subgroup of $GL_n(k)$, so that $L = \text{Lie}(G)$ is a subalgebra of $gl_n(k)$ this representation can be written as

$$Ad(g)(x) = gxg^{-1}, \quad x \in L, g \in G \tag{7.2}$$

(One needs to prove of course that gxg^{-1} is again in L.) In the more general and more abstract setting of 6.2 above, this goes as follows. Let G be a Lie group and L its Lie algebra, i.e. the Lie algebra of left invariant smooth vectorfields on G. Again let $L_g: G \to G$ be defined by $L_g(x) = gx$ and define $i_h: G \to G$ by $x \mapsto hxh^{-1}$. Now observe that

$$L_g i_h = i_h L_{h^{-1}gh}$$

It follows immediately that if the vectorfield $i_h^* X$ is defined by

$$(i_h^* X)(f) = X(i_h^* f)$$

that then i_h^* is left invariant if X is left-invariant.

In case $G = GL_n(\mathbf{R}) \subset \mathbf{R}^{n \times n}$, we can identity the tangent space $T_g G$ at $g \in G$ with the space of $n \times n$ matrices $\mathbf{R}^{n \times n}$. The left invariant vectorfields then are of the form

$$g \mapsto gA, \quad g \in G$$

for some fixed $A \in \mathbf{R}$. If X is a vectorfield then $i_h^* X$ is the vectorfield defined by

$$(i_h^* X)_x = J(i_h)(i_h^{-1}(x))(X_{i_h^{-1}(x)})$$

Now $i_h(x) = hxh^{-1}$ which is linear in x. So $J(i_h)(y)(v) = hvh^{-1}$ for all $y \in G$, $v \in \mathbf{R}^{n \times n} = T_y G$. So if X is the left invariant vectorfield $g \mapsto gA$, then i_h^* is the vectorfield

$$\begin{aligned}(i_h^* X)_g = J(i_h)(i_h^{-1}(g))(X_{i_h^{-1}(g)}) &= J(i_h)(i_h^{-1}(g))(h^{-1}ghA) \\ &= h(h^{-1}ghA)h^{-1} = ghAh^{-1}\end{aligned}$$

which is indeed again of the same form, hence left invariant, and we see that the induced action on $T_e G = \mathbf{R}^{n \times n}$ is indeed $A \mapsto hAh^{-1}$ This also serves to prove that if $g \in G \subset GL_n(\mathbf{R})$, $x \in L \subset gl_n(\mathbf{R})$, $L =$ Lie(G) then indeed $gxg^{-1} \in L$.

For $G = GL_n(k)$, $L = gl_n(k)$ there is a local diffeomorphism of a neighborhood of 0 in $gl_n(k)$ to a neighborhood of $e = I_n \in GL_n(k)$ given by $A \mapsto \exp(A) = I + \frac{A}{1!} + \frac{A^2}{2!} + \cdots$ and more generally exp takes a neighborhood in the subalgebra L of G into one around $e \in G$.

The particular case of (7.2) where g is of the form $e^A, A \in L$, is of importance. One has *'the adjoint representation formula'*

$$e^A B e^{-A} = B + \frac{[A,B]}{1!} + \frac{[A,[A,B]]}{2!} + \cdots = \sum_{n=0}^{\infty} (n!)^{-1} ad(A)^n(B) \tag{7.3}$$

(From which it is clear that indeed $e^A B e^{-A} \in$ (a completion of) the Lie algebra generated by A and B). This formula is easily proved by induction. One needs $ad(A)^n(B) = \sum_{i+j=n} (-1)^j \frac{1}{i!} \cdot \frac{1}{j!} A^i B A^j$.

Occasionally in the literature this formula occurs under the name Campbell-Baker-Hausdorff formula (or Campbell-Hausdorff formula). This name, however, belongs more properly to the far deeper result that if $A, B \in L$ then $e^A e^B$ is the exponential of some element in (a completion of) the sub-Lie-algebra of L spanned by A and B. This element can be expressed as an infinite sum (the $C-B-H$ formula)

$$A + B + \frac{1}{2}[A,B] + \frac{1}{12}[[A,B],B] - \frac{1}{12}[[A,B],A] + \cdots$$

(This result can be extended to the most general case: that of a free Lie algebra, either in terms of formal series identities, or by means of suitable completions.)

8. Postscript

The above is a very rudimentary introduction to Lie algebras. Especially the topic "Lie algebras and Lie groups" also called "Lie theory" has been given very little space, in spite of the fact that it is likely to become of some importance in filtering (integration of a representation of a Lie algebra to a representation of a Lie (semi)group). The books [1, 4, 5, 6, 8, 9] are all recommended for further material. My current personal favourite is [9] with [4] as (a far more difficult) close runner-up; [6] is a classic and in its present incarnation very good value indeed.

1. A. Bourbaki (1960). Groupes et Algèbres de Lie, Chap. 1: Algèbres de Lie, Hermann.
2. M. Demazure (1967). Classification des algèbres de Lie filtrés, Sém. Bourbaki 1966/1967, Exp. **326**, Benjamin.
3. M. Hazewinkel. Tutorial on Manifolds and Calculus on Manifolds. This volume.

4. S. HELGASON (1978). Differential Geometry, Lie Groups and Symmetric Spaces, Acad. Pr..
5. J.E. HUMPHREYS (1972). Introduction to Lie Algebras and Representation Theory, Springer.
6. N. JACOBSON (1980). Lie Algebras, Dover reprint.
7. A.J. KRENER (1973). On the Equivalence of Control Systems and the Linearization of nonlinear Systems, SIAM J. Control **11**, 670-676.
8. J.P. SERRE (1965). Lie Algebras and Lie Groups, Benjamin.
9. V.S. VARADARAJAN (1984). Lie groups and Lie algebras and their representations, Springer (orginally published by Prentice-Hall, 1974).

Printed in the United States
By Bookmasters